Springer Series in Advanced Manufacturing

Series editor

Duc Truong Pham, University of Birmingham, Birmingham, UK

The **Springer Series in Advanced Manufacturing** includes advanced textbooks, research monographs, edited works and conference proceedings covering all major subjects in the field of advanced manufacturing.
The following is a non-exclusive list of subjects relevant to the series:

1. Manufacturing processes and operations (material processing; assembly; test and inspection; packaging and shipping).
2. Manufacturing product and process design (product design; product data management; product development; manufacturing system planning).
3. Enterprise management (product life cycle management; production planning and control; quality management).

Emphasis will be placed on novel material of topical interest (for example, books on nanomanufacturing) as well as new treatments of more traditional areas.

As advanced manufacturing usually involves extensive use of information and communication technology (ICT), books dealing with advanced ICT tools for advanced manufacturing are also of interest to the Series.

Springer and Professor Pham welcome book ideas from authors. Potential authors who wish to submit a book proposal should contact Anthony Doyle, Executive Editor, Springer, e-mail: anthony.doyle@springer.com.

More information about this series at http://www.springer.com/series/7113

Pingyu Jiang

Social Manufacturing: Fundamentals and Applications

Springer

Pingyu Jiang
State Key Laboratory for Manufacturing
Systems Engineering
Xi'an Jiaotong University
Xi'an
China

ISSN 1860-5168 ISSN 2196-1735 (electronic)
Springer Series in Advanced Manufacturing
ISBN 978-3-030-10295-1 ISBN 978-3-319-72986-2 (eBook)
https://doi.org/10.1007/978-3-319-72986-2

Printed on acid-free paper

This Springer imprint is published by the registered company Springer International Publishing AG part of Springer Nature
The registered company address is: Gewerbestrasse 11, 6330 Cham, Switzerland

Preface

Internet-based connecting and communicating behaviors with each other are becoming basic requirements for human beings, not only in daily life but also during product manufacturing activities both inside an enterprise and across enterprises. They also influence in depth the dynamically organizational shape and runtime logic of current manufacturing industry from the angles of business interactions, resource utilization, manufacturing flows, systems integration, etc. In fact, product manufacturing activities in the whole stages of product life cycle work inside a huge socio-technical system and one of the most available trends in the near future is to create a sustainably ecological enterprise circle under the philosophy of social manufacturing.

Social manufacturing is a kind of service-oriented manufacturing paradigm. This term was originally from news report at ***BusinessWire*** and follow-up ***Economist*** about in the later 2011 on the basis of predicting that 3D printing would motivate a new-round industrial revolution and make anyone in principle become a manufacturer besides a consumer. Different from the original term description mentioned in the above news reports, since the end of 2011, my research team at state key laboratory for manufacturing systems engineering, Xi'an Jiaotong University, China has focused mainly on generic manufacturing environment and been doing some pioneer studies on defining concepts, designing organizational architecture and runtime, identifying key enabled technologies, verifying theory through various prototypes, doing case studies under the context of social manufacturing. This book just sums up our initial research outcomes in the last 6 years based on our related publications in various international journals and conferences. Through case studies, we can also know recent initial explorations concerning social manufacturing in China manufacturing industry.

In order to explain what social manufacturing is, why it can enable manufacturing industry, and how it works, in this book, we will present our research outcomes concerning social manufacturing with 12 chapters. In Chap. 1, we first discuss driving factors and trends of creating new manufacturing paradigms from the angles of socialization, digitalization, servitization, and intelligentization, and indicate that social manufacturing will play an important role in manufacturing industry in the near future. And then, we propose the fundamental implementing architecture and runtime logic of social manufacturing, and identify the correspondent-enabled technologies and computing methods to support the above technologies after giving a series of concept definitions in Chap. 2. In Chaps. 3–9, we respectively describe the following key enabled technologies in detail under the context of social manufacturing:

- socialized manufacturing resources and interconnections,
- social business relationship and organizational network,
- open product design for social manufacturing,
- RFID, social sensors, and extended cyber-physical system,
- social factory and interconnections,
- product service systems for social manufacturing, and
- blockchain models for cyber-credits of social manufacturing.

On the basis of the contents of the above chapters, we further describe how to configure a social manufacturing system and run it respectively in Chaps. 10 and 11. In order to verify theory, methods, models, and key enabled technologies, etc., we use three industrial cases from Haier, RepRap open-source 3D printers manufacturer network, X-part manufacturers so as to demonstrate three types of preliminarily distributive implementing mechanisms for social manufacturing, such as partially centralized control, centerless self-control, and completely centralized control mechanisms, in Chap. 12.

It is very clear that social manufacturing meets the needs of some new manufacturing philosophies like sharing, collaborating and competing, socialized resource utilization, enterprise minimalization and microlization, outsourcing and crowdsourcing, crowdfunding, product services systems, etc., under the support of Internet-based connecting and communicating behaviors in business. Current industrial practices have also indicated that this paradigm would be one of the most important manufacturing configuration and runtime modes. Original 3D-printing-driven manufacturing manner, in which someone holds his/her own 3D printers, can become a manufacturer in principle, and is a node inside a huge social network, proposed in the early news reports is just an extreme example of social manufacturing paradigm presented in this book.

What I want to mention here is that it is very happy for us to at last finish this book with the great help of a lot of financial sources and individuals. So I would like to take this opportunity to give my thanks to Natural Science Foundation of China (NSFC). In fact, NSFC issued the first nationwide research project (Grant No.: 71571142) to me starting from 2016 in the title of *social manufacturing*. I also thank Ministry of Science and Technologies (MOST) of China to give me the financial support through project *Manufacturing-cluster-driven innovation methods for micro-and-small-scale manufacturing enterprises* (Grant No.: 2016IM010100). The project issued by MOST lets us have a big chance to do case studies in Haier Groups, Jerui Groups, etc. I need to give many thanks to my current and graduated students who are not listed as authors of this book but they make different contributions for the research outcomes, such as, Dr. Fuqiang Zhang, Dr. Wei Cao, Dr. Peilu Sun, Dr. Kai Ding, et al. It is also good news that one of the authors, Dr. Jiewu Leng, has gotten his first early-career financial support from NSFC on the basis of our research outcomes in social manufacturing when he went to Guangdong University of Technology as a lecturer in 2017 and further got the young talent-fund from his new host university. I also sincerely thank Springer staffs who give me many kind helps during planning, writing, and publishing the book.

Finally, I would like to give my most sincere thanks to my wife, Zi Liu, who completely supported me to work for the book, especially in the period from January to March of 2018.

Xi'an, China
May 2018

Pingyu Jiang

Contents

Chapter 1
Introduction

Pingyu Jiang

1.1 Background and Problems

Connecting and communicating behaviors with each other are top requirements for human beings besides eating and sleeping. There is no exception even for product manufacturing activities during thousands of years. Looking this situation back to the history of product manufacturing, we know that human beings originally used hand-made manner to make goods so as to satisfy their living needs, which lasted the whole agriculture era. In that time, product manufacturing activities of human beings were essentially organized in families, which were basic organizational units, and main connecting and communicating media of the society.

It was changed because the first industrial revolution happened in United Kingdom in the 18th century. One of the biggest impacts to human beings was to extend connecting and communicating behaviors of human beings besides ones mainly in families. The role of human beings began their varieties not only in daily life but also in product manufacturing activities. Such varieties especially in product manufacturing activities may be described as a kind of evolution from ***family-centered, enterprise-centered, cross-enterprise-centered to Internet-based connecting and communicating behaviors*** along with changing of product types and production volumes. It means the organizational structure of product manufacturing, in fact, has to fit with the evolution of social relationship and development besides satisfying the needs in aspects of various product types and production volumes.

The objective of setting up manufacturing enterprises is to produce products not only for consuming in daily life but also for using in industrial productions. Referring to various product types and production volumes, the production modes of manufacturing enterprises have gone through huge changes from ***hand-made fabrication, mass production, mass customization to individualized production*** so as to satisfy the needs of increasing the production efficiency, especially since the first industrial revolution happened. Besides factors of the social relationship and development mentioned above, such changes also depend on innovations of production organizations,

P. Jiang, *Social Manufacturing: Fundamentals and Applications*, Springer Series in Advanced Manufacturing, https://doi.org/10.1007/978-3-319-72986-2_1

developments of manufacturing technologies themselves, and progresses on correlative science and technologies like automation and computer science, information technologies, industrial engineering, etc.

Today the development of Internet, Internet of Things (IoT) and cyber-physical systems (CPS), together with fantastical Internet-based connecting and communicating behaviors in business, makes the organizational structure and business interactions inside a manufacturing enterprise and among different manufacturing enterprises face a very big challenge more over than before. Sometimes big data are taking the place of traditional methods to help the manufacturing enterprise improving its various capabilities. It means the manufacturing enterprise has to answer how it can fit with the challenge as an Internet-based enterprise, what it must change in its product manufacturing activities during the product life cycle and how it does. Those problems must be solved by any manufacturing enterprise living in the 21st century.

1.1.1 Product Orders and Production Modes

One of the most important purposes for human beings to set up manufacturing enterprises is to produce products. In fact, the first business task for a manufacturing enterprise is to get product orders in which product types and production volumes to be needed are declared in detail. At least, there are two types of product orders:

- ***inventory-prediction-based product order***, which often represents to need a big volume of products based on producers' prediction and deals with a large number of potential customer groups, and
- ***customer-requirement-driven product order***, which often stands for needing one-of-a-kind, small batch, or middle and big volume of products and comes from specific customers who sign contracts in advance with producers as constraints in law for both sides.

These two types of product orders are correspondent with different financial issues. For the first type of product order, producers who are the representative of a manufacturing enterprise need to prepare the finance by themselves. Actually, this type of product order is a kind of pseudo-order. For the second type of product order, producers who are the representative of a manufacturing enterprise produce products according to customers' requirements and prepare the partial or the whole finance on the basis of contracts between producers and customers.

At present, there has been the third way to prepare the finance if a manufacturing enterprise even an individual has a good product design scheme but no money to produce the products. This way is called as crowdfunding. Actually, crowdfunding implies a kind of product order in which customers prepay the whole money to the party of holding the product design scheme and have a time delay to get the products they want. From the angle of classification, a product order created by crowdfunding mechanism also belongs to the customer-requirement-driven one.

In addition, a kind of new product-driven order, entitled as product-services order, appears along with the practical applications of product service systems.

It must be pointed out that product orders, which provide the information related to product types, production volumes, etc., hint such a fact that product manufacturing activities can be organized in the mode of mass production, mass customization, or individualized production. This means that the product types and production volumes would influence the organizational structure of product manufacturing activities. The fact will also be presented in depth under the consideration of Internet-based connecting and communicating behaviors related to a new manufacturing paradigm in this book.

1.1.2 "Smiling Curve" and Value-Added Activities in Product Life Cycle

Generally speaking, the term "*manufacturing*" covers different domains depending on the contexts of problems we discuss. It is necessary to view product manufacturing activities from the angle of the whole life cycle of the product. In this case, the term "*manufacturing*" is generic, not limited in a range of machining parts and assembling products, and covers the whole stages of product life cycle. While the product life cycle consists of ***requirement analysis stage, design stage, production stage, selling stage, running and maintaining stage, and recycling and abandoning stage.*** Here, product manufacturing activities indicate requirement analysis activities, design activities, production activities, delivery activities, running and maintaining activities, and recycling and abandoning activities which are respectively correspondent with different stages of a product life cycle.

What we must emphasize on firstly is that Bill of Materials (BOM) of products is just a kind of index to do product manufacturing activities in different stages of a product life cycle. Figure 1.1 shows BOM-driven self-made, outsourced and purchased relationships and their connections with different enterprises in the production stage of a product life cycle. It becomes a typical enterprise alliance if the core enterprise holding products assigns and outsources its production tasks to partner co-enterprises in "*acquaintance model*" [1]. It can also be said that shaping a supply chain needs to understand the evolutional processes of BOM in different stages of a product life cycle. Here, the outsourcing mechanism is an important driver.

What we must point out secondly is that value creations and benefit shares in different stages of a product life cycle are quite different. In fact, the "*smiling curve*" shown in Fig. 1.2 just gives us a glimpse on creating values and sharing benefits in different stages of a product life cycle.

It is quite evident that product manufacturing activities get much more benefits in both design stage and running & maintaining stage, and relatively smaller benefits in production stage. In addition, increasing value-added benefits in different stages of a product life cycle by means of using outsourcing and crowdsourcing mechanisms must be taken into consideration in depth.

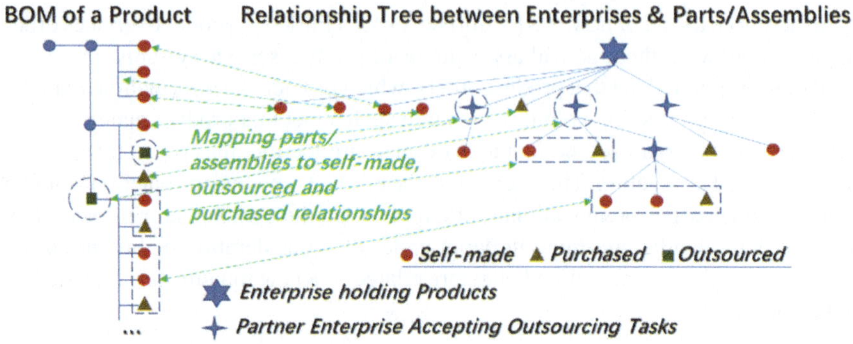

Fig. 1.1 BOM-driven production relationships related to enterprises and their parts/assemblies

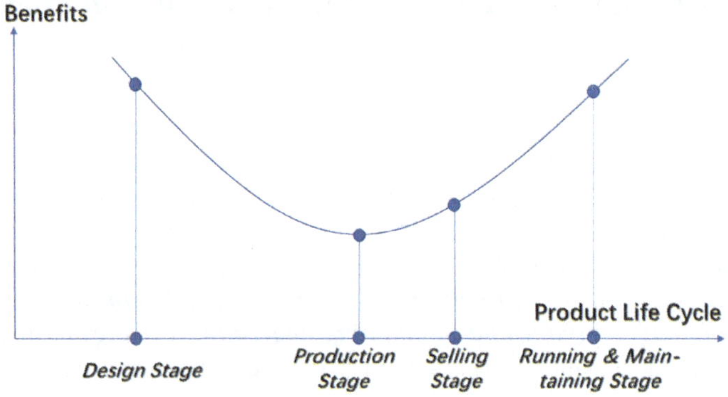

Fig. 1.2 Smiling curve related to relationship between product life cycle and its obtained benefits [2]

In product design stage, using the crowdsourcing mechanism to cluster socialized design capabilities is a new procedure for a manufacturing enterprise to solve its own design difficulties. Comparing with its own original designing workflow, this is a value-added procedure and can let the correspondent product design activities implement the reduction of the design costs. In addition, outsourcing product design tasks is also a kind of value-added activities. Therefore, it means that the organizational structure and the correspondent workflow of product design activities, which can fit with integrating and using socialized design resources, need to be reconstructed.

In product production stage, it is very clear for a manufacturing enterprise to outsource its partial and even the whole production tasks of products to the outside on the basis of considering its own production capabilities and cost reductions. Although sometimes those production tasks can also be finished by itself, outsourcing them can reach much lower costs. The purpose to do as this is to extend its own production capabilities and get value-added benefits. In this way, the manufacturing enterprise works in a "*dumbbell*"-like organizational structure. Outsourcing production tasks, of course, is a value-added procedure through enhancing production capabilities and reducing production costs, and also a result of social division of labors. By the way, as a new value-added procedure of assigning production tasks competitively in the production stage of a product life cycle, furthermore, crowdsourcing mechanism can be also involved in the competition of production tasks. It means that the organizational structure and the correspondent workflow of product production activities, which can fit with integrating and using socialized production resources, need to be reconstructed.

In product selling stage, a manufacturing enterprise can select to sell either its products or services attached to the products to customers according to product orders. Here, selling services attached to the products, which is also called as selling product service systems, provides a new value-added way through finishing both selling activities and running and maintaining activities to get more benefits. This point will be discussed below in detail. Another key point is how to implement effective product logistics. Similar to using the outsourcing and crowdsourcing mechanisms mentioned above, a value-added procedure related to product logistics can also be realized accordingly. It means that the organizational structure and the correspondent workflow of product selling activities, which can fit with integrating and using socialized logistic resources, need to be reconstructed.

In product running and maintaining stage, traditional way is to get value-added benefits from repairing of products, changing of spare parts, etc., which is realized mainly in terms of selecting suitable repairing-type micro-enterprises and outsourcing repairing tasks to them. A new way to get more value-added benefits is to run and maintain product service systems, which fuse tangible products and intangible services together. Here, service providers concerning the product service systems are also able to consist of socialized running and maintaining resources. Both selecting service providers and being in charge of running and maintaining these systems may depend on either outsourcing or crowdsourcing mechanisms. It means that the organizational structure and the correspondent workflow of product running and maintaining activities, which can fit with integrating and using socialized running and maintaining resources, need to be reconstructed.

To sum up, both outsourcing and crowdsourcing mechanisms play a key role in realizing value-added procedures with the help of correspondent product manufacturing activities during a product life cycle and influence the organizational structure and runtime logic of manufacturing enterprises when a lot of socialized manufacturing resources are integrated and used.

1.1.3 Key Factors to Create and Impact a New Manufacturing Paradigm

On the basis of analysis mentioned above, we know that the factors of influencing the socialized manufacturing resource collection, organizational structure and correspondent runtime logic for a manufacturing enterprise include:

- connecting and communicating behaviors in business, especially dealing with cross-enterprise-centered and Internet-based behaviors,
- product manufacturing activities in a product life cycle, especially being outsourcing or/and crowdsourcing activities related to bill of the materials of the product in the context of supply chain, and
- product orders related to product types and production volumes, which are concerned respectively with mass production mode, mass customization mode, or individualized production mode.

The different roadmaps to satisfy above three factors would guide us to create new different manufacturing paradigms. For example, current existing manufacturing paradigms, like agile manufacturing, enterprise alliance, cloud manufacturing, e-manufacturing, service-oriented manufacturing, etc., just are solutions which satisfy the above three factors. But the further question we have to answer right now is what is the correct roadmap to create a new manufacturing paradigm.

This feasible roadmap is visible through analyzing the correlation among the above factors. In fact, product orders and product manufacturing activities in a product life cycle deal mainly with technology-driven implementations. Connecting and communication behaviors in business are just related to socialization-driven practices and cover the whole procedure concerning finishing product orders and product manufacturing activities. From the angle of socio-technical systems, therefore, fusing both of them would find a roadmap to create a new manufacturing paradigm. Accordingly, in Fig. 1.3, we can identify the key domain the roadmap focuses on. This roadmap depends strongly on Internet-based connecting and communicating behaviors in business, covers product manufacturing activities in the whole stages of a product life cycle, and enables three types of production modes (mass production, mass customization, and individualized production).

It is obvious that following characteristics related to Internet-based connecting and communicating behaviors in business would continuously impact how to plan the roadmap:

- role-based interactions in communities,
- microlization and minimalization of resources,
- self-organization,
- virus-like propagation,
- collaboration and share,
- big-data-driven decision-making and execution, and
- distributive overall infrastructure.

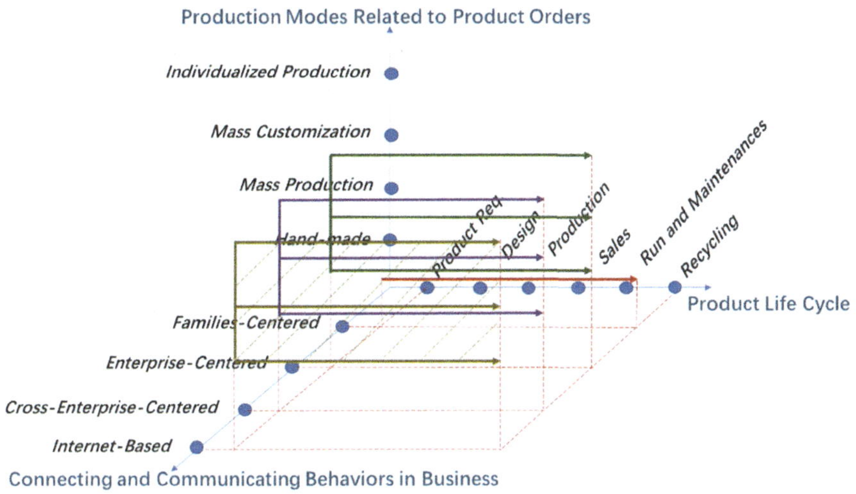

Fig. 1.3 Key domain identification of creating a new manufacturing paradigm

The descriptions related to Internet-based connecting and communicating behaviors in business just give us a basic view on socialization issue in manufacturing. In addition, digitalization, servitization, and intelligentization, together with socialization, are also key points to impact how to create a new manufacturing paradigm and run through the whole procedure of planning its roadmap.

1.2 Related Work Analysis

Construction of a new manufacturing paradigm is on the basis of a huge number of research outcomes which have been done by lots of scholars and engineers. Here, we will further analyze socialization, digitalization, servitization and intelligentization issues in manufacturing so as to guide us to propose our solution related to a new manufacturing paradigm.

1.2.1 Socialization in Manufacturing

In fact, social issues have been discussed for decades depending on the development level of social contexts and media. Gertler thought that the socio-political context surrounding machinery production in regional areas is socially distant from one another because of human beings' different backgrounds [3]. However, this starting point considering the social issues is quite different from using the concept and framework of socio-technical systems to study product manufacturing activities, which often would reveal new social phenomena [4]. For example, Kwahk and Ahn adopted the socio-technical systems to analyze the procedure of using an ERP system so as to obtain much better organizational methods for ERP applications [5]. Also through socialized ERP applications, the social issues connected closely with the manufacturing context were discussed in a BusinessWire release and a term "*social manufacturing*" was first put forward [6]. Almost at the same time, the concept of social manufacturing related only to 3D-printing-driven individualized production mode was declared as being the third industrial revolution in an Economist release in 2012. In addition, the recent development of social media or platforms on Internet, such as *Facebook*, *Twitter*, *WeChat*, etc., has changed our daily communicating manners and deeply influenced business interactions. This means that it's the time to reconsider the role of social issues in the sense of fusing with technical issues in manufacturing, especially human beings' role. This is "*socialization*" in manufacturing.

Viewing social issues in manufacturing at much larger ranges, at least following research and practical topics are concerned with the term "*socialization*":

- cloud manufacturing paradigm [7, 8],
- Application-Service-Provider-based CAX/PDM/MES/ERP usages [9],
- global and local production network [10],
- searching and using of socialized manufacturing resources [11],
- supply chain management [12],
- social business interactions [13], and
- outsourcing and crowdsourcing mechanisms [14, 15], etc.

In a word, we find that such current progresses on socialization in manufacturing have provided us many feasible and efficient references, typically in the aspects of global production network, social business interactions, crowdsourcing mechanism, etc., when we reconsider social issues in manufacturing under the promise of Internet-based connecting and communicating behaviors in business.

1.2.2 Digitalization in Manufacturing

Digitalization in manufacturing has become an indispensable premise and engineering fundamentals to any manufacturing paradigms. Along with the development of CNC technologies and CAX/PDM/MES/ERP systems, the basic digitalization related to manufacturing facilities and various workflows which deal with the whole stages of a product life cycle has been realized.

However, Internet-based connecting and communicating behaviors of human beings in business are bringing some new troubles such as unimaginable increases of manufacturing data, habit changes of using the above software, virus-like-propagation-driven business interaction methods and workflows, new embedded systems, etc. For example, APP-like software access would gradually replace multi-layer-menu-based system access, shaping of a production network would be changed in terms of new business interaction models, and embedded systems would be commonly attached to facilities for real-time monitoring and maintaining [16]. In addition, a huge amount of manufacturing data need new computing technologies to implement data-driven decision-making, controlling and optimizing. It is necessary to introduce big data and digital twins technologies to enhance the capability of digitalization in manufacturing [17, 18].

To sum up, we find that such current progresses on digitalization in manufacturing can provide us many feasible and efficient references, typically in the aspects of APP-like software model, big data, digital twins, embedded systems attached to various facilities and products, etc., when we reconsider digital issues in manufacturing under the promise of Internet-based connecting and communicating behaviors in business.

1.2.3 *Servitization in Manufacturing*

Servitization in manufacturing is often correlated closely to social issues in the whole stages of a product life cycle because any service activities happen among people under the business contexts. Traditionally, manufacturing services depend mainly on either direct services from a manufacturing enterprise or outsourced third-party services provided by partner co-enterprises from an enterprise alliance [19].

Similar to the above, the Internet-based connecting and communicating behaviors of human beings in business are also bringing some troubles for servitization in manufacturing, like changing of outsourcing service workflows, using of new crowdsourcing mechanism [20], etc.

In fact, one of biggest progresses on manufacturing services is to introduce the concept, implementing the framework and key enabled technologies of product service systems (PSS) into manufacturing reality [21]. It is inspiring that the applications of PSS make manufacturing services able to closely fuse with social issues. Especially, using socialized resources to run and maintain PSS can fit completely with the needs of socialization in manufacturing [22].

In a summary, we find that such current progresses on servitization in manufacturing can provide us many feasible and efficient references, typically in the aspects of crowdsourcing mechanism, product service systems, etc., when we reconsider service issues in manufacturing under the promise of Internet-based connecting and communicating behaviors in business.

1.2.4 *Intelligentization in Manufacturing*

Different from scientific fields like mathematics, most of product manufacturing activities are experiment-oriented and cannot be modeled very well in the form of precisely mathematical descriptions most of the time. It implies such a fact that intelligentization in manufacturing plays a very important role in realizing various functions and achieving their good performances related to each stage of a product life cycle. Here, two key intelligent handling steps which consist of manufacturing data sampling and intelligent computing for decision-making, optimizing and controlling are often used. Typically, following researches and practices have or are being done:

- configuration and runtime logic of CPS/IoT for data sampling [23], and
- construction of various manufacturing models covering different product manufacturing activities, which can be solved with correspondent intelligent algorithms, like cutting parameter optimization [24], production planning and scheduling [25], socialized manufacturing resources searching [11], social business relationship extracting [26], etc.

It must be pointed out the Internet-based connecting and communicating behaviors of human beings in business are also bringing some troubles for intelligentization in manufacturing, like changing of data sampling modes, introducing of big data, intelligent methods combined with social computing [27], etc.

In a word, we find that such current progresses on intelligentization in manufacturing have provided us many feasible and efficient references, typically in the aspects of big data and intelligent methods combined with social computing, etc., when we reconsider intelligent issues in manufacturing under the promise of Internet-based connecting and communicating behaviors in business.

1.3 Proposed Solutions for Theory and Applications

On the basis of discussions mentioned above, we believe that human beings essentially play one of the most important roles in product manufacturing. Although someone thought automation without people's participations was the future of product manufacturing, this point of view has been proved to be not correct. Nowadays, it is very clear that human-beings-centered manufacturing paradigm is being influenced deeply by means of putting emphasis on Internet-based connecting and communicating behaviors of human beings in business. It can also be seen through identifying a key domain of constructing a new manufacturing paradigm in Fig. 1.3. Here, virus-like propagation, self-organization, collaboration and share, social business interactions, crowdsourcing/outsourcing-driven value-added workflows, microlization and minimalization of enterprises, makers and correspondent working space, socialized manufacturing resource fusion and uses, etc., present new characteristics of product manufacturing in the near future. Therefore, it can be said that socialization of an advanced manufacturing paradigm is the first problem that needs to be solved. At the same time, digitalization, servitization, and intelligentization as attachments of the socialization in manufacturing would make the new manufacturing paradigm much more powerful.

As illustrated in Fig. 1.3, this new manufacturing paradigm will support product manufacturing activities during the whole stages of a product life cycle and cover different product types and production volumes. Especially, it is the best choice for individualized production.

This kind of new manufacturing paradigm is called as "***social manufacturing***" in the book and quite different from its initial concept mentioned in both *BusinessWire* and *Economist*.

1.4 Concluding Remarks

In this chapter, human-beings-centered product manufacturing activities are analyzed from the three dimensions, that is, product life cycle, product orders and connecting and communicating behaviors in business on the basis of the current manufacturing background related to how to implement Internet-based enterprises. Accordingly, the following conclusions can be drawn that:

- value-added activities covering the whole stages of a product life cycle come from suitable using of outsourcing and crowdsourcing mechanisms,
- three key factors mentioned in Sect. 1.1.3 influence how to construct a new manufacturing paradigm,
- one of the most important factors with which we can identify the key domain of constructing a new manufacturing paradigm is the Internet-based connecting and communicating behaviors in business, and
- socialization in manufacturing, together with digitalization, servitization, and intelligentization, plays the core role in constructing a new manufacturing paradigm.

The above conclusions also imply that the term "*social manufacturing*" stands for a kind of suitable presentation of the new manufacturing paradigm.

References

1. Jiang PY (2004) Chapter 1: introduction. In: Jiang PY et al (eds) E-service theory and technologies for networked manufacturing. China Science Press, pp 1–29 (in Chinese)
2. NN (2018) Smiling curve. Wikipedia. Available at: https://en.wikipedia.org/wiki/Smiling_curve
3. Gertler MS (1993) Implementing advanced manufacturing technologies in mature industrial regions: towards a social model of technology production. Reg Stud 27(3):665–680
4. Geels FW (2004) From sectoral systems of innovation to socio-technical systems: insights about dynamics and change from sociology and institutional theory. Res Policy 33(6–7):897–920
5. Kwahk KY, Ahn H (2010) Moderating effects of localization differences on ERP use: a socio-technical systems perspective. Comput Hum Behav 26(2):186–198
6. NN (2011) Kenandy delivers social manufacturing application on Force.com, bringing social, mobile, and open cloud computing technologies to global, distributed manufacturing. BusinessWire. Available at http://www.businesswire.com/news/home/20110901006181/en/Kenandy-Delivers-Social-Manufacturing-Application-Force.com-Bringing

7. Ren L, Zhang L, Wang LH, Tao F, Chai XD (2017) Cloud manufacturing: key characteristics and applications. Int J Comp Integr Manuf 30(6):501–515
8. Adamson G, Wang LH, Holm M, Moore P (2017) Cloud manufacturing—a critical review of recent development and future trends. Int J Comp Integr Manuf 30(4–5):347–380
9. Jiang PY et al (2007) e2-MES: an e-service-driven networked manufacturing platform for extended enterprises. Int J Comp Integr Manuf 20(2–3):127–142
10. Holt TV, Weisman W (2016) Global production network mapping for transforming socio-ecological systems. Curr Opin Environ Sustain 20(2016):61–66
11. Wu L, Yang C (2010) A solution of manufacturing resources sharing in cloud computing environment. In 7th International conference on cooperative design, visualization, and engineering, Calvia, Mallorca, Spain, 19–22 Sept 2010
12. Cousins PD, Menguc B (2006) The implications of socialization and integration in supply chain management. J Oper Manage 24(5):604–620
13. Aalst W, Song M (2004) Mining social networks: uncovering interaction patterns in business processes. In: Desel J, Pernici B, Weske M (eds) Business process management, vol 3080. Lecture notes in computer science. Springer, Berlin, Heidelberg
14. McCarthy I, Anagnostou A (2004) The impact of outsourcing on the transaction costs and boundaries of manufacturing. Int J Prod Econ 88(1):61–71
15. Brabham DC (2008) Crowdsourcing as a model for problem solving—an introduction and cases. Int J Res New Media Technol 14(1):75–90
16. Foster C, Graham M (2017) Reconsidering the role of the digital in global production networks. Glob Networks 17(1):68–88
17. Lee J, Lapira E, Bagheri B, Kao HA (2013) Recent advances and trends in predictive manufacturing systems in big data environment. Manufact Lett 1(2013):38–41
18. Rosen R, Wichert G, Lo G, Bettenhausen KD (2015) About the importance of autonomy and digital twins for the future of manufacturing. IFAC-PapersOnLine 48(3):567–572
19. Dickson PH, Weaver KM (2011) Institutional readiness and small to medium-sized enterprise alliance formation. J Small Bus Manage 49(1):126–148
20. Vukovic M (2009) Crowdsourcing for enterprises. In: 2009 world conference on services, Los Angeles, CA, USA, 6–10 July 2009
21. Mont O, Tukker A (2006) Product-Service Systems: reviewing achievements and refining the research agenda. J Clean Prod 14(17):1451–1454
22. Overholm H (2017) Alliance formation by intermediary ventures in the solar service industry: implications for product eservice systems research. J Clean Prod 140(2017):288–298
23. Lee J, Kao HA, Yang S (2014) Service innovation and smart analytics for industry 4.0 and big data environment. Procedia CIRP 16(2014):3–8
24. Yildiz AR (2013) Cuckoo search algorithm for the selection of optimal machining parameters in milling operations. Int J Adv Manuf Technol 64(1–4):55–61
25. He N, Zhang DZ, Li Q (2014) Agent-based hierarchical production planning and scheduling in make-to-order manufacturing system. Int J Prod Econ 149(2014):117–130
26. Bhuiyan T (2010) A survey on the relationship between trust and interest similarity in online social networks. J Emerg Technol in Web Intell 2(4):291–299
27. Sheth A, Anantharam P, Henson C (2013) Physical-cyber-social computing: an early 21st century approach. IEEE Intell Syst 28(1):78–82

Chapter 2
Social Manufacturing Paradigm: Concepts, Architecture and Key Enabled Technologies

Pingyu Jiang

2.1 Concepts of Social Manufacturing

The development history of human beings proves such a fact that product manufacturing activities must be synchronous with social, scientific and technological developments in correspondent eras, and need to run and be managed under the context of socio-technical systems.

In fact, impacts in scientific and technological developments were much more than ones in social development connected with the above socio-technical systems during the period from the first industrial revolution happened in United Kingdom of the 18th century to world-wide web creation in the early 1990s of the 20th century. Such impacts enhanced mechanization, automatization and informatization characteristics of the socio-technical systems and generated a variety of available manufacturing paradigms like mass production, enterprise alliance, etc.

However, especially in recent years, progresses on Internet and IoT/CPS technologies have been making human beings live in a data-driven connecting and communication environment they never met before. Impacts in social development related to the above socio-technical systems are being enlarged to extremes that human beings cannot imagine. Under the guidance of Internet-based connecting and communicating behaviors in business, for example, value-added manufacturing service activities covering the whole stages of a product life cycle based on outsourcing and crowdsourcing mechanisms, handling methods of product orders, etc., would deeply influence functions and performances of the socio-technical systems. It becomes the important premise to introduce social manufacturing into reality so as to transfer current manufacturing enterprises to Internet-based ones today and in the near future from the angle of not only organizational structure but also runtime logic [1].

P. Jiang, *Social Manufacturing: Fundamentals and Applications*, Springer Series in Advanced Manufacturing, https://doi.org/10.1007/978-3-319-72986-2_2

2.1.1 Definitions and Clarifications

The term "***social manufacturing***" in this book is defined as a kind of Internet-based and service-oriented advanced manufacturing paradigm covering the whole stages of a product life cycle [2, 3]. Under the consideration of the social media and the context that enable Internet-based connecting and communicating behaviors in business, this paradigm organizes socialized manufacturing resources in communities and further in social manufacturing network, makes use of product-order-driven runtime logic based on outsourcing and crowdsourcing mechanism as well as product service systems, and depends on digital supports of a type of new social-media-like industrial software model. It is also a dynamically changeable socio-technical system.

Furthermore, the following concepts have to be defined so as to explain the social manufacturing paradigm in detail.

Under the framework of social manufacturing paradigm, the term "***socialized manufacturing resources***" is defined as a kind of combination of tangible and untangible objects, which consists of sites, facilities, manpower, manufacturing capabilities, etc., has independent financial and running powers, depends on the link of specific capabilities, and is geographically distributive on sites. In a generic meaning, they can also be presented further in the form of either micro-and-small-scale manufacturing enterprises or individuals like makers, individual designers, etc., through considering their organizational structure and runtime logic. Here, the term "***micro-and-small-scale manufacturing enterprises (MSMEs)***", also called as "XiaoWei" in Chinese, indicates that their amounts of facilities and manpower are relatively smaller except that manufacturing capabilities, organizational structure and runtime are agile enough to satisfy the needs of market changes and challenges. It must be pointed out that "***manufacturing***" here stands for a generic term which is not limited to the production stage and covers the whole stages of a product life cycle. While the term "***makers***" is defined as a kind of individual who does for some creative tasks concerning new conceptual products, services or software in personalized or freestyle mode.

It has been proved that big-and-middle-scale manufacturing enterprises would make their decision-making procedures become too slow and inefficient to respond market changes quickly and agilely. As a trend, therefore, ***minimalization and microlization of manufacturing resources*** in an existing big-and-middle-scale manufacturing enterprise are very important for its agility and efficiency and can be defined as a procedure of decomposing and organizing its manufacturing resources into a group of MSMEs and individuals like makers, and at least reconstructing a transitional software platform in which correspondent web portals are used as the front-ends of original CAX/PDM/MES/ERP systems that belong to this existing big-and-middle-scale manufacturing enterprise and finally changing the current software platform to the social-media-like software model. It must be declared that this doesn't mean the existing big-and-middle-scale manufacturing enterprise is replaced with a group of completely independent MSMEs and individuals, but implies the existing big-and-middle-scale manufacturing enterprise has a new organizational structure

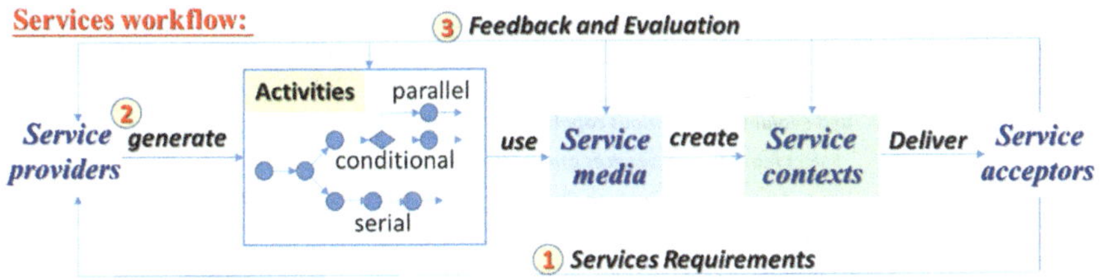

Fig. 2.1 Services workflow to demonstrate services concept

and runtime and can still dominate these MSMEs and individuals although it assigns partial powers, especially including independent financial power, to them. As a comparable reference, for example, "*Amoeba mode*" that came initially from *Kyocera* company of Japan can also do such an original decomposition and reorganization work but is not enough for social manufacturing applications because "*Amoeba mode*" mainly emphasizes collaborative tasks inside a manufacturing enterprise [4].

One of the most important characteristics of social manufacturing paradigm is service-oriented. Generally speaking, the term "***services***" is defined as a series of activities running in the form of serial, parallel or/and conditional connections, which are started by ***service providers*** who create ***service contexts*** according to services requirements, delivered these contexts to ***service acceptors*** who put forward service requirements, accept services and send feedback information if needed to the service providers through ***service media*** [5]. Figure 2.1 just shows a workflow of the services.

Under the framework of social manufacturing paradigm, several service-oriented concepts need to be declared further based on the definition of the services. These concepts include producers, customers/users, procumers, manufacturing services, manufacturing outsourcing, manufacturing crowdsourcing, and product service systems.

Here, the term "***producers***" is defined as a kind of people's role in product manufacturing activities covering the whole or partial stages of a product life cycle, who are representatives or owners of socialized manufacturing resources, run correspondent product manufacturing activities and produce products. Corresponding to the term "***producers***", the term "***customers/users***" is also defined as a kind of people's role, who buy products made by producers, use products under the support of services concerning running and maintaining mechanism of the products provided by producers or the third-party service providers. While the term "***procumers***" is defined as a kind of people's dual role, who play either producers' role or customers' one in different product manufacturing activities.

Furthermore, the term "***manufacturing services***" is defined as a kind of service concerning product manufacturing activities in the whole stages of a product life cycle. In fact, manufacturing services can be implemented in different mechanisms like outsourcing, crowdsourcing and product service systems.

The term "***manufacturing outsourcing***" is defined as a kind of productive service mechanism to implement manufacturing services in which producers/procumers who

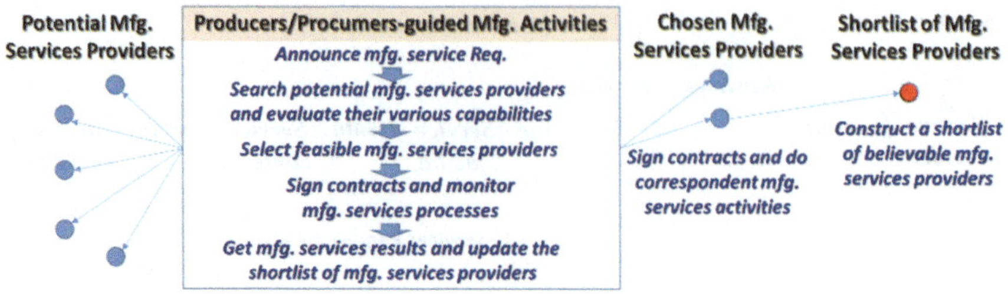

Fig. 2.2 Producers-/procumers-guided activity flow of manufacturing outsourcing services

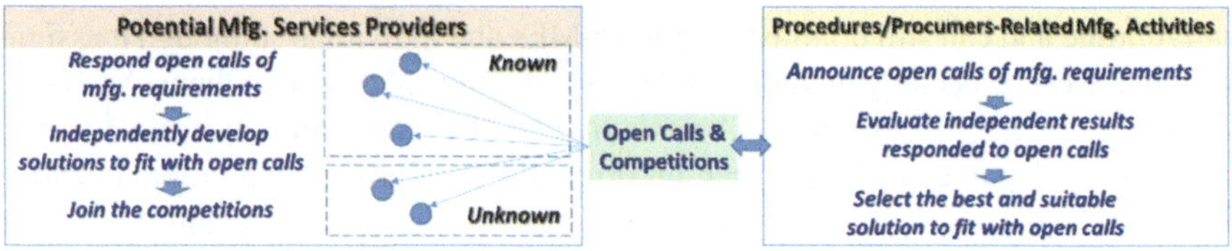

Fig. 2.3 Collaborative activity flow of manufacturing crowdsourcing services

act as service acceptors announce requirements of manufacturing tasks, search and evaluate potential manufacturing services providers, select feasible manufacturing services providers, sign contracts with manufacturing services providers, monitor manufacturing services processes, get manufacturing service results, and update the shortlist of available manufacturing service providers [6]. Figure 2.2 just illustrates such a producers-/procumers-guided activity procedure of implementing manufacturing outsourcing mechanism. It must be pointed out that manufacturing outsourcing can be done extremely by means of assigning manufacturing services providers directly from the above shortlist.

Different from the manufacturing outsourcing mechanism that depends on capability judgements in advance for producers/procumers to obtain manufacturing services, manufacturing crowdsourcing mechanism is result-driven and hopes to use a lot of known and unknown socialized manufacturing resources to respond open calls of manufacturing requirements announced by producers/procumers. Here, the term "***manufacturing crowdsourcing***" is defined as a kind of productive service mechanism to implement manufacturing services in which producers/procumers who act as service acceptors announce their open calls of requirements, manufacturing service providers who are interested in doing the correspondent manufacturing tasks independently respond these open calls and develop correspondent candidate results for upcoming competitions, and then producers/procumers evaluate all the results that are responses to open calls and select the best and suitable solution for their uses [7]. Figure 2.3 just shows a workflow of using the manufacturing crowdsourcing mechanism. What we want to emphasize here is that the contents of open calls especially in the product production stage may deal with one or several machined and assembled samples instead of providing a batch of machined parts and assembles.

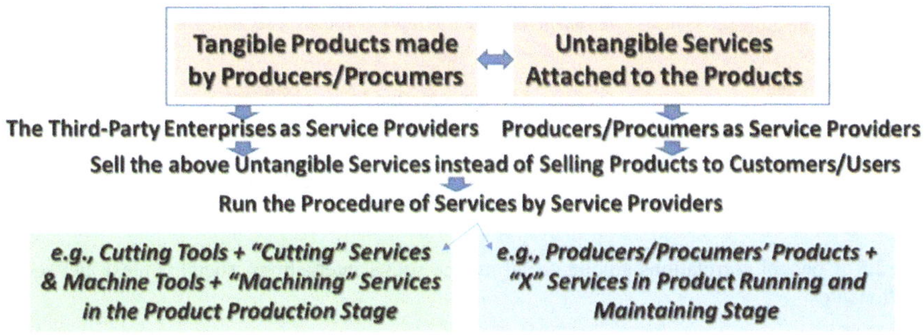

Fig. 2.4 Attributes of product service systems and their applications in the whole stages of a product life cycle

During the whole stages of a product life cycle, some of product manufacturing activities can be outsourced by means of using business models called as product service systems. On the one hand, for example, "cutting services" related to cutting tools and "machining services" concerned with machine tools can be provided by the domain-specific third-party service providers so as to decrease producers/procumers' production costs in the product production stage. On the other hand, either producers/consumers or the third-party service providers can use producers/prosumers' products as tangible basements to sell the product-related untangible services instead of selling products in the product running and maintaining stage. Here, the term "***product service systems***" is defined as a kind of service-driven business model in which untangible services are attached to tangible products and provided to customers/users by either producers/consumers or the third-party service providers, and concurrently selling services attached to the products is in place of selling the products to customers/users [8], as Fig. 2.4 shows.

Another important characteristic of social manufacturing paradigm is Internet-based, which also implies that social media, social interactions, social context, and social relationships of enabling Internet-based connecting and communicating behaviors in business play a key role in enabling this paradigm. Generally speaking, the term "***social media***" is defined as a kind of connecting and communicating toolset or platform for human beings, which includes *Facebook*, *WeChat*, *Twitter*, *QQ* and *QQ Space*, etc. The term "***social interactions***" is defined as a kind of dialog among people in terms of using the above social media. The term "***social context***" is defined as a kind of goal-driven combination of different social interactions according to specific domain-related meanings. While the term "***social relationships***" is just defined as a kind of correlative connection among role-driven agents or people, which depend on the domain-specific semantic information coming from the social interactions and social context. Here, the term "***role***" means that an agent or person standing for different attributes needs to focus respectively on different tasks. Figure 2.5 just demonstrates the correlations among the above concepts.

Under the framework of social manufacturing paradigm, several extended Internet-based social business concepts need to be declared further based on the

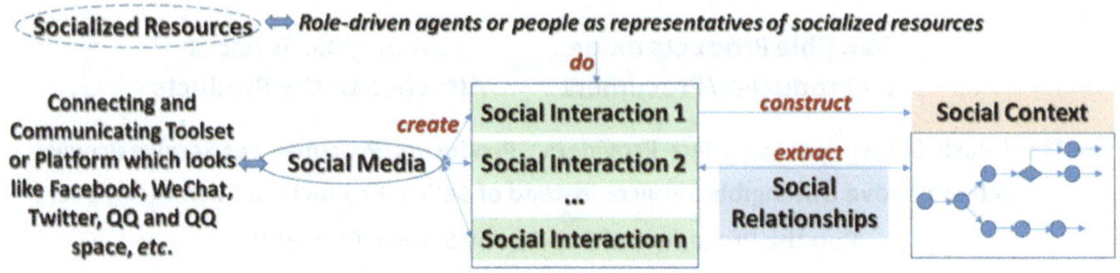

Fig. 2.5 Concept correlations among social media, social interactions, social context and social relationships

definition of the social media, the social interactions and the social context. These concepts include social manufacturing media, social business interactions, social business context, social business relationships, manufacturing communities, organizational structures of social manufacturing, social manufacturing network, and social manufacturing runtime.

Referring to the definitions mentioned above, the term "***social manufacturing media***" is defined as a kind of new social-media-like toolset or platform with which a variety of role-driven digital operations related to product manufacturing activities in the whole stages of a product life cycle, connecting and communicating behaviors, manufacturing data computing and transferring, etc., can be done correctly. The term "***social business interactions***" is defined as a kind of dialog in business under social manufacturing paradigm. The term "***social business context***" is defined as a kind of social context in business under social manufacturing paradigm. While the term "***social business relationships***" is just defined as a kind of social relationship in business under social manufacturing paradigm. Here, the role-driven producers/procumers act as representatives of socialized manufacturing resources. What we would like to mention here is that social business relationships can be represented as a classifying-tree according to focusing on different, similar, and supply-chain-like semantic information demands. Such a classifying-tree is a fundamental to generate upcoming manufacturing communities, organizational structure of social manufacturing, and social manufacturing network.

Accordingly, the term "***manufacturing communities***" is defined as a kind of domain-specific sub-network that uses socialized manufacturing resources as nodes. Here, selecting of the nodes depends on leaf or low-level nodes of the classifying-tree of the social business relationships. This means that there exist common benefits and shares among the members of a manufacturing community. In order to make any manufacturing community work well, in fact, we need to use the correspondent social manufacturing media fitting with the community. Such a specific social manufacturing media is also called as a toolset or platform attached to the manufacturing community. One kind of the toolset or platform implementation is in the form of role-driven webAPPs. It must be pointed out that all the platforms or toolsets attached respectively to correspondent manufacturing communities construct the social manufacturing media.

On the basis of the definition of the manufacturing communities, the term "***organizational structures of social manufacturing***" is defined as a kind of distributive mode of interconnecting different manufacturing communities as well as independent socialized manufacturing resources. There are two kinds of main distributive organizational structures, that is, centralized-control-like and centerless-self-control structures. Here, the centralized-control-like organizational structure means that a core manufacturing enterprise which is also a socialized manufacturing resource will govern the procedure of constructing the sustainably ecological enterprise circle, and control the most or partial power of running this circle through managing key product manufacturing activities. Haier's case just shows this situation. The centerless-self-control organizational structure means that all the related communities and independent socialized manufacturing resources are interlinked according to dynamic needs of product manufacturing activities and work collaboratively and equally. The case for producing *RepRap* open-source 3D printers on the *TaoBao* e-commerce platform just demonstrates this situation. In fact, the output of the organizational structure of social manufacturing just is a kind of configuration of generic social manufacturing network.

The term "***generic social manufacturing network***" is defined as a kind of network in which the related manufacturing communities and independent socialized manufacturing resources are dynamically interlinked according to the selected organizational structure of social manufacturing. As soon as a product order arrives, a ***product-order-based social manufacturing network***, which is called simply as ***social manufacturing network*** and oriented from the above generic social manufacturing network, is created. In this book, social manufacturing network implies a kind of product-order-based network.

In order to run both the generic social manufacturing network and the product-order-based social manufacturing network, the term "***runtime***" is defined as a kind of planning, scheduling, monitoring, tracking and tracing mechanism to run, explain and manage the correspondent network.

In fact, nodes in a social manufacturing network represent socialized manufacturing resources which respectively cover the whole stage of a product life cycle. In order to make the nodes work well in the form of data-driven mechanism and meet the needs of Internet/IoT/CPS, and Internet-based connecting and communication behaviors in business, a data sampling and computing environment has to be built. Here, the term "***social sensors***" is defined as a kind of sensor which receives, computes and transfers Internet-based interactive data among human-to-human, human-to-machine, machine-to-human, and machine-to-machine communications. Any social sensor may be attached to a node in social manufacturing network and also a social media to enable relaying functions for the virus-like propagation of social data.

The term "***extended CPS***" is defined as a kind of enhanced Cyber-Physical-System in which RFID/social sensors are attached to the original system in hardware and at least social computing modules are increased in software. Normally, an extended CPS is used as the front-end of a node in a social manufacturing network [9].

There are many different types of nodes in this social manufacturing network including designing node, production node, running and maintaining node, etc. Here, one of the most important nodes is social factory which belongs to a type of production node. Here, the term "***social factory***", actually also as a type of socialized manufacturing resource used only for finishing production tasks, is defined as a kind of either part machining or product assembling node in the social manufacturing network, in which social-data-driven product manufacturing activities focusing on the production stage of a product life cycle are planned, run and are scheduled by means of using extended CPS as the front-end of social factory.

In addition, credit and security assurance mechanism needs to cover all the product manufacturing activities in the whole stages of a product life cycle. Traditionally, contracts between socialized manufacturing resources are used as the credit and security assurances of correspondent product manufacturing activities. In this way, a contract network related to the correspondent social manufacturing network is gradually shaped according to the assigning methods of product manufacturing tasks such as outsourcing, crowdsourcing or product service systems. It is obvious that such a credit and security assurance mechanism is subjective, changeable and revisable. In order to meet the needs of Internet-based connecting and communicating behaviors in business, therefore, block chain can be used as a substitution of the contract network mentioned above. Actually, the block chain is an ideal choice as the credit and security assurance mechanism because it is objective, unchangeable and unmodifiable [10]. The term "***block chain for social manufacturing***" is defined as a kind of objective, unchangeable and unmodifiable credit and security mechanism attached to both generic and product-order-driven social manufacturing networks so as to support all the product manufacturing activities in the whole stages of a product life cycle.

The whole view of defining the concept "***social manufacturing paradigm***" that uses the above concepts can be summed up in Fig. 2.6.

2.1.2 Characteristics of Social Manufacturing

Social manufacturing paradigm, as a kind of next generation manufacturing technology covering the whole stages of a product life cycle, is closely concerned with Internet-based connecting and communicating behaviors in business. It is obvious that following characteristics would decide how far the social manufacturing paradigm goes and which kind of its roadmap is planned, and will also continuously influence how to configure and run it.

The first characteristic of the social manufacturing paradigm is microlization and minimalization of manufacturing resources. It is clear that a big manufacturing enterprise often has a rigid organizational structure in making decisions concerning its product line, human power, capitals, resources, etc. This trouble has been identified actually decades ago. One of the early solutions was "*Amoeba Mode*" that came initially from *Kyocera* Company of Japan and relied on independent financial checks. At present, a new round of macrolization and minimalization of manufacturing resources inside a very big manufacturing enterprise like Haier has happened.

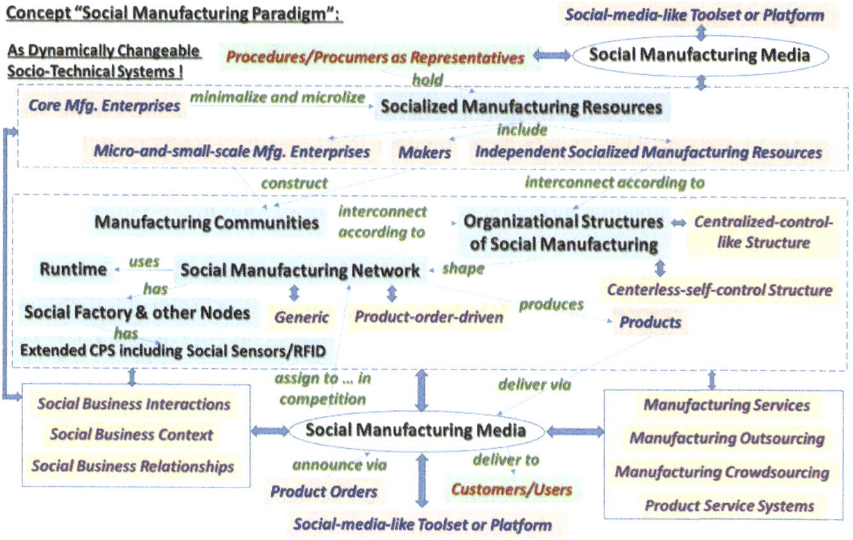

Fig. 2.6 Definition of concept "social manufacturing paradigm"

These decomposed micro-and-small-scale manufacturing enterprises, together with a variety of socialized manufacturing resources and individuals, are being integrated into a kind of new sustainably ecological enterprise circle so as to let this original big manufacturing enterprise have more flexible and agile and meet the needs of dramatic changes of markets. The obvious difference between "*Amoeba Mode*" and this round of macrolization and minimalization locates connecting and communicating behaviors in business. In fact, "*Amoeba Mode*" is a kind of organizational change limited inside a manufacturing enterprise. Of course its connecting and communicating behaviors in business are also limited in small ranges according to the roles and working tasks of "*Amoeba*" groups or units in the enterprise. While the macrolization and minimalization of manufacturing resources will create new and relatively independent micro-and-small-scale manufacturing enterprises. Their connecting and communicating behaviors in business happen not only inside a manufacturing enterprise but also across different enterprises. In addition, their roles and working tasks are basically equal to ones of the above manufacturing enterprise that runs "*Amoeba Mode*".

The second characteristic of the social manufacturing paradigm is the self-organization of socialized manufacturing resources. Generally, both socialized manufacturing resources and micro-and-small-scale manufacturing enterprises decomposed from big manufacturing enterprises will be integrated into different supply chains according to their roles and specialized capabilities to undertake working tasks. The method to reach such an integration in the context of social manufacturing paradigm just depends on Internet-based connecting and communicating behaviors in business among the above socialized manufacturing resources. Here, role-based interactions among the representatives of these socialized manufacturing resources would look for "*win-win*" points, share business benefits and avoid business risks. Depending on their role-based interactions, of course, the socialized manufacturing resources will also shape the correspondent manufacturing communities in the form of self-organization so as to obtain maximal business benefits for any manufacturing resource inside the community during either the competitions or the shares related to product orders, manufacturing outsourcing and crowdsourcing, product services, etc.

The third characteristic of the social manufacturing paradigm is virus-like propagation of organizational structure. Comparing with current Internet-based connecting and communicating behaviors of human beings by using *Facebook*, *WeChat*, *QQ*, etc., interaction relationships are dynamically grouped according to their common interests. An individual can belong to different groups and actively or passively join different groups. Such a situation looks like role-driven behaviors. Furthermore, grouping people is so fast to work like the propagation of virus. This means that the social organizational structure of human beings driven with Internet-based social media depends on people's common interests and demonstrates the nature of dynamical changes, role-driven and virus-like propagations. Similar to facts mentioned above, Internet-based connecting and communicating behaviors in business in the context of social manufacturing paradigm have the same attributes. It means that socialized manufacturing resources are grouped into different manufacturing communities on the basis of dynamical changes, role-driven and virus-like propagations of organizational structure.

The fourth characteristic of the social manufacturing paradigm is of sharing both capabilities and business benefits of socialized manufacturing resources inside a manufacturing community or among different manufacturing communities. The purpose to do as this is to avoid business risk and get business benefits in "w*in-win*" mode. It is obvious that doing business interactions in group is much better than ones in individual. Here, business interactions deal with sharing product orders, manufacturing outsourcing and crowdsourcing services, and product services inside a manufacturing community or across manufacturing communities which depend on the relationships created with the mechanism of role-driven and virus-like propagations and related to people's common interests.

The fifth characteristic of the social manufacturing paradigm is dynamically distributive infrastructure. In fact, the social manufacturing paradigm focuses on trying to create a kind of new sustainably ecological enterprise circle, which uses manufacturing communities as basic clustering blocks. The backbone of such a circle can be built by either a core manufacturing enterprise or a group of dynamically changeable manufacturing enterprises which are able to shape a supply chain through enabling different capability complementation or a big benefit-shareable cluster through integrating similar capabilities. Extremely, such a backbone can also be created on the basis of open source product philosophy. While the manufacturing communities are just utilized to accumulate socialized manufacturing resources in the forms of self-organization, role-driven and virus-like propagations, and capability and business benefit shares. It is almost impossible for any manufacturing enterprise to completely govern the whole sustainably ecological enterprise circle. It is because such a sustainably ecological enterprise circle has to be shaped as a dynamically distributive infrastructure. Partial manufacturing communities in the infrastructure can only be controlled respectively by different leading enterprises that are transferring their business into "*platform*" strategies and decomposing their organizations into micro-and-small-scale manufacturing enterprises that are also connected with their own platforms to be created. It means that the platforms act as a set of social front-end tools of enabling manufacturing communities.

The sixth characteristic of the social manufacturing paradigm is big-data-driven decision-making and performance optimization. The uses of IOT and extended CPS, together with the Internet-based connecting and communicating behaviors in business, in the context of social manufacturing would generate a huge number of datasets which are often represented as videos, images, signals, texts, and numeric data. It becomes very important to make decisions and implement performance optimizations by using such datasets. It also implies that social computing, services computing, cloud computing, intelligent computing such as machine learning, neural network, swarm intelligence algorithms, etc., which will be discussed in the next section, are all the core computing techniques to handle the above datasets.

The seventh characteristic of the social manufacturing paradigm is industrial software model to be used. Basically, the industrial software model is correspondent with the configuring mechanism and the runtime logic of social manufacturing. It is also dependent on a distributive, social-media-like and service-oriented architecture in which a lot of platforms as the front-ends of manufacturing communities are interlinked and each platform delivers the correspondent webAPPs to suitable users according to their roles. It may be predicted that traditional CAX/PDM/PLM/ERP/MES systems will be replaced with this kind of new industrial software model. It is very clear that the complexity of using the industrial software is greatly going down although its development would become more complicated.

2.1.3 Comparisons Between Social Manufacturing and Other Manufacturing Paradigms

Social manufacturing, as a kind of new next generation manufacturing paradigm, actually is a kind of service-oriented manufacturing and inherits all the natures of the service-oriented manufacturing. In order to sum up the same and different points of social manufacturing paradigm with other manufacturing paradigms, we will do an analysis from the following several catalogues:

- background and start points,
- organizational structure,
- runtime,
- resources,

- business interactions,
- spaces to enable connecting and communicating behaviors,
- main core principles to change organizational structure and runtime, and
- industrial software model.

Table 2.1 just lists some key attributes of different manufacturing paradigms like intelligent manufacturing, agile manufacturing, service-oriented manufacturing, cloud manufacturing and social manufacturing. It can be seen that the social manufacturing paradigm has an obvious difference with the other manufacturing paradigms.

2.2 Basic Architecture and Runtime Logic of Social Manufacturing

2.2.1 Basic Architecture of Social Manufacturing

The basic architecture of social manufacturing paradigm is a distributive, social-media-like and service-oriented, and works under the environment of the socio-technical system. It is easy to understand this architecture through comparing it with popular social interaction tools like *Tencent*'s *QQ* [11].

Here, a node in the right side of the Fig. 2.7 which is related to a *QQ* member stands for a socialized manufacturing resource which is presented in the form of either a micro-and-small-scale manufacturing enterprise or an individual like maker. Any manufacturing community circled with the dash line in the right side of the Fig. 2.7, which is concerned with a *QQ* group, expresses a group of socialized manufacturing resources which have common business interests and are linked with each other in "*win-win*" mode. While the linkages among nodes depend on Internet-based connecting and communicating behaviors based on social business interactions, which are comparable with *QQ* interactions happened among different *QQ* members inside a *QQ* group.

2.2.2 Basic Runtime Logic of Social Manufacturing

A good product spectrum which meets the needs of customers is the premise for an enterprise to live a long time. While the goal of any basic runtime logic of advanced manufacturing paradigm covering the whole stages of a product life cycle is to make the manufacturing efficiency increase, the various costs be cut down, and the quality of product and services become high. In order to reach the goal mentioned above, the basic runtime logic of social manufacturing in each stage of the product life cycle just follows seven steps referring to Fig. 2.8:

Table 2.1 Comparison among different manufacturing paradigms

	Intelligent Mfg.	Agile Mfg.	Service-Oriented Mfg.	Cloud Mfg.	Social Mfg.
Background and start points	Efficiency and quality enhancement through artificial intelligence	Dynamic and changeable enterprise alliance construction to response markets quickly	Value-added creation in the whole product life cycle via service-driven runtime	A kind of service-oriented mfg. via integrating and controlling socialized resources in cloud computing philosophy	A kind of service-oriented mfg. via self-organizing and using socialized resources in communities
Organizational structure	Available for any a kind of organizational structure	Core-enterprise-centered alliance and correspondent resource reorganization to response markets quickly	Service-driven resource reorganization	Organization of socialized resources in cloud philosophy	Self-organization of socialized resources in communities and their interconnections
Runtime	Either centered or decentralized mechanism	Core-enterprise mechanism	Either centered or decentralized mechanism by using IaaS/PaaS/SaaS	Cloud-platform-centered mechanism	Self-organized -community-based distributive mechanism
Resources	Any forms	Core-enterprise plus assigned resources via limited competition	Resources as services	Socialized resources organized as pools in cloud	Socialized resources in communities by using role-driven and virus-like propagations
Business interactions	Any forms	Core-enterprise-centered interactions inside the alliance	Business interactions mainly in the name of services	Business interactions mainly among cloud platform and users	Business interactions among resources inside a community and/or across communities
Spaces to enable connecting and communicating behaviors	Any spaces	Fixed space inside an alliance	Spaces covering the whole product life cycle	Cloud-media space	Social-media-based space
Main core principles to change organizational structure and runtime	Artificial intelligence techniques	Supply chain theory	Production outsourcing, crowdsourcing, product service systems, manufacturing services, etc.	Production outsourcing, crowdsourcing, manufacturing services, etc. in the form of cloud services	Production outsourcing, crowdsourcing, product service systems, manufacturing services, etc. in the form of community-driven services
Industrial software model	Any intelligent industrial software model	Traditional CAX/PDM/PLM/ERP/MES systems	Any service-oriented industrial software model	Cloud-based service platforms	Social-media-like and role-driven webAPP delivery and platforms to end-users

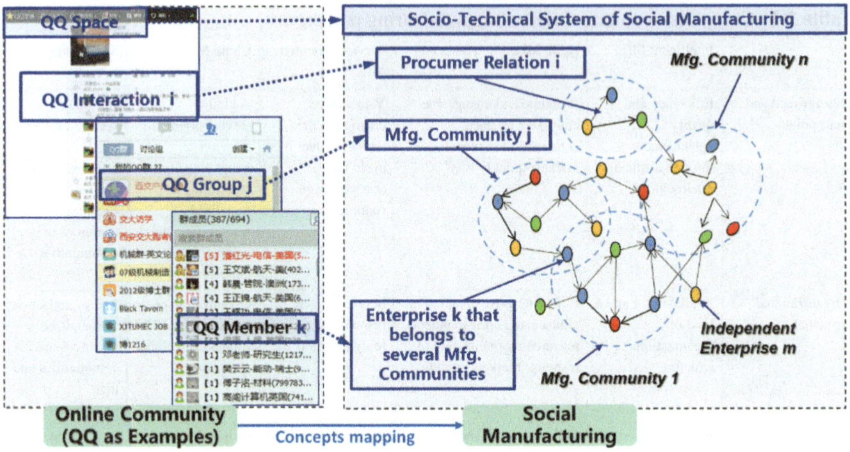

Fig. 2.7 Basic architecture of social manufacturing paradigm comparing with popular social interaction tool tencent's "QQ"

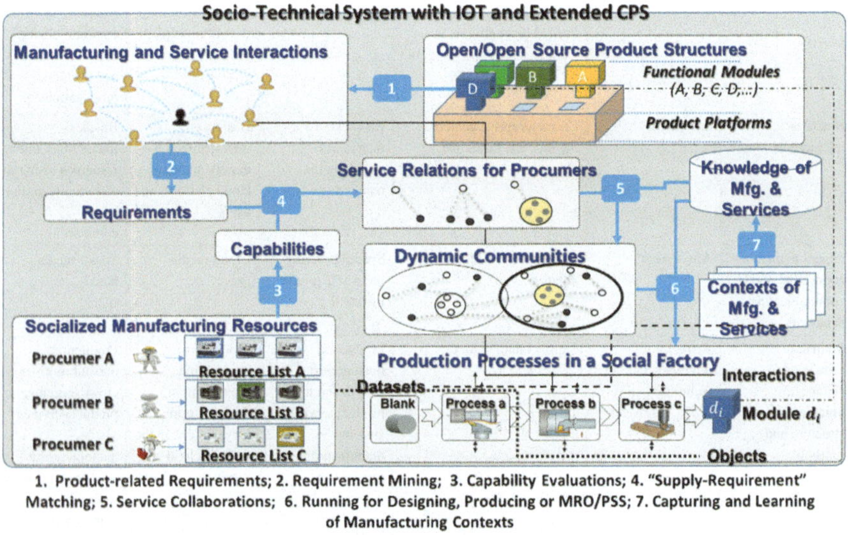

Fig. 2.8 Basic runtime logic of social manufacturing paradigm

- delivering product-related requirements,
- mining the requirements,
- evaluating the capabilities of socialized manufacturing resources,
- matching "*supply-requirement*" pairs to enable services,
- planning, scheduling and collaborating service activities,
- executing product manufacturing activities, and
- capturing and learning the manufacturing knowledge.

Here, three typical runtime procedures respectively in product designing stage, product production stage, and product running and maintaining stage are described in brief.

In the product designing stage, requirements are firstly delivered to or found by different manufacturing communities, social business interactions related to such requirements happen and then the correspondent requirements are correlated to feasible socialized manufacturing resources on the basis of mining the social business interactions in depth. At the same time, product designing capabilities of different socialized manufacturing resources are evaluated. In an iterative way of using the specific algorithms to handle the datasets from the requirements, mine the social business interactions, and get the results of the capability evaluations, useful "*supply-requirement*" pairs can be matched successfully and product designing service relationships are created. Furthermore, product designing services can be scheduled and run collaboratively under crowdsourcing and outsourcing mechanisms inside a dynamic manufacturing community and/or among several manufacturing communities after dong the service planning. Concurrently manufacturing service knowledge can be acquired from the context of product designing service activities.

Similar to the description mentioned above, the same procedures can be executed in both the product production stage and product running and maintaining stage. But the different points mainly locate that:

- requirements in the product production stage are how to finish part machining and product assembling tasks in the form of high production efficiency, low production costs, and high part and product quality. In addition, production activities are done in one or several social factories mainly in terms of using outsourcing and industrial equipment service system mechanisms, and
- requirements in the product running and maintaining stage are how to run and maintain the products by using product service system and MRO mechanisms.

2.3 Key Enabled Technologies

In order to implement the social manufacturing paradigm, at least following key enabled technologies are needed:

- self-organization of socialized manufacturing resources,
- manufacturing service relationship modeling, coordinating and generating,
- open product design and crowdsourcing design,
- RFID, social sensors and extended cyber physical systems,
- social factory and interconnections,
- product service system for social manufacturing, and
- credit and security mechanism, and IP protection under social manufacturing.

These key enabled technologies will be described briefly in the following subsections and upcoming chapters in detail.

2.3.1 *Self-organization of Socialized Manufacturing Resources*

Along with the increasing product and service demands in finer-grained markets, many micro-and-small-scale manufacturing enterprises (*e.g.*, micro-and-small-sized firms, start-ups, small factories, workshops, and even individuals) spring up with socialized manufacturing resources (SMRs) and provide specialized manufacturing services for prosumers. For example, many SMRs possessing the characteristics of decentralized, self-adaptive, and self-organization begin to cluster as manufacturing community for providing specialized manufacturing services in the different stages of a product life cycle.

Initially, SMRs self-organize them to perform various specialized product-related or production-related manufacturing service capabilities to satisfy customers' requirements. This brings advantages of flexibility and responsiveness over traditional core manufacturers-centered type. Gradually, to achieve bargaining power competitiveness and collaboration efficiency during the production interaction and collaboration with core manufacturers, SMRs with similar interests and capabilities begin to aggregate into dynamic manufacturing communities through social networking and sharing to organize their capabilities autonomously. As shown in Fig. 2.9, extremely, the self-organization mechanism makes community-based decentralized manufacturing resources have become a new shape of social manufacturing network that is different from traditional platform-driven centralized management [12].

This self-organization shape of social manufacturing network calls for new decision-supporting methods and coordination technologies. Firstly, SMRs should be defined and encapsulated into the manufacturing communities and social manufacturing network, so as to depict production capability that prosumers provide to accomplish the parts or product manufacturing tasks from product manufacturers. It helps to achieve value-added manufacturing services and rapid dynamic responses to market. Secondly, based on the established connection via social-media-like connecting and communicating behaviors, various communities of SMRs are formed as complex, dynamic autonomous systems to co-create products and manufacturing services especially for mass individualization mode. Since a disorganization of these decentralized SMRs will hinder the coordination decision-making and result in an inferior position to prosumers when bargaining with the core manufacturer, a better matchmaking and coordination mechanism should be developed for shaping of SMRs network. Thirdly, the group intelligence in decision making should be addressed for organization of SMRs in the product life cycle. With these technologies, SMRs can thrive and reorganize themselves into more-effective interacting and collaborating shapes to accomplish the whole product manufacturing activities driven by different demands respectively related to mass production, mass customization, or mass individualization.

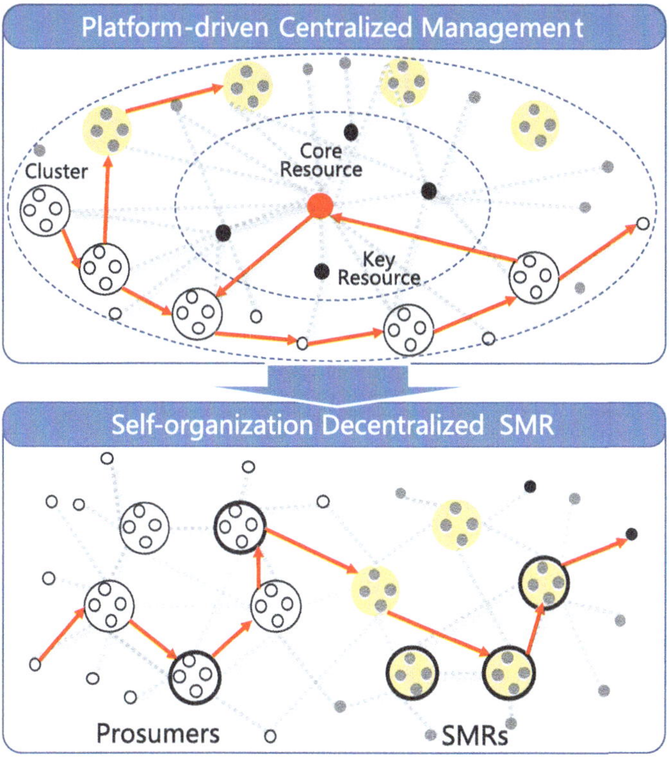

Fig. 2.9 Self-organization of socialized manufacturing resources

2.3.2 *Manufacturing Service Relationship Modeling, Coordination and Generation*

The above demands on product and manufacturing services have driven producers/prosumers, who either own the correspondent socialized micro-and-small-scale manufacturing enterprises or are individuals like makers, clustering into various manufacturing communities. Lots of producers/prosumers spring up to provide product-services to satisfy customers' requirements. Traditional giant product manufacturers are becoming dumbbell-shaped and build social business relationships with producers/prosumers. The product manufacturing activities are not limited in the core product manufacturers but in multiple producers/prosumers, forming a product-oriented social manufacturing network. Unfortunately, limited by the lack of an effective manufacturing service platform that integrates SMRs and shares manufacturing information and capabilities through a distributed credit and security mechanism, most of producers/prosumers are suffering from the dilemma that only manufacturing services provided by few dependable partners in the same area can be obtained without trembling.

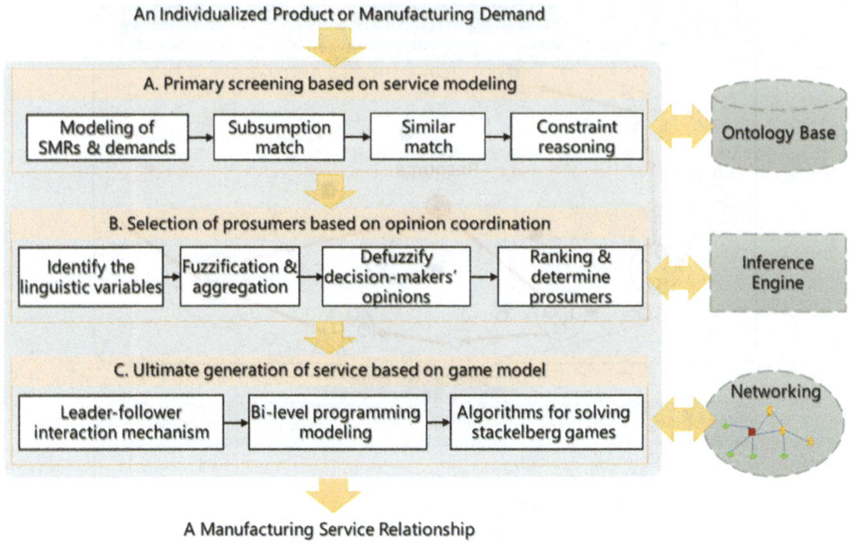

Fig. 2.10 Procedures of service modeling, coordination and generation

Outsourcing and crowdsourcing services related to specialized product manufacturing activities are becoming one of the most significant manufacturing services and value-added activities during the collaborative production procedures. It is necessary to have a decision-supporting tool to model, coordinate and generate manufacturing services. As shown in Fig. 2.10, the specific implementing procedure is correspondent with several steps.

With respect to the modeling of manufacturing service relationships, as producers/prosumers are in large amounts, manufacturing service interactions among them become vital and need to be well managed, because they affect customer experiences and expectations much. In fact, manufacturing service interactions derive from the concept of services and are defined as "*the direct interactions between service providers and customers to provide customers with timely and relevant information to enable them to make informed decisions, complete their work easily, and co-create added value*" [13]. The modeling method for manufacturing service interactions are always empirical and lack a unified graphical way to solve it.

With respect to the coordinating of manufacturing services, through online public platforms and interfaces of enterprise information systems, producers/prosumers could achieve real-time interaction and communication information such as collaborative design, progress monitoring, quality feedback and so on. Thus, online-offline integrated manufacturing service coordination among producers/prosumers should be comprehensively considered. With different product order coordination strategies on producers/prosumers, different manufacturing communities are established and there are more than one communities can meet the performance requirements of customers. The production cost and delivery time are chosen to represent the performance of the manufacturing community.

With respect to the generating of manufacturing services, the purpose is to provide a systematic solution for producers/prosumers to integrate the manufacturing with services so as to realize the added-value of product manufacturing activities. Through the intelligent integration of SMRs, manufacturing service matching and finding, manufacturing service running and monitoring in a certain manufacturing community, the producers/prosumers have a better product order relation and creditworthiness to the manufacturing service acceptors. While massive producers/prosumers relationships and manufacturing service relationships exist in diverse social media, which integrates and virtualizes plenty of SMRs, aggregates producers/prosumers into manufacturing communities by recommendation and self-organization to provide manufacturing services, promotes intelligent business and all-around product order management by using social networking tools [14].

2.3.3 Open Product Design and Crowdsourcing Design

As with other manufacturing paradigms, product design for social manufacturing is one of the most important starting points. Here, open product design is often used as a kind of typical methodology to enable the correspondent product manufacturing activities dealing with the designing stage of a product life cycle and also influence deeply the runtime logic in the other stages. It is clear that such a design methodology is a kind of service-driven mode and quite different from traditional one in many aspects, including idea creations, designers operations, designing activities, design workflow, design result evaluations, etc.

In fact, open product design is a kind of generic design thinking methodology which meets closely with the needs of social manufacturing paradigm. From the angle of products themselves, both open-source and closed-source products can be designed under the open product design environment. Here, closed-source products imply that the intellectual property (IP) of the products belongs to either manufacturing enterprises or individuals. From the angle of design service philosophy which is also the baseline of the open product design, both crowdsourcing and outsourcing mechanisms are often used for supporting the open product design activities and workflow. On the basis of descriptions mentioned above, a kind of open product design method powered with crowdsourcing design services is studied in depth in the follow-up chapter. Some core problems and the correspondent key enabled technologies are summed up as follows.

The first problem that needs to be solved is designer roles in social manufacturing. Under the context of social manufacturing, on the one hand, open product design is able to openly integrate tremendous socialized designers, actively participate into an interactive product design process, and organize these socialized designers into design communities (DCs) concerning customer-centric markets. On the other hand, customers can also join open product design activities if needed. It leads to the situation that the role of the above customers is changing swiftly to prosumers who will be involved into correspondent design activities. The role changing is very important for the further understanding of all the concepts related to this proposed open product design method integrated with the crowdsourcing mechanism.

The second problem that needs to be solved is to configure the conceptual architecture of crowdsourcing-driven open product design services and run the open design activities under the support of the correspondent design service workflow and key enabled technologies. As mentioned before, the whole open product design workflow is different from traditional one. Here, the conceptual architecture is dependent on a self-organization-driven and service-based DC which consists of socialized designers, customers, traditional designers who belong to specific manufacturing enterprises. Depending on types of products to be designed, that is, open-source or closed-source products, extremely, the DC is only composed of socialized designers besides customers as designing participants. The design service workflow is constructed in the form of activity diagrams and presents the idea of both competitions and shares. In order to enable the above conceptual architecture, three key enabled technologies, that is, customer requirements analysis, design community construction and design result evaluations, need to be studied in depth.

Customer requirements analysis for crowdsourcing-driven open product design is the first key enabled technology. As with other design methods, open product design activities are also carried out based on the customer requirements (CRs). Due to its specific nature, however, the CR acquirement under the environment of open product design is hugely different from traditional one at least in the aspects of formal description, analyzation, and decision making. Hence, the specific techniques, including deep learning, random forest algorithm, etc., are often used for solving the problems related to the CR acquirement so as to make sure that the correspondent solution can be fused into the self-driven and self-organized design service activities and workflow.

Shaping and analyzing for design community is the second key enabled technology. The designers in an open product design environment deal with a large number of socialized designers, designers who belong to specific manufacturing enterprises, and customers who are interested in participating some design tasks. They are from different fields and with different backgrounds. Not all the designers have enough design abilities for the design tasks and they don't even know each other before they join in the same DC. Therefore, deep learning and complex network analysis theory are used for solving some problems related to community generation, operation and management, decision making, communication and interaction, trust and security issues, etc.

Evaluation for crowdsourcing-driven open product design outputs is the third key enabled technology. The open and free characteristics of open product design bring it with both advantages and disadvantages. One of the most crucial disadvantages is unreliable. Hence, three basic assessment approaches are proposed to provide initial support for open product design practices, including DC resilience evaluation, project feasibility evaluation, and design result evaluation.

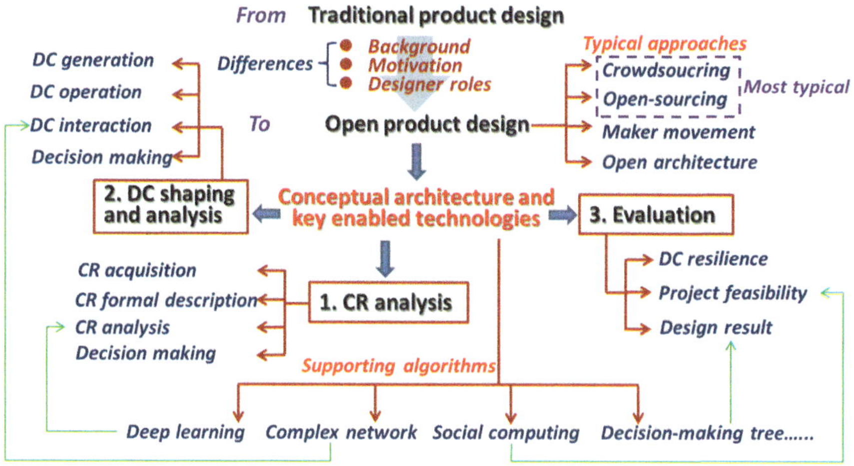

Fig. 2.11 Concept architecture and the correspondent key enabling technologies for open product design

The conceptual architecture and the correspondent key enabled technologies for open product design can be illustrated in Fig. 2.11.

2.3.4 RFID, Social Sensors and Extended Cyber Physical Systems

RFID, social sensors and CPS play a significant role in ubiquitous interconnections, manufacturing and business interactions, and mass collaborations that belong to the fundamental requirements for social manufacturing paradigm [15, 16]. They are also key enabled technologies to implement the big-data-driven decision-making and performance optimization covering the whole stages of a product life cycle in the context of social manufacturing. Here, the relationship among RFID, social sensors, and extended CPS can be illustrated in Fig. 2.12.

In fact, RFID is mainly used for detecting and reacting the dynamic changes of state, time, position and other attributes of physical objects attached with RFID tags, and ensuring that visibility and controllability of production and logistics processes in both the intra-enterprise level and the inter-enterprise level can be achieved. Typically, such physical objects include rude materials, work-in-progress(WIP), finished parts/assemblies/products, various physical resources, etc. There are four types of basic RFID application scenarios which are described respectively with four types of enhanced state blocks so as to present the behaviors of RFID-tagged objects inside a limited controllable space. Furthermore, these enhanced state blocks are developed for constructing an RFID-based event-state-position graphical deduction

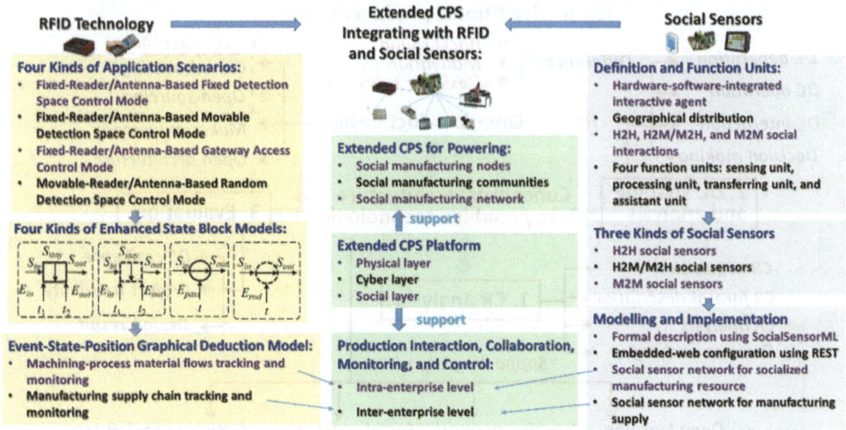

Fig. 2.12 Correlations among RFID, social sensors and extended CPS

model (rfid-GDM) so as to track and monitor WIP material flows, supply chains and other various processes in the context of social manufacturing. Here, the construction procedure of rfid-GDM includes four steps, that is, multi-granularity process decomposition, automatic events generation, RFID-enabled process flow modelling, and final rfid-GDM construction. It is clear that the rfid-GDM provides a guideline to help us to deploy RFID devices, and track and monitor a variety of processes in both the intra-enterprise level and the inter-enterprise level.

Social sensors are mainly used for sampling, computing and transferring datasets from social business interactions in the context of social manufacturing. With the help of SocialSensorML which is an improved SensorML-based description language, accordingly, we develop a social sensor model which can be seen as a type of hardware-software-integrated interactive agent, consists of sensing, processing, transferring and assistant units, and deals with enabling four types of the communication and interaction modes, that is human-to-human, human-to-machine, machine-to-human and machine-to-machine. Here, the sensing unit captures the input dataset which deals with physical sensor data from the embedded and add-on sensors, human interactive data from the human's texts, signals, images, voices, gestures, etc., and crawls data from other social and physical sensors. The processing unit merges the above input dataset into machine-readable format. The transferring unit transfers the formatted data to other sensors or cloud database for sharing or further processing. The assistant unit provides the assistant services such as power, screen, storage cache, etc. In addition, social sensors can be enabled physically through tiny embedded webservers, like *Raspberry Pi*, with the *REST* architecture. Through invoking *REST* operations, the functions of social sensors, such as crawler, chatbot and data computing, etc., are reached. Furthermore, social sensors can be added into the CPS.

Taking into the consideration of social manufacturing philosophy, both RFID and social sensors will be integrated into the CPS to meet the needs of social business interactions. The new CPS, also called as extended CPS, plays a role of the hardware-software-integrated mediators that have capabilities of perception, communication, interaction and control feedback for the autonomous operations related to not only single machine but also multi-machine collaborations. Here, an extended CPS node model, which includes RFID, sensors, actuators, human-machine-interface (HMI), network, etc., is developed and attached only to a single machine or facility. Similar to the social sensor model, this node model is configured with an URL, and different extended CPS nodes can interact with each other for autonomous business coordination. It should be pointed out that RFID devices can be viewed as special sensors so as to monitor and track a variety of processes and social sensors act as specifically function-added HMIs so as to enable four types of communication and interaction modes.

In this way, a three-layer extended CPS framework is built, which includes physical layer, cyber layer and social layer. Corresponding with the social manufacturing network, an extended CPS network will also be shaped by using lots of extended CPS nodes.

2.3.5 Social Factory and Interconnections

As mentioned above, social factory is a kind of production node used for finishing either part machining or product assembling tasks in the context of social manufacturing [17]. The interconnections among all the social factories will shape a sub-network of social manufacturing network. This sub-network is a dynamically changeable production environment for producing a product, a batch of products or a product spectrum. As the same as the traditional manufacturing factory, social factory still requires itself being flexible, reconfigurable and intelligent besides presenting its Internet-based connecting and communicating behaviors in business through various mechanisms related to social business interactions and transferring, autonomous product order sharing, extended-CPS-driven dataset handling, etc.

Generally, a product order is firstly decomposed into sub-orders which depend on the product bill of materials (BOM) in the level of sub-network mentioned above. And then one or some of the sub-orders together with customer requirements are respectively obtained and shared by a group of suitable social factories on the basis of using outsourcing and crowdsourcing service mechanisms. In this way, a group of social factories build the ubiquitous connections with each other and customers, and make the product order production process transparent to the whole social manufacturing network in the form of either competitions or shares. This situation is based on the interconnections of social factories.

Only for inside a social factory, furthermore, there are two basic things to need to do too. One is to finish the sub-orders it gets and another to share its own unoccupied capability temporarily and autonomously with the outside production nodes if needed. These two things still need the support of the above mechanisms related to social business interactions and transferring, autonomous product order sharing,

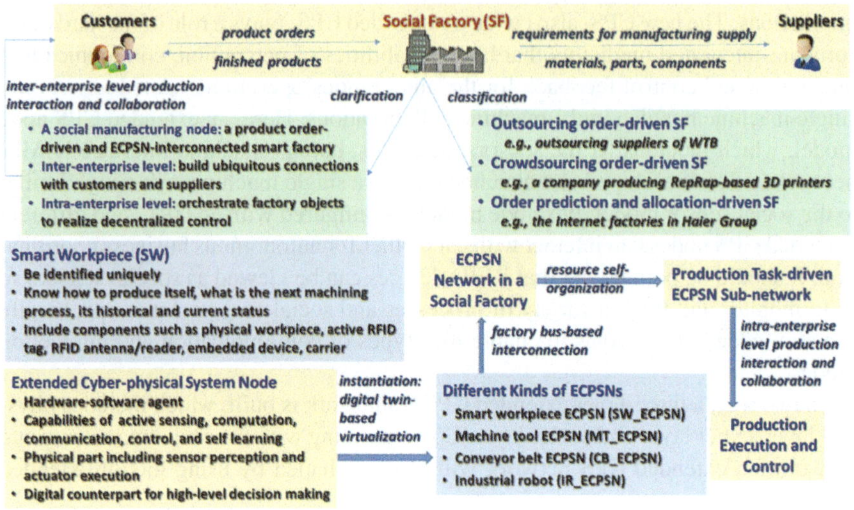

Fig. 2.13 Key concepts of social factory and their correlations

extended-CPS-driven dataset handling, etc. In fact, extended CPS node (ECPSN) and the correspondent ECPSN-network inside this social factory is the basic middleware to enable the above mechanisms. Under the control of them, as shown in Fig. 2.13, the expected/unexpected production events are captured and handled in real time, the production decisions are made concurrently, and humans, smart machines and workpieces/assemblies, physical processes, and software systems are orchestrated to fulfil the production tasks via the ubiquitous interconnections, social business interactions, and mass collaboration.

It must be emphasized that ECPSN and the correspondent ECPSN-network make it possible to capture and handle a huge number of data. Especially an ECPSN is often used as a kind of hardware-software agent attached to a single machine or facility, this means that ECPSN equips the machine or facility with capabilities of active sensing, computing, communicating, controlling, and self-learning. In order to model the phenomenon, a digital twin referring to the ECPSN and the attached machine/facility is created. The digital twin model of the social factory can also be shaped in terms of interconnecting all the ECPSN digital twins. By using digital twin technology, the physical space and cyber space are seamlessly connected. For example, the physical ECPSN updates the current status to its digital twin hosted in the embedded device when expected/unexpected production events occur. The digital twin handles the sensory information and then sends commands to the physical space for feedback control. To enhance the intelligence and autonomy of the ECPSN, some intelligent algorithms or reasoning methods such as machine learning algorithms, rule-based reasoning, and event trigger conditions are integrated into its digital twin. The input data of this ECPSN come from the physical sensors, humans, and other ECPSNs. The output data of this ECPSN, which are also from its digital twin, include production commands, interaction information, and manufacturing data that can be sent respectively to the physical facilities, other ECPSNs, and backend database.

Different types of ECPSNs, such as ECPSN for smart workpiece (SW_ECPSN), ECPSN for machine tools (MT_ECPSN), ECPSN for conveyor belt (CB_ECPSN), and EPSN for industrial robot (IR_ECPSN) are instantiated with the digital-twin-based virtualization. Here, smart workpiece is a special ECPSN, can be uniquely identified, knows how to produce itself, knows its next machining process, knows its historical and current status, and can communicate with other ECPSNs for dynamic and decentralized production control. All the ECPSNs are interconnected via factory bus and form an ECPSN-network. Driven by the specific production tasks, the partial ECPSNs in this ECPSN-network self-organize into a sub-ECPSN-network that is dynamically formed and resolved. Within the sub-ECPSN-network, operators, managers and different ECPSNs interact and collaborate autonomously with each other to realize the production execution and control related to the above specific production tasks.

2.3.6 Product Service System for Social Manufacturing

Currently, it has become a common sense to use services attached to specific products for getting more business profits and satisfying customers' requirements. Product service system (PSS) is just a sustainable and powerful tool for reaching such a goal. It also means that manufacturing enterprises where want to enable the PSS configuration and runtime need to change their business model firstly. However, transforming to product-service providers (PSPs) would face various challenges and obstacles such as internal barrier inside an enterprise, increasing of labor costs and service complexity, customers' culture and misapprehend and so on. In order to respond the challenges, we integrate the social manufacturing paradigm with the PSS philosophy to realize a kind of collaborative, flexible and socialized PSS which is also called PSS for Social manufacturing.

PSS for social manufacturing is a kind of order-driven and multi-PSP-based service mode in which socialized service resources (SSRs) are self-organized into service communities to collaboratively finish all the tasks related to the same PSS order. The runtime logic and the key enabled technologies are illustrated in Fig. 2.14. Here, four enabled technologies include service capability modeling, service flow modeling, service monitoring and scheduling, and service quality evaluating [18]. These four enabled technologies cover the whole running processes of any PSS for social manufacturing, including service contract establishing (service capability), product-service providing (service flow), service process controlling (monitoring and scheduling) and service resulting (service evaluating).

The runtime logic of PSS for social manufacturing can be described in detail as follows. The distributed PSPs release their SSRs to cluster into different service communities by the similarity of service capabilities. The service communities match with the customer requirements and suitable communities are selected to establish a

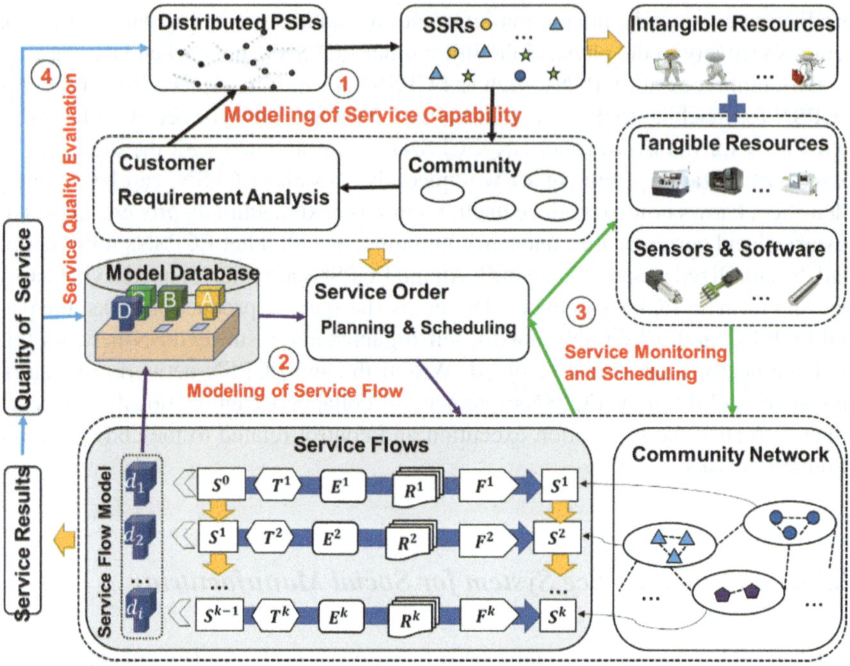

Fig. 2.14 Runtime logic and key enabled technologies of PSS for social manufacturing

service contract with the specific customers. According to the contract, the service flows are modeled with the details of service triggering condition T^k service event E^k socialized service resources R^k and the service flows F^k own in Fig. 2.14. On the basis of the service flows, PSPs provide their product-services to the customers. In order to provide lean services, the service contents of a PSP are scheduled and the service flows are also monitored. Here, the embedded sensors and corresponding software are developed to schedule and monitor the PSPs so as to control them to stick with the modeled service flows. After completing the product-services, the service results should be evaluated with the evaluation criterions so as to determine the service quality and efficiency, and act as a reference for future services. Supporting by four key enabled technologies mentioned above, the core functions of PSS for social manufacturing are realized.

2.3.7 *Credit and Security Mechanism, and IP Protection Under Social Manufacturing*

Social manufacturing network is a distributive network in which the adaptation toward partial and complete decentralization begins to make the system more robust, flexible, secure and efficient. Therefore, the credibility, security and intellectual property (IP) protection of product manufacturing activities under the environment of cross-enterprises are very difficult to ensure because there are no trusted third parties to supervise them.

The credibility, security and intellectual property protection is the foundation of mass collaboration for cross-enterprises. On the one hand, crediting the competence and honesty of socialized manufacturing resources (SMRs) is a vital factor for prosumers to select trustworthy SMRs to cooperate with. And SMRs must have the credit and security assurance they can afford for their manufacturing services. On the other hand, the credit and security assurance involves the profit distribution among SMRs. Apparently, the establishment of credit and security mechanism is an important aspect for implementing cross-enterprise product manufacturing activities under the context of social manufacturing, which is crucial to keep manufacturing operations smoothly and reliably. Generally speaking, the credit and security establishment among SMRs and customers may depend on the trusted third party. However, there is a gap among SMRs and customers in establishing the initial credit and security assurance because they know nothing about each other. Hence, how to build and guarantee the credit and security of SMRs becomes an important problem. At the same time, the increasing personalized demands require a lot of cross-enterprise product manufacturing activities to achieve high manufacturing flexibility. This requires more trustworthy technologies to guarantee the balance of profits among SMRs and the quality of such activities.

The credit and security mechanism acts as the trusted third party to prevent defaults and frauds during manufacturing service interactions, and bridges the credit and security gap among SMRs. Customers, as initiators of product orders, must publish reasonable crowdsourcing/outsourcing demands and bear the manufacturing service charges as contracts. SMRs are the providers to satisfy the crowdsourcing/outsourcing demands, and their processing capability, product quality, product delivery time must be reliable. A distributive social manufacturing platform can be used as the social media to make the credit and security assurance operate smoothly and responsibly, including supervising product manufacturing activities, rating the credit and security grades of SMRs based on their historical transaction data, balancing interests among SMRs and so on.

On the basis of credit and security requirements during the mass collaboration among SMRs and customers, we find that blockchain technology has a huge potential to address the interoperability challenges. Under the blockchain-based credit and security mechanism, as shown in Fig. 2.15, a unified agreement of economic incentives and participations in the form of stable consensus of blockchain can be established among SMRs and customers [10]. The consensus process of the blockchain is a crowdsourcing process of manufacturing tasks. It is an important prerequisite for smooth operations of the blockchain how to design a reasonable crowdsourcing mechanism so as to facilitate SMRs to initiatively manufacturing supply services. Also, the smart contracts will be triggered and the interests distribution will be implemented automatically based on the service tasks of the SMRs, which greatly reduce profit disputes among the SMRs. Here, participants draw their own smart contracts referring to the correspondent interests distribution and use a voting mechanism based

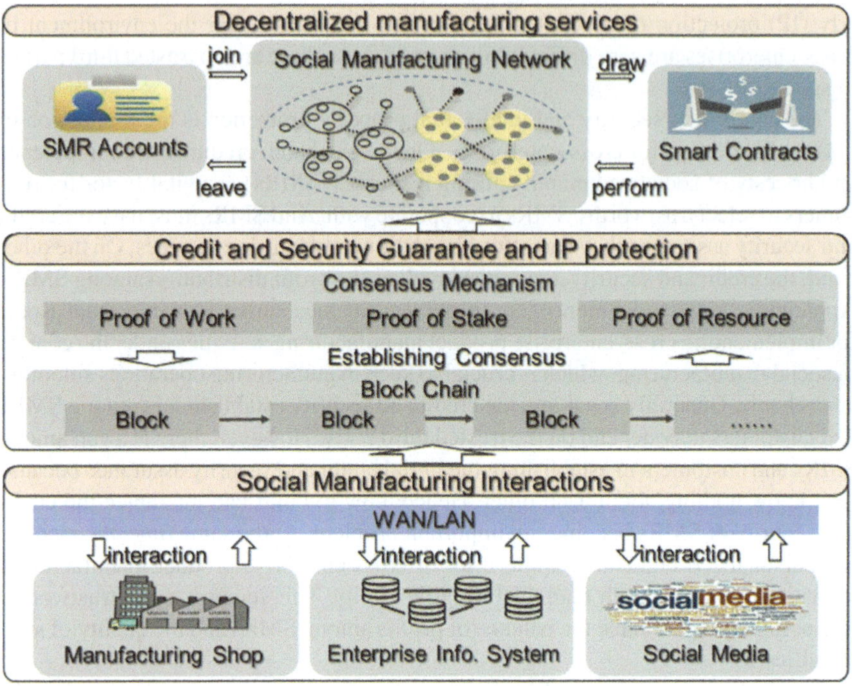

Fig. 2.15 Blockchain-based credit mechanism for IP protection

on Proof of Work or other proofs. Once a SMR accomplishes its manufacturing task, it publishes a proposal about manufacturing information, and other members will verify the proposal. In this way, a new blockchain-driven credit and security mechanism for social manufacturing can be reached.

2.4 Computing, Decision-Making and Evaluating in Social Manufacturing

It is necessary to use a number of computing, decision-making and evaluating methods to support all the product manufacturing activities covering the whole stages of a product life cycle in the context of social manufacturing. Here, typical methods include:

- reasoning, machine learning and neural networks,
- swarm intelligence, genetic algorithm and immune algorithm,
- decision-making algorithms like random forest algorithm,
- social computing and complex network analysis, and
- services computing including cloud computing.

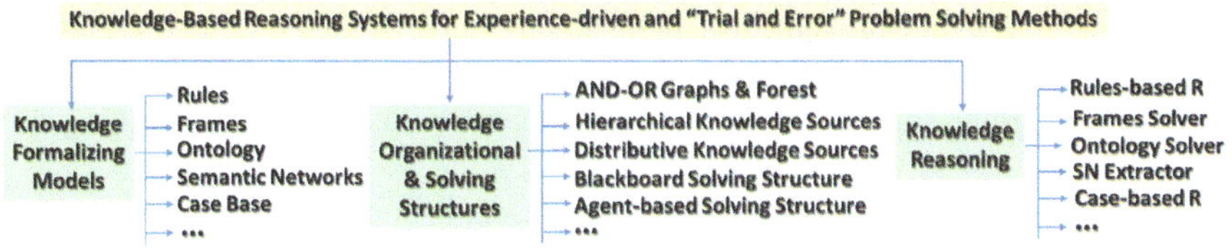

Fig. 2.16 Symbol-based knowledge expressions, reasoning and problem-solving structures

2.4.1 *Reasoning, Machine Learning and Neural Networks in Social Manufacturing*

There exist lots of experience-driven and "*trial-and-error*" problem-solving methods in manufacturing engineering. It has also no exception to social manufacturing. This means that a variety of explicit and implicit manufacturing knowledge and datasets are accumulated from time to time during the above problem solving procedure. In general, such a kind of knowledge is often presented as semantic contexts which stand for the symbol abstraction of manufacturing experiences. While the correspondent datasets are concerned with either labeled "*input–output*" pairs or unlabeled "*input*" datasets which respectively illustrate some kinds of either classifying or clustering relationships among manufacturing status, entities or attributes in the form of "*black box*". Furthermore, the results of these classifying and clustering relationships can be used for predicting changes of the correspondent manufacturing status, entities and attributes in the future.

In order to acquire and use the symbol-based knowledge from manufacturing experiences during problem solving, on the one hand, it is very important to choose a suitable way to express and organize the formalized knowledge, and then use it by means of reasoning. Figure 2.16 just shows some useful symbol-based knowledge expression models which deal with rules, frames, ontologies, semantic networks and case bases, organizational and solving structures which are concerned with AND-OR graphs, hierarchical or distributive knowledge sources, blackboard solving structure and agent-based solving structure, reasoning mechanisms which mainly include foreword-based rule reasoning, frame and ontology solvers, case-based reasoning algorithms, etc.

Looking those back to the social manufacturing paradigm, we know that typical symbol-based knowledge deals with:

- carrying out various either "*if-then*"-based or "*cause–result*"-driven judgements,
- describing a huge number of either correlative "*entity-attribute-value*" relationships such as products and socialized manufacturing resources or manufacturing contexts like assembly and logistic flows,
- formalizing and reusing successful and unsuccessful cases from both casebase and FEMA database of an enterprise node, etc.

Here, "*if-then*"-like judgements can be mainly implemented with rule-based reasoning systems, entity-centered relationships and social manufacturing contexts with ontology and frame solving systems, and case reuses with case-based reasoning systems.

On the other hand, a variety of original datasets in social manufacturing can be sampled in the forms of videos, images, signals in either time sequence or frequency spectrum, text data and numeric data. Typical examples can be found as follows:

- capturing on-site monitoring videos such as videos related to socialized manufacturing resources and collaborative work,
- taking real-time images such as images from monitoring wearing conditions of cutting tools,
- recording real-time signals such as continuously electrical power consumption curves of machine tools, time sequence signals for fault diagnosis,
- collecting a large number of text data such as social interaction texts in business,
- sampling, measuring or collecting numeric data such as a huge number of data from sensors/RFID in extended CPSs or an industrial IoT, quantity-changeable data from product-related measurements in which either big or small samples are obtained depending on product production volumes, etc.

The above original datasets need firstly to be pre-handled in terms of using either outlier tests in statistics or experience-driven hand-made deletions, which are concerned with the types of the datasets. Unfortunately, this task is very complicated because of the complexity of the datasets.

Furthermore, the unlabeled datasets which only have "*input*" data can be used for clustering, that is, automatically for generating the "*input-output*" relationships by means of using unsupervised learning models like self-organization mapping (SOM) neural network. While the labeled datasets which have certain "*input–output*" relationships can be used for either classifying all the types of the datasets with the help of supervised learning models like back propagation neural network, deep convolutional neural network specifically for very big samples and supported vector machines (SVM) models specifically for small samples, or regressing the formulas of the numeric datasets with the help of statistics. Figure 2.17 just lists some available regression, machine learning and neural network models.

It must be mentioned here that it is very difficult to acquire labeled datasets which are presented in videos, images and signals in the context of social manufacturing. We often use the hand-made method to mark a label for a video segment, an image, a signal fragment, or a text sentence.

It is also be pointed out that lots of datasets from the context of social manufacturing have "*Volume*", "*Velocity*", "*Variety*" and "*Veracity*" characteristics of big data. Therefore, we can use big data analytics including a variety of computing models and algorithms to handle the above datasets. Figure 2.18 just lists some available modeling techniques and algorithms.

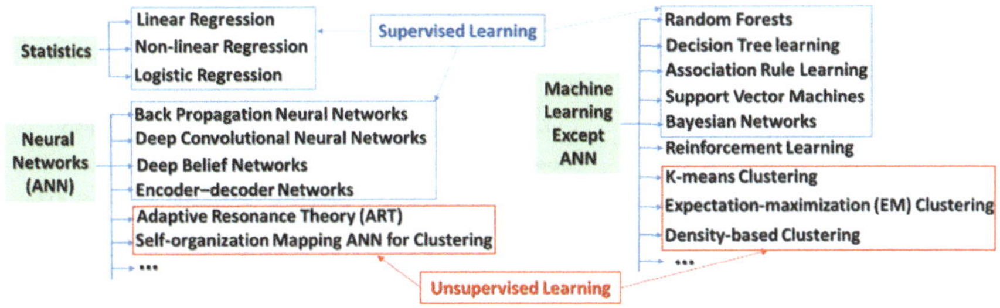

Fig. 2.17 Statistics, machine learning and neural network methods for handling various datasets

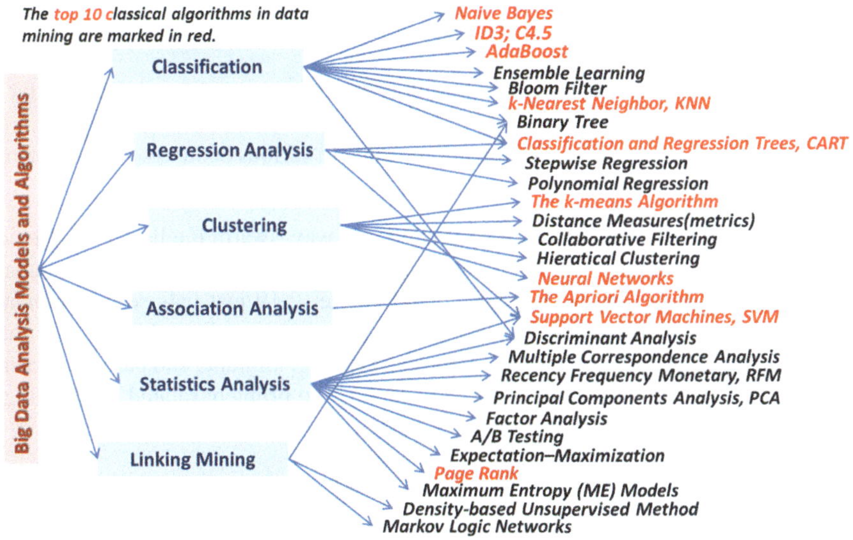

Fig. 2.18 Modeling techniques and algorithms for big data analytics

2.4.2 *Swarm Intelligence and GA/IA in Social Manufacturing*

Optimization is always a key topic for problem solving especially in the area of manufacturing engineering. Because of both the NP-hard and the mathematically ineffable characteristics for most of manufacturing engineering problems, it is very difficult even impossible to use traditional optimal models like gradient descent optimization model to solve such problems. Similarly, the same situations exist in problem solving related to the social manufacturing paradigm. For example, there are following optimal problems typically in social manufacturing:

- selecting a short-cut from all the feasible service or logistic routines,
- generating an optimal layout for socialized manufacturing resources or extended CPSs,

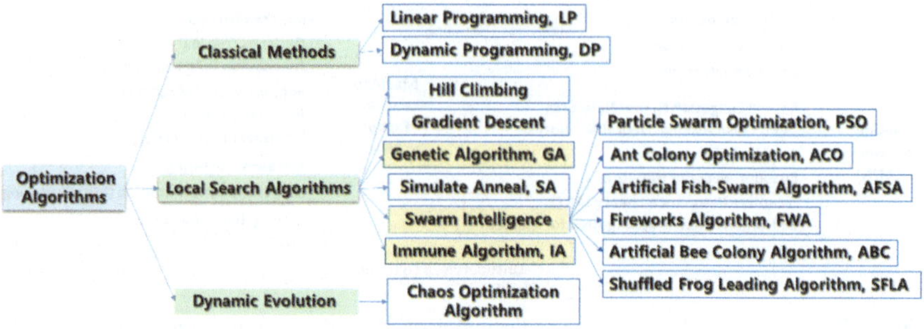

Fig. 2.19 Swarm intelligence and GA/IA algorithms for optimization

- searching an optimal planning or scheduling schema in running a social manufacturing system, and
- implementing a parameters optimization related to various contexts of social manufacturing, etc.

Specially in social manufacturing, as a feasible way, swarm intelligence, genetic and immune algorithms (GA/IA) can of course be used for reaching the optimal goal of solving the above engineering problems. Figure 2.19 just lists several typical optimal algorithms. Here, ant colony algorithm is much more suitable to solve the problem concerning path selection, layout, etc. Genetic, immune, and particle swarm optimization algorithms are often used for solving the problem related to planning, scheduling, parameter optimization, etc.

2.4.3 Decision-Making Methods in Social Manufacturing

Decision-making is technologically taken into consideration as a kind of methodology to get the final solution of a domain-specific problem from a group of feasible candidate solutions. Either evaluation or optimization algorithms play a key role in a decision-making procedure. The researches have indicated that decision-making of the human beings works in the environment of "*socio-technical systems*" and possesses multi-disciplinary attributes related to psychology, sociology, operational research, science, domain-specific engineering technologies, etc.

In fact, there are also a dozen of decision-making problems in social manufacturing, which cover all the product manufacturing activities in the whole stages of a product life cycle and partially include:

- selecting suitable physical objects such as manufacturing communities, collaborative partners, socialized manufacturing resources, machine tools, cutting tools, and so on, from a group of feasible candidates,
- selecting suitable schemes/plans like multi-stage machining and assembling processes, rescheduling solutions, rough production planning, and so on, and

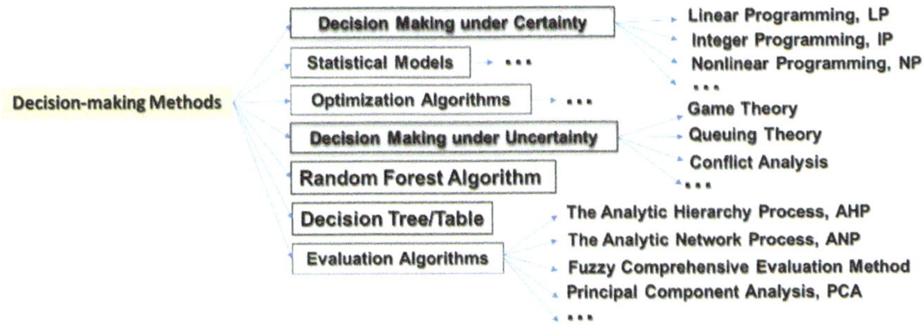

Fig. 2.20 Frequently used methods for decision-making

- obtaining a variety of control strategies which are concerned with fault diagnosis, quality control, orientation of machining and assembling error variation sources, and so on.

Figure 2.20 gives us a glimpse on different decision-making methods. Here, one of the simplest decision-making methods is decision tree or decision table. For more complicated cases, both game theory and random forest algorithm can be used for determining a variety of collaborative relationships such as outsourcing and product services in social manufacturing, depending on different application contexts. In addition, AHP, ANP or fuzzy comprehensive evaluation methods can also be used for choosing a final solution from a group of feasible ones through integrating with the other methods of generating feasible candidate solutions. Typical applications include planning and scheduling issues like multi-routes of machining and assembling process planning and scheduling, product order sharing, socialized manufacturing resource configuring, etc.

2.4.4 Social Computing and Complex Network Analysis in Social Manufacturing

Social computing is one of the most important computing methods for social manufacturing because "*it refers to* ***systems*** *that support the gathering, representation, processing, use, and dissemination of information that is distributed across social collectivities such as teams, communities, organizations, and markets*" [19]. Here, the term "***systems***" also implies the linkages of people in society. So it is obvious that social network is the core computing goal the social computing must be focused on.

In fact, social manufacturing network is a kind of social network and makes a huge number of socialized manufacturing resources connect with each other and work together. It becomes much more significant how to measure the performance of this network with the help of complex network analysis theory which is also a part

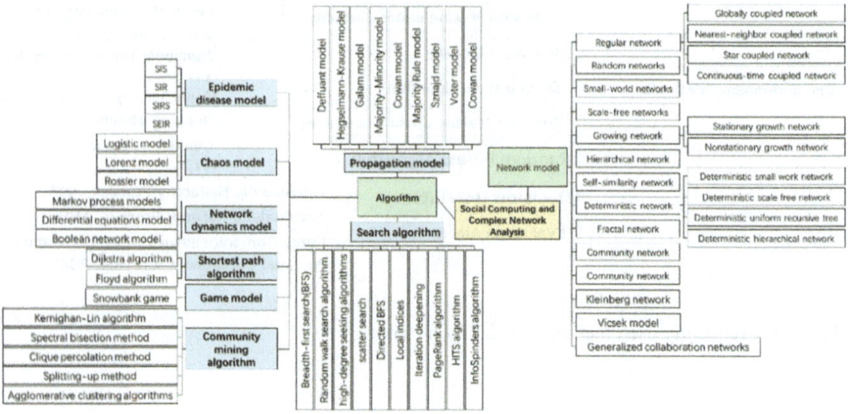

Fig. 2.21 Some models and algorithms related to social computing and complex network analysis theory

of social computing. Here, the performance of the network deals with node attributes, weight distributions, node degrees, robustness, survivability, dynamics attributes, etc.

In addition, at least the following problems concerning the social manufacturing network should be solved by using social computing:

- searching various paths that possess specific physical meanings,
- constructing manufacturing communities via mining the relationships of social business interactions,
- finding the propagating evidences and roadmaps,
- researching the phenomena of chaos, and
- discussing the sharing, gaming and collaborating mechanisms of people inside and across teams, manufacturing communities, organizations, etc.

Figure 2.21 just shows some models and algorithms related to social computing and complex network analysis theory only for the reference.

2.4.5 Services Computing in Social Manufacturing

Social manufacturing is a kind of service-oriented manufacturing paradigm. It is very clear that services computing plays an important role in enabling the above manufacturing paradigm. The purpose of introducing services computing into social manufacturing is to enable IT services and this computing technology to perform correspondent business services more efficiently and effectively. In general, services computing is taken into consideration as a series of theoretical models, key enabled technologies, and algorithms in a kind of multi-disciplinary and computing-driven mode so as to bridge the gap between business services and IT services. Basically, services computing includes a scientific, technological and social terminology catalogue in cross domains, which deals with Web services and service-oriented architecture (SOA), cloud computing, business consulting methodology and utilities, business process modeling, transformation and integration, etc. Here, services computing covers "*the whole life-cycle of services innovation research that includes business componentization, services modeling, services creation, services realization, services annotation, services deployment, services discovery, services composition, services delivery, service-to-service collaboration, services monitoring, services optimization, as well as services management*" [20].

Services computing appears almost in all the product manufacturing activities of social manufacturing along with the mechanisms of outsourcing, crowdsourcing and product service systems. It would influence some software-using infrastructures such as industrial software platforms as the front-ends of manufacturing communities. Here, generic applications deal with:

- setting up and running the suitable industrial software front-ends and the fundamental computing infrastructures for social manufacturing network, manufacturing communities, socialized manufacturing resources, or individuals, and
- enabling the mechanisms of outsourcing, crowdsourcing, and product service systems during different product manufacturing activities in the whole stages of a product life cycle like service description, service search and correlation, service use and service evaluation, etc.

Figure 2.22 lists the core techniques and models of services computing. It is very useful as fundamental supporting computing methods to enable manufacturing services in the context of social manufacturing. Here, one of the biggest contributions of services computing is to use its SaaS (Software as a Service) infrastructure for enabling the industrial software front-ends of manufacturing communities. In addition, service-oriented architecture (SOA) from *IBM* is also very useful for enabling the correspondent manufacturing services which cover their searching, finding, matching, executing, evaluating and ending stages.

What we would like to mention here is the role of cloud computing in services computing although the cloud computing is often thought as an independent classification in some cases. The essential characteristics, deployment models, key enabled technologies, platforms and providers of the cloud computing are listed in Fig. 2.23. It is very obvious that cloud computing as an extended part of services computing will take a key place in generating the context of social manufacturing.

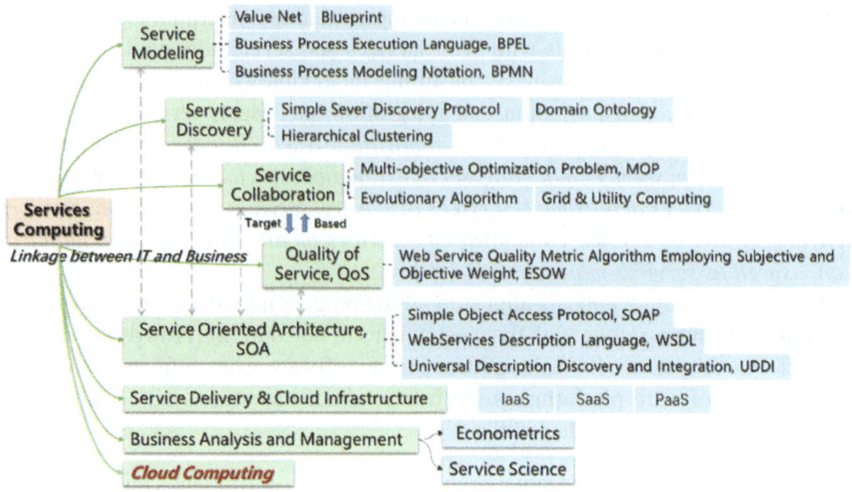

Fig. 2.22 Core techniques and models of services computing

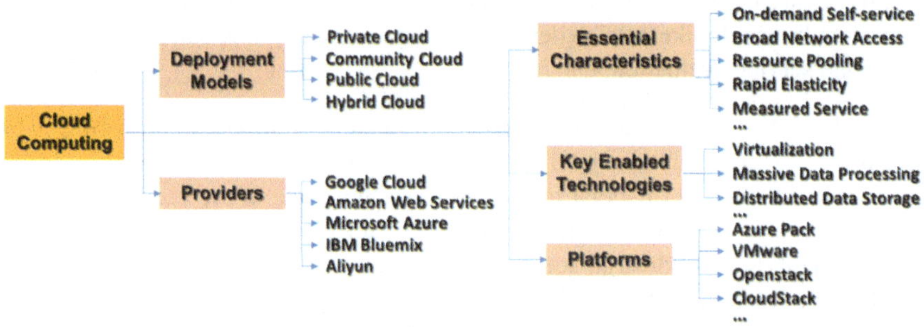

Fig. 2.23 Key techniques, models and platforms for cloud computing

2.5 Conclusions

In summary, we mainly discuss the key enabled technologies and computing methods in this chapter after introducing the concepts, characteristics, basic architecture and runtime mode, which deal with the social manufacturing paradigm. Here, the concepts concerning socialized manufacturing resources, social business relationships, manufacturing communities, manufacturing services, social manufacturing networks, etc., are emphasized on in depth. At the same time, a comparison among the different manufacturing paradigms proves such a fact that social manufacturing possesses obvious characteristics which are quite different from others like its resources, business interactions, organizational structure, runtime, working principles, and so on. It is also clear that at least seven key enabled technologies discussed in the above sections, together with computing methods like social computing and services computing, play a core role in constructing the basic architecture and implementing the runtime logic specifically for the social manufacturing paradigm.

References

1. Jiang PY, Ding K (2012) Social manufacturing: a new way to enable outsourcing productions. In: International conference on innovative design and manufacturing, Tapei, China, 12–14 Dec 2012
2. Jiang PY, Ding K, Leng JW (2016) Towards a cyber-physical-social-connected and service-oriented manufacturing paradigm: social manufacturing. Manuf Lett 7(2016):15–21
3. Jiang PY, Leng JW, Ding K, Gu PH, Koren Y (2016) Social manufacturing as a sustainable paradigm for mass individualization. Proceedings of IMechE, Part B: J Eng Manuf 230(10):1961–1968
4. Urban W (2017) Amoeba management system transformation in the light of organizational change literature. Manage Prod Eng Rev 8(1):16–23
5. Jiang PY (2004) E-service theory and applications for networked manufacturing. China Science Press, Beijing, China, pp 273 (in Chinese)
6. Ehie IC (2001) Determinants of success in manufacturing outsourcing decisions: a survey study. Prod Invent Manag J 42(1):31–39
7. Evans RD, Gao JX, Mahdikhah S, Messaadia M, Baudry D (2015) A review of crowdsourcing literature related to the manufacturing industry. J Adv Manag Sci 4(3):224–321
8. Baines TS, Lightfoot HW, Evans S et al (2007) State-of-the-art in product-service systems. Proceedings of IMechE, Part B: Journal of Engineering Manufacture 221(10):1543–1552
9. Liu C, Jiang PY (2016) A cyber-physical system architecture in shop floor for intelligent manufacturing. Procedia CIRP 56(2016):372–377
10. Liu JJ, Jiang PY, Leng JW (2017) A framework of credit assurance mechanism for manufacturing services under social manufacturing context. In: The 13th IEEE international conference on automation science and engineering (IEEE CASE 2017), Xi'an, China, 20–23 Aug 2017
11. Leng JW (2016) Mass individualization demands driven social manufacturing network configuration and decision methods research. Doctoral dissertation, Xi'an Jiaotong University, Xi'an, China, p 143
12. Ding K, Jiang PY et al (2016) Modeling and analyzing of an enterprise relationship network in the context of social manufacturing. Proceedings of the IMechE, Part B: Journal of Engineering Manufacture 230(4):752–769
13. Lewis B, Entwistle T (1990) Managing the service encounter: a focus on the employee. Int J Serv Ind Manag 1(3):41–52
14. Leng JW, Jiang PY (2017) Mining and matching relationships from interaction contexts in a social manufacturing paradigm. IEEE Trans Syst Man Cybern: Syst 47(2):276–288
15. Jiang PY, Cao W (2013) An RFID-driven graphical formalized deduction for describing the time-sensitive state and position changes of work-in-progress material flows in a job-shop floor. Trans. of the ASME: J Manuf Sci Eng 135(3):1–14
16. Ding K, Jiang PY et al (2017) RFID-enabled physical object tracking in process flow based on an enhanced graphical deduction modeling method. IEEE Trans Syst Man Cybern: Syst 47(11):3006–3018
17. Jiang PY, Liu C (2017) Discussion on the implication, architecture and operation logic of social factory. Aeronaut Manuf Technol 2017(7): 26–30. (in Chinese, Cover Story)
18. Ding K, Jiang PY, Zheng M (2017) Environmental and economic sustainability-aware resource service scheduling for industrial product service systems. J Intell Manuf 28(6):1303–1316

19. Erickson T (2013) Chapter 4: social computing. In: The encyclopedia of human-computer interaction, 2nd edn. (also available at: https://www.interaction-design.org/literature/book/the-encyclopedia-of-human-computer-interaction-2nd-ed/social-computing)
20. NN (2018) Services computing. Wikipedia. Available at: https://en.wiki-pedia.org/wiki/Services_computing

Chapter 3
Socialized Manufacturing Resources and Interconnections

Jiewu Leng and Pingyu Jiang

3.1 Concepts and Classification

Under the current industrial economy trend of evolving into experience economy and socialnomics, both manufacturers and their manufacturing activities are heading towards socialization, servitization, and mass individualization [1]. Consumers emphasize more focus on their manufacturing participation and value embodiment [2], and their demands on specialized manufacturing services. Such a focus is becoming complex, individualized, and service-oriented. Small modular infrastructure for manufacturing is therefore burgeoning and stimulating big prosumers to "*think small*" for flexibility benefits [3]. Thus, the role of customers is changing from buyer to "*prosumer*" (producer + consumer) [4], who widely interact with manufacturers and service providers and involve in the product lifecycle to transform their demands into individualized products and services [5]. They widely infiltrate in the product lifecycle activities and collaborate with each other to improve capabilities related to the development, manufacturing and usage of products. Social media such as *Facebook*, *Twitter*, *LinkedIn*, and *Kenandy*'s cloud ERP provide them with viable solutions [6]. These trends are changing the manufacturing paradigm into a socialized and collaborative one.

On the other hand, along with the explosive demands in finer-grained markets, many small-and-medium-sized producers, e.g., micro-and-small-sized firms, start-ups, small factories, workshops, and even individuals who have 3D printers [7], spring up with socialized manufacturing resources (SMRs) and involve in different stages of market segments. They provide various specialized product-related or manufacturing service capabilities to satisfy customers' individualized requirements, act as SMR service providers, and bring advantages of flexibility and responsiveness over traditional big manufacturers. Moreover, SMR service providers aggregate and self-organize themselves into dynamic social communities (SCs) to win bargaining power and efficiency in the interaction with core manufacturers.

P. Jiang, *Social Manufacturing: Fundamentals and Applications*, Springer Series in Advanced Manufacturing, https://doi.org/10.1007/978-3-319-72986-2_3

3.1.1 Manufacturing Resources and Socialization

Definition 3.1 Socialized manufacturing resources (SMRs) in a narrow meanings are defined as all kinds of the property-type and consumable-supply-type manufacturing resources in this chapter. They are geographically distributed across the prosumers and could support the product manufacturing activities by providing manufacturing services [8–10].

Definition 3.2 Property-type manufacturing resources include core manufacturing equipment (e.g., lathe, milling machine, boring machine, etc.). As social manufacturing deals especially with the individualized machining demands, machining resources involved are limited to property-type manufacturing resources, such as turning machines, milling machines, machining devices, and CNC centers.

Definition 3.3 Consumable-supply-type manufacturing resources include materials and its auxiliary device suits (e.g., cutting tools, measuring tools, etc.).

Definition 3.4 Manufacturing resource services (MRSs) are defined as the service encapsulation of the SMRs and production capability prosumers provide so as to accomplish the manufacturing tasks, including parts machining and product assembling, from product manufacturers. It helps to achieve value-added services and rapid dynamic responses to market.

The linkages between these concepts are described as follows. Core manufacturers release MRS demands in the social manufacturing platform, while prosumers release their SMRs and MRS capabilities in the platform [11]. Through initial clustering and self-organization, the prosumers aggregate into different kinds of groups and communities to enlarge their bargain power and common profits. To integrate the SMRs distributed all over the society, unified modeling technique is one of the most effective socialization tools and is hence introduced to realize the virtual access of the SMRs. For the first step of unified modeling technique, the dynamic production capability modelling of the SMRs is significant. For example, it is considered as a core of the machining system in the production stage of a product lifecycle since a well-built capability model can provide accurate information to support the subsequent retrieval and matching. According to the demands of both of SMR providers and demanders, two capabilities including machining capability and production capability are extracted from the viewpoint of machining targets, delivery date, quality, and cost.

Taking into consideration of modelling SMR capabilities in the production stage of a product life cycle further, as shown in Fig. 3.1, machining function and machining performance are defined to evaluate the machining capability of SMRs. To describe the model clearly, some detailed concepts involved should be clarified at first.

Machining function refers to a set of machining features an SMR can handle. It is determined by the basic physical properties of an SMR, including machine type, number of machines, machine attributes and basic parameters. For instance, a vertical milling machine has the function of milling a surface. *Performance* is a

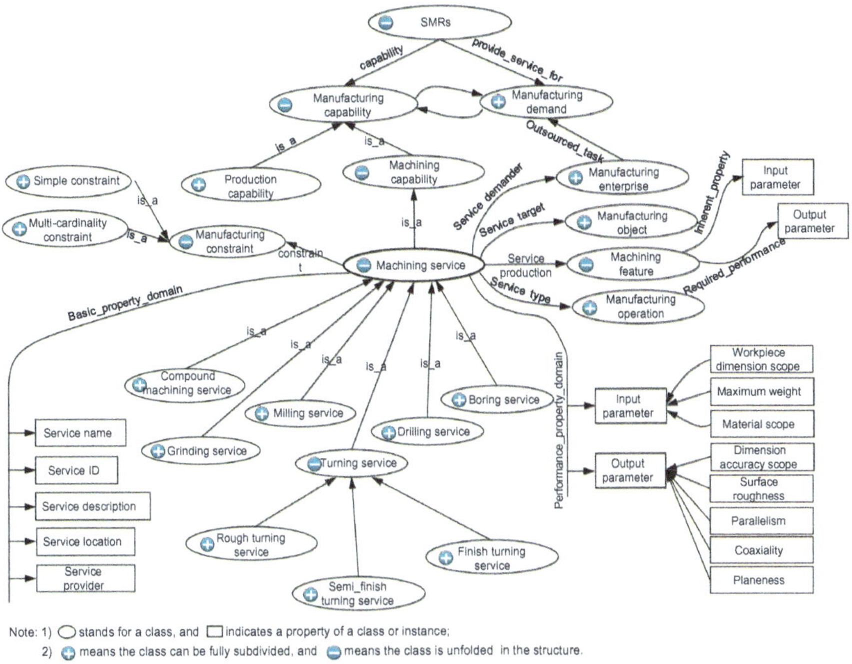

Fig. 3.1 Capability model of a single SMR

metric of evaluating the highest accuracy and surface quality the SMR can reach in terms of its machining function. It includes machining type, machining precision, machining material and machining features of the SMR. Machining performance directly determines whether an SMR can complete the task or not. *Performance quality* is used to evaluate the system stability of an SMR when fulfilling a machining function, which usually is reflected by the tolerance distribution of machining a batch of workpieces. *Machining capability* denotes which machining functions an SMR have, what performance it can reach, and how about the stability of the SMR when machining a batch of workpieces. *Production capability* is a conventional concept that mainly deals with the number of workpieces that an SMR can fulfill in a unit time interval. It somehow relates to the delivery date of an outsourced task. Besides the production capabilities of an SMR, the ancillary service is essential for assisting the SMR to complete the production task.

Generally, *machining function* is implied by the machining features an SMR can machine. The classification of all the machining features can refer to STEP AP224 Standard. Performance can be considered as a constriction of the *machining function*, that is, the highest accuracy and surface quality. Accuracy and surface quality are very relative to one or several machining features. Therefore, different *machining functions* usually have different performances. As the states of SMRs in the social manufacturing environment are changing by time, a dynamic information model is

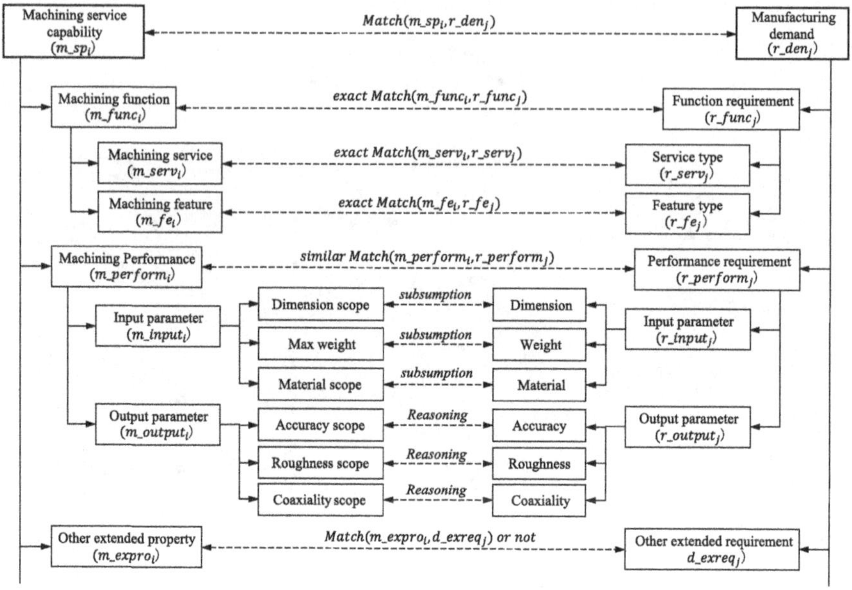

Fig. 3.2 Descriptive model of SMRs

also necessary for scheduling and monitoring the SMRs timely. Here, the dynamic information mainly includes the running state of each SMR together with its current working-load, which is demonstrated by a kind of Gantt chart. What's more, price and cost are also very important information for describing an SMR. Although physical SMRs can be connected by the above unified modeling procedure, we still need to establish a descriptive model for transforming these physical SMRs into virtual ones and subsequently digitalized them in the network. Therefore, SMR descriptive model is presented from the viewpoints of resource identification, basic information, capability information, and mapping relationships, as shown in Fig. 3.2.

After the descriptive model has been established, a formalized description of the SMRs can be realized by using set theory and relational algebra. Here, a five-element-array is built for this purpose as follows:

$$SMR = \{ID, Intro, BasicInfo, CapaInfo, MapInfo\} \tag{3.1}$$

where *ID* represents the unique identification of an SMR in the manufacturing community. *Intro* enotes the introduction of the SMR, which shows its brief and original information for both service demanders and providers. *BasicInfo* s the basic attribute of the SMR, including name, type, manufacturer, version, and ownership, to provide basic information for the management, scheduling, and analyzing of SMRs. *CapaInfo* escribes the capabilities of the SMR such as machining capability, production capability, cost and other dynamic information related to its capabilities. *MapInfo* epresents the mapping relationship between the physical SMR and the cor-

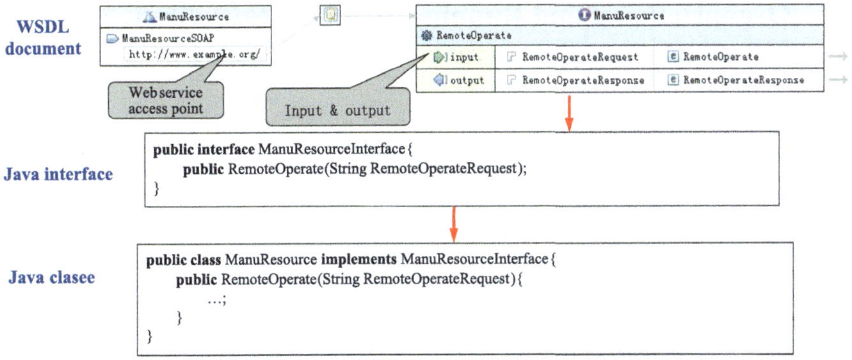

Fig. 3.3 Virtual access of MRSs based on web service

responding digital SMR. When the SMRs are described formally, the SMR pool can be formed as a supergiant set that comprises of all the SMRs and their information.

For the other stages in a product life cycle, the capability models of SMRs can also be created in the similar way.

3.1.2 Manufacturing Resource Services

According to Definition 3.4, the virtualization of MRSs means encapsulating the MRSs and putting the virtual MRSs into the social manufacturing communities and network. A standard and open encapsulating method that provides unified interfaces to connect distributed MRSs is the base of sharing and configuring SMRs optimally. Here, the encapsulation of MRSs is realized by employing Web Services Description Language (WSDL) based on ontology philosophy [12]. Simple Object Access Protocol (SOAP) is also applied to shield the diversity among different software platforms. Diverse kinds of web services are related to diverse kinds of MRSs, and usually one MRS connects only one web service. Prosumers visit these web services through their access points and call their remote methods to control and command the corresponding physical SMRs. Because of the diversity of the communication methods and the needed information among different MRSs, the encapsulation procedures of different MRSs need to define different methods and operations according to the intrinsic functions of the MRSs themselves. Figure 3.3 reveals the realization procedure of a web service. We firstly use WSDL to define the visiting operations and port address of each web service according to a related MRS, secondly develop a web service interface for each MRS according to the WSDL document, and finally develop software classes to realize these interfaces.

3.2 Organizational and Running Shapes of Socialized Manufacturing Resources

The mass individualized demands and SMRs stimulate manufacturing industry to thrive, and also force core manufacturers to re-organize themselves into small-modular-infrastructure or platform-typed prosumers such as *Haier*, which is a household electrical appliances manufacturer in China. Such core manufacturers are transforming their organizing structures into more-effective shapes of adapting into the mass individualization demands, and integrating outside SMRs to involve in the finer-grained markets. Based on that, a socialized manufacturing resources network made up of prosumers, SMRs and their social relationships can be built for certain individualized manufacturing tasks. Within the network, all the participants interact and collaborate to accomplish the product manufacturing activities in the whole stages of a product life cycle.

3.2.1 Shaping Manufacturing Resources into Manufacturing Communities Under a Social Network

From the evolutionary view, manufacturing organizing structure evolves from the vertical organization (i.e., traditional manufacturer) to the project/product-oriented virtual organization (i.e., virtual enterprise), and finally to the service-oriented social organization. The flexibility, complexity, and collaboration scope of the above three types of organizing structures range from low to high. In the social manufacturing paradigm, the shape of manufacturing resources organizations is heading to manufacturing communities (MCs), which are defined as associations where prosumers collaborate for unified MRSs to improve their common profits and bargaining powers. From the viewpoint of bargain power, there are two kinds of MCs, that is, horizontal MC and vertical MC. Horizontal MCs consist of prosumers specializing at one kind of MRS in the certain phase of a product's manufacturing chain, and are formed through initial clustering and afterwards self-organization. There is a committee elected in each horizontal MC to represent the rights and profits of its member prosumers. While vertical MCs consist of prosumers during all the phases of a certain product's manufacturing chain. Additionally, there are social spaces with social networking tools in both horizontal MCs and vertical MCs to support enterprise interactions and industry trends analysis.

In the manufacturing community, prosumers can easily disseminate service relationships across an entire community or from different level services across the community, and be aware of problematic situations before they occur. In our early studies, which mainly focus on the production stage of a product life cycle, Ding et al. proposed a method to configure customized community space of an enterprise, which can search for best partners for the enterprise [13]. Three key enabling technologies were discussed to support its implementation, including socialized manufacturing resources configuration, social manufacturing community organization and collaboration and manufacturing outsourcing service searching and matching. Cao and Jiang proposed a cloud machining community to integrate the distributed resources and provide on-demand services for service requestors who need them. Some key enabling technologies are detailed, including virtual access and matching of machining resources and machining demands, and generation of machining groups and creditworthiness evaluation mechanism [14]. Details of the discussion will not be included here for concise reason.

3.2.2 Analyzing Organizational Shapes of Socialized Manufacturing Resource Network

From the viewpoint of elements in MC, there are two kinds of MCs, namely, prosumer MC and SMR MC (e.g., design MC, manufacturing MC, and transportation MC) [13]. Both are dynamic, changeable, and led by one or more members with higher leaderships. Prosumers or SMRs can opt in and opt out multiple MCs autonomously according to the similarity of their interests/capabilities. Each pair of members in an MC establishes a social relationship. The relationship tightness can be high or low, determined by their business overlaps.

As shown in Fig. 3.4, prosumers, SMRs, MCs, and their relationships make up a big socialized manufacturing resources network (SMRN), which is open, stochastic, self-organized, and stimulated by societal policies, and dynamic prosumer requirements. The self-organizing mechanism, autonomy mechanism, decision-making mechanism and others ensure that the SMRN evolves into a dynamical-steady system. Based on the complex network theory, the SMRN can be described as a complex network:

$$G = (V, E, W) \tag{3.2}$$

where $V = \{v_1, v_2, \ldots, v_n\}$ is the node set of prosumers and SMRs; $E = \{e_1, e_2, \ldots, e_m\}$ is the edge set of social relationships among nodes; $W = \{w_1, w_2, \ldots, w_p\}$ represents the weights indicating the relationship tightness of e_i, and $w_i \geq 1$. Define $w_i = 1$ if two nodes just have social relationships, and $w_i = k$ $(k > 1)$ if two nodes have service relationships (i.e., there are manufacturing interaction between them). The more interaction between them, the bigger k is.

3.2.3 Building a Running Shape of Socialized Manufacturing Resource Network

Networked SMRs are self-organized into multi-modal MCs in an SMRN. The dynamics of the SMRN is determined by their interactions, which is crucial to understand the structural and functional properties of the complex system. To build a running shape of socialized manufacturing resource network, an *EAGLE* algorithm for extracting both the hierarchical and overlapping properties of community can be applied [15]. Its main procedure includes two steps:

Step 1: Generate a dendrogram based on the similarity M between two communities MC_1 and MC_2 (initially two nodes).

$$M = 1/2m \sum_{v \in MC_1, w \in SM, v \neq w} (A_{vw} - k_v k_w / 2m) \tag{3.3}$$

where A_{vw} is the element of adjacency matrix of the network, $A_{vw} \in [0, 1]$, $m = 1/2 \sum_{v,w} A_{vw}$ is the total number of edges in the SMRN, k_v is the degree of node v determined by its relationship tightness with others.

Step 2: Choose an appropriate cut to break the dendrogram into MCs based on the modularity EQ.

$$EQ = 1/2m \sum_i \sum_{v \in SC_i, w \in SC_i} 1/O_v O_w (A_{vw} - k_v k_w / 2m) \tag{3.4}$$

where O_v is the number of MCs included node v.

The dendrogram and its cut are depicted in Fig. 3.5. The revealed MCs are the hidden modules of the SMRN, which are helpful for rapid SMRN matching and

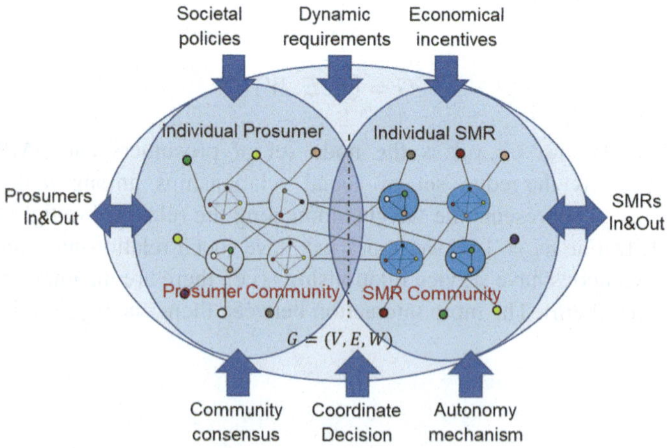

Fig. 3.4 Relationships between SMRs and MCs in social manufacturing

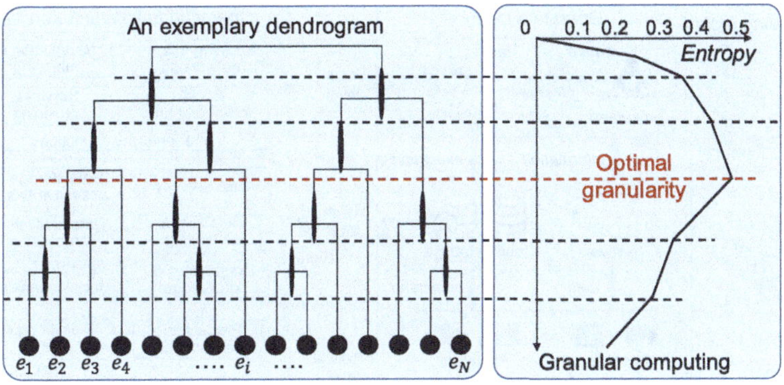

Fig. 3.5 A dendrogram and its appropriate cut

efficient collaboration. The network dynamics and network coordination should be further studied so as to reveal the SMRs self-organizing disciplines and shapes, and to promote the market moving towards ordering [16, 17].

3.3 Management of Social Manufacturing Resources

The wide interconnection of prosumers, SMRs, and other participants is the core for SMRN management. There are two main types of SMRN management patterns, which are platform-driven centralized management and self-organization decentralized management [18].

3.3.1 Platform-Driven Centralized Management

Platform-driven centralized management is the basis for efficient interaction, collaboration, and sharing in an SMRN that has leading prosumers or core manufacturers. A framework for platform-driven centralized management of the SMRN is depicted in Fig. 3.6. Cyber-Physical-Social System (CPSS) is the supporting technology and is built upon cyber-physical systems by adding social factors [19, 20]. From the organizing logic view, an SMRN can be viewed as a CPSS network, MCs as CPSS units, and SMRs as CPSS nodes.

The SMRN deals with three levels, which are physical interconnection, cyber interconnection, and social interconnection. At the cyber level, the network infrastructure is cloud-based, and the gathered data, information, and knowledge are stored in the public or private cloud [21]. Cloud computing and social computing are applied to explore the industrial and social big data for supporting decision-making. At the physical level, intelligent sensors and actuators are integrated in equipment, vehi-

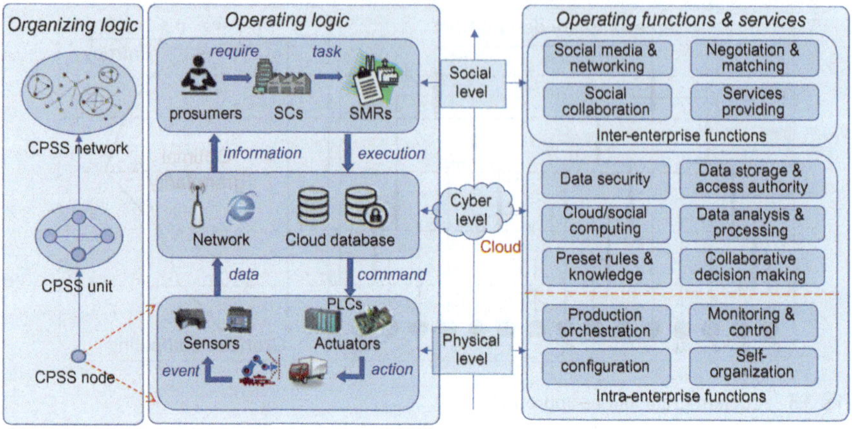

Fig. 3.6 The CPSS for platform-driven centralized management of SMRN

cles and other CPSS nodes, making them become smart objects. Sensors monitor dynamic events and gather real-time operating data, actuators receive commands from the cyber level and execute actions, and CPSS nodes collaborate with each other to execute manufacturing operations. At the social level, prosumers, SMRs, and other participants interact with each other via social networking and social media tools, which facilitate the achievement of social business and crowd intelligence. Prosumer's demands are mapped into product functions and lifecycle tasks via social communications with SMRs.

The above three levels ensure the cyber-physical-social interconnection among prosumers, SMRs, manufacturers, and other participants. Based on that, real-time manufacturing data can be gathered and shared among them to realize dynamic and transparent inter-enterprise manufacturing management. Thus, the SMRN can respond in real time to the prosumer needs and the changing conditions in the SMRs. Note that the cyber level is the link of the social level and the physical level because the latter two levels rely on the data gathering, processing, and analyzing at the cyber level [22].

With the help of the platform, the well-thought-of members undertake the daily management of the community, such as allow-in of members, status updates, and topics initiation. The SMR MC applies outsourced/crowdsourced tasks by taking the aggregated capabilities of its members as chips. The members collaborate with each other to finish the tasks and the social media help them to collaborate efficiently.

Because the equipment in an SMR is cyber-physical-social connected as a CPSS node, real-time manufacturing data can be gathered with radio frequency identification (RFID) devices and sensor network. Thus, material flow of each outsourced task can be transparently monitored. On the one hand, if disturbances (e.g., machine breakdown, quality defects, and task delay) occur, the correspondent SMR can perceive them in time and monitor the self-adjustment of the CPSS nodes until it needs manual interventions. On the other hand, if prosumer's demands change, the SMR will rapidly analyse them, re-allocate the changed tasks, and update the manufacturing plans. All the SMRs are cyber-physical-social connected too, thus real-time manufacturing data from each SMR can be shared among prosumers and other SMRs via authorized data interfaces. Besides, real-time transportation states between two partners can be achieved by RFID and GPS. Thus, the total physical flows including material flow and product flow can be monitored transparently.

Based on the real-time inter-enterprise manufacturing data, dynamic integrated production and transportation planning and scheduling (IPTPS) can be addressed. IPTPS is important to improve inter-enterprise collaboration efficiency in social manufacturing. Because prosumer's partners are scattered around the world, the transportation processes of physical objects among them need to be well planned together with the manufacturing processes to save cost and time. The constraints of the IPTPS problem are the quality satisfaction and the due date satisfaction of each order. The objective of the IPTPS problem is to minimize the total manufacturing and transportation cost. Social interaction and mass collaboration are the key ways to realize the above two aspects. With the increasing applications of social networking and social media tools in manufacturing industry, the synchronous and asynchronous communications among prosumers become easier. Under *Web 3.0*, semantic web and instant messaging are used to handle prosumers' daily business. For example, prosumers can discuss a product CAD/CAM model with SMRs and SCs through online meeting, shared document editing, and live-streaming technologies. The inter-enterprise collaboration ranges are from the whole product life cycle activities. For example, designers from design MCs can participate in the production stage, interacting with SMRs in the production MCs via social media to co-decide the production process planning [23, 24].

3.3.2 Self-organization Decentralized Management

In the self-organization decentralized management of an SMRN, prosumers can interact with each other via social media and make suggestions or demands to manufacturers to improve product conceptual design [25]. Then manufacturers response to such suggestions or demands rapidly and transform them into engineered features. *Fiat* 500 car and *Xiaomi* cell phone are the examples utilizing crowd intelligence of prosumers to improve product design. There are two types of SMRN self-organization decentralized management modes, which are SMRN with core manufacturer and SMRN without core manufacturer.

As shown in Fig. 3.7, prosumers, core manufacturer, SMR MCs and SCs make up an SMRN. Here, SCs mean Prosumer MCs. Prosumers propose individualized demands in the form of context, video, and other forms of non-structural data in the online prosumer SC. Core manufacturer analyzes their demands via contextual mining or big data analytics, and maps them into functions via deep learning and big data analytics methods [26, 27]. Then, it crowdsources product design tasks to the design MCs or just to the prosumer SC, utilizing crowd intelligence to rapidly develop the

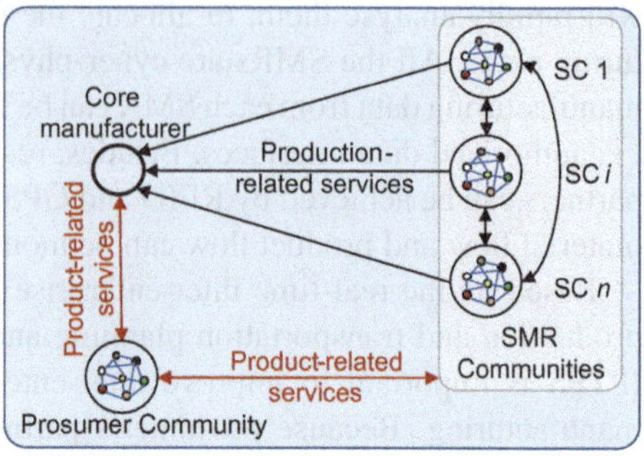

Fig. 3.7 SMRN with core manufacturer

required products [28]. After the product design is accomplished, core manufacturer decomposes production tasks according to the product bill of materials (BOM) and outsources them to production MCs that specialize at various part production tasks. The negotiation and supplier selection mechanism ensure the selected MCs are optimal for the tasks. The selected MC allocates the outsourced tasks to its members based on their capabilities, which ensure that each SMR can win its profit according to its contribution. Meanwhile, the selected MCs provide real-time manufacturing data, based on which core manufacturer can synthesize them into comprehensive information to prosumers. Besides, some MCs will undertake production assistance for others in the form of product-service system (PSS). For example, machine tool service providers from PSS MCs will assist core manufacturer to schedule the machine tools they provide and give operating suggestions to them. In some sense, core manufacturer acts as the system integrator to aggregate different SMRs and PSS providers for individualized manufacturing.

As shown in Fig. 3.8, prosumer utilizes SMRs and SCs to develop individualized products without core manufacturer dealing with system integration. That is, an active prosumer will act as the system integrator. This mode is previously called "*Do It Yourself*" (DIY). All the product lifecycle activities are outsourced to SMRs and are managed by this kind of prosumer. This special prosumer proposes the ideas or demands to the designer SCs, and then the correspondent SMRs with design intelligence will apply for them and upload their design schemes as solutions for that special prosumer. The optimal SMR with its scheme will win its rewards. Then, the special prosumer will outsource production tasks according to the product BOM, and optimal SMRs from different production MCs are selected. Other SMRs like transportation service providers from transportation MCs also participate in correspondent product manufacturing activities. Thus, kinds of manufacturing-related services and product-related services are provided to the special prosumer. The special prosumer and these designer SCs, producer SCs and transporter SCs, together with their SMRs related respectively to design MCs, production MCs and transportation MCs make up the final SMRN for interactions and collaboration. So this final SMRN is used for realizing the goal of "*co-innovation, co-manufacturing, and co-operating*." In this way, distributed SMRs can be efficiently organized and utilized.

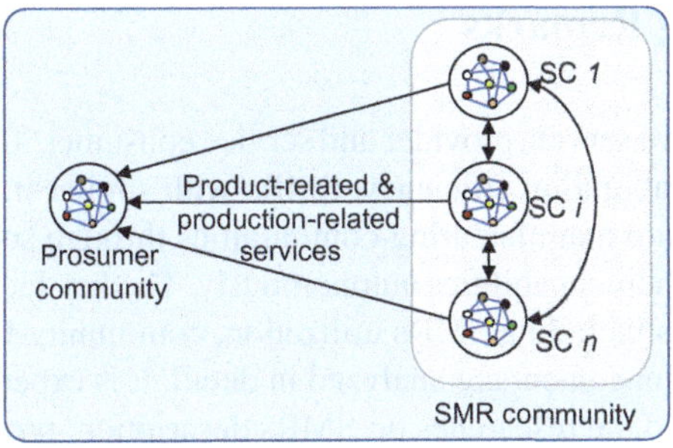

Fig. 3.8 SMRN without core manufacturer

Table 3.1 Mapping between evolutionary theory and social manufacturing

Evolutionary theory		Social manufacturing	
Name	Example	Name	Example
Cell	Neuron	SMR	Design workshop
Tissue	Nervous tissue	SC	Crowdsourcing designer community
Organ	Brain	SMRN	Product design network
Multicellular organism	Human	SM	Product development eco-system

Currently, this kind of SMRN mainly lies in the fast-moving consumer goods or DIY products manufacturing. When SMRN applied broadly in industrial products manufacturing, it will stimulate social manufacturing to move towards a mature stage. Note that prosumers in this kind of SMRN can be traditional manufacturers too. To some sense, the second kind of SMRN is an extension of the first kind.

SMRNs are the extractions of social manufacturing system (SMS), different correlated SMRNs compose the final big system. From the view of evolutionary theory, single cells are the fundamental unit of structure and functions in all living organisms, and they specialize into different cell types that are adapted to particular functions. Cells that are similar to each other in appearance and have the same function aggregate into tissues to act the specific functions. Multiple tissues form the organs of the multicellular organism by the functional grouping. Analogously, SMRs are just like the single cells, MCs are the tissues, SMRNs are the organs, and the big SMS is the multicellular organism, as shown in Table 3.1. Thus, the manufacturing organizing of SMRNs is conformed to the development of industry towards a stronger one. Future work should be devoted to explore the evolution-centric manufacturing organizing mechanism.

From another point of view, social big data generated from intertwined social interactions and mass collaboration are valuable assets. Applying big data analytics, important information such as market trends and prosumer preference can be explored. Social big data analysis helps to carry out social product development and predictive manufacturing [29, 30].

3.4 Concluding Remarks

SMRs act dual-roles of service provider and service consumer. To enhance their competitiveness and collaboration efficiency, SMRs with similar interests and capabilities are aggregated into manufacturing communities through social networking and sharing to organize their capabilities autonomously. The key factors in the resources configuration process including SMRs utilization, community-based resources self-organization, and management are analyzed in detail. It is expected that this chapter would be beneficial to the researches on SMRs description, prosumers organization and SMRs configuration under the context of social manufacturing.

References

1. Koren Y, Shpitalni M, Gu P, Hu SJ (2015) Product design for mass-individualization. Procedia CIRP 36(2015):64–71
2. Mourtzis D, Doukas M (2014) Design and planning of manufacturing networks for mass customisation and per-sonalisation: challenges and outlook. Procedia CIRP 14(2014):1–13
3. Dahlgren E, Goecmen C, Lackner K, van Ryzin G (2013) Small modular infrastructure. Eng Econ 58(4):231–264
4. Ritzer G, Dean P, Jurgenson N (2012) The coming of age of the prosumer. Am Behav Sci 56(4):379–398
5. Xie C, Bagozzi RP, Troye SV (2008) Trying to prosume: toward a theory of consumers as co-creators of value. J Acad Mark Sci 36(1):109–122
6. Andreadis G (2015) A collaborative framework for social media aware manufacturing. Manuf Lett 3(2015):14–17
7. Errasti A, Rudberg M, Corti D (2015) Managing international operations: configuration of production network for SMEs. Prod Plann Control 26(9SI): 691–692
8. Jiang PY, Leng JW, Ding K, Gu P, Koren Y (2016) Social manufacturing as a sustainable paradigm for mass individualization. Proc Inst Mech Eng [B] 230(10):1961–1968
9. Jiang PY, Leng JW (2017) The configuration of social manufacturing a social intelligence way toward service-oriented manufacturing. Int J Manuf Res 12(1):4–19
10. Jiang PY, Ding K, Leng JW (2016) Towards a cyber-physical-social-connected and service-oriented manufacturing paradigm: Social Manufacturing. Manuf Lett 7(2016):15–21
11. Cao W, Jiang PY, Jiang K (2015) Demand-based manufacturing service capability estimation of a manufacturing system in a social manufacturing environment. Proc Inst Mech Eng [B] 38(4):242–252
12. Maedche A, Staab S (2001) Ontology learning for the semantic web. IEEE Intell Syst 16(2):72–79
13. Ding K, Jiang PY, Leng JW, Cao W (2016) Modeling and analyzing of an enterprise relationship network in the context of social manufacturing. Proc Inst Mech Eng [B] 230(4):752–769
14. Liu JJ, Jiang PY, Leng JW (2017) A framework of credit assurance mechanism for manufacturing services under social manufacturing context. In: 13th IEEE conference on automation science and engineering (CASE), Xi'an, China, 20–23 Aug 2017
15. Shen H, Cheng X, Cai K, Hu M (2009) Detect overlapping and hierarchical community structure in networks. Physica A 388(8):1706–1712
16. Leng JW, Jiang PY, Ding K (2014) Implementing of a three-phase integrated decision support model for parts machining outsourcing. Int J Prod Res 52(12):3614–3636
17. Leng JW, Jiang PY, Zheng M (2017) Outsourcer—supplier coordination for parts machining outsourcing under social manufacturing. Proc Inst Mech Eng [B] 231(6):1078–1090

18. Mourtzis D, Doukas M, Psarommatis F (2012) Design and planning of decentralised production networks under high product variety demand. Procedia CIRP 3(11):293–298
19. Wolf W (2009) Cyber-physical systems. Comput 42(3):88–89
20. Hussein D, Park S, Han SN, Crespi N (2015) Dynamic social structure of things: a contextual approach in CPSS. IEEE Internet Comput 19(3):12–20
21. Åsa E, Johansson C, Nergård H (2015) Manufacturing knowledge: going from production of things to designing value in use. Intell Decis Technol 9(1):79–89
22. Frazzon EM, Hartmann J, Makuschewitz T, Scholz-Reiter B (2013) Towards socio-cyber-physical systems in production networks. Procedia CIRP 7(2013):49–54
23. Leng JW, Jiang PY (2017) Granular computing based development of service process reference models in social manufacturing contexts. Concur Eng Res Appl 25(2):95–107
24. Leng JW, Chen Q, Mao N, Jiang PY (2018) Combining granular computing technique with deep learning for service planning under social manufacturing contexts. Knowl-Based Syst 143(2018):295–306
25. Tseng MM, Jiao RJ, Wang C (2010) Design for mass personalization. CIRP Ann—Manuf Technol 59(1):175–178
26. Leng JW, Jiang PY (2017) Mining and matching relationships from interaction contexts in a social manufacturing paradigm. IEEE Trans Syst Man Cybern: Syst 47(2):276–288
27. Leng JW, Jiang PY (2016) A deep learning approach for relationship extraction from interaction context in social manufacturing paradigm. Knowl-Based Syst 100(2016):188–199
28. Wang FY, Zeng D, Carley KM, Mao WJ (2007) Social computing: from social informatics to social intelligence. IEEE Intell Syst 22(2):79–83
29. Jiang PY, Leng JW, Ding K (2016) Social manufacturing: a survey of the state-of-the-art and future challenges. In: IEEE international conference on service operations and logistics, and informatics, Beijing, China, 10–12 July 2016
30. Leng JW, Jiang PY, Zhang F, Cao W (2013) Framework and key enabling technologies for social manufacturing. Appl Mech Mater 312(2013):498–501

Chapter 4
Social Business Relationship and Organizational Network

Jiewu Leng, Wei Guo and Pingyu Jiang

4.1 Enterprise Relationships and Social Business Relationship

Considering nowadays' distributed manufacturing industries, small and medium service-oriented enterprises (SMSEs) with socialized manufacturing resources (SMRs) are flourishing to provide manufacturing services [1]. The traditional giant product manufacturers are becoming dumbbell-shaped. They start to outsource and build social business relationships with SMSEs. The manufacturing activities of a product are not limited within core product manufacturers but in multiple SMSEs, forming a product-oriented social manufacturing network. Unfortunately, it lacks an effective service platform that can integrate SMRs and share machining information and capabilities through a sophisticated trust mechanism [2]. At present, most manufacturing enterprises encounter a dilemma that only manufacturing services provided by few dependable partners in the same area can be obtained without troubling. In this context, there are massive enterprise relationships and social business relationships in diverse social media. It integrates and virtualizes plenty of SMRs, aggregates SMSEs into manufacturing communities (MCs) by recommendation and self-organization to provide manufacturing services, and promotes intelligent business and all-around order management by using social networking tools.

For example, *Kenandy* has built a cloud-based platform for social manufacturing management and collaboration, integrating globally distributed companies to work together to produce a product.

P. Jiang, *Social Manufacturing: Fundamentals and Applications*, Springer Series in Advanced Manufacturing, https://doi.org/10.1007/978-3-319-72986-2_4

4.1.1 Enterprise Relationship Network

Massive SMSEs need to be well organized to find the optimal one for outsourced manufacturing services [3]. Driven by diverse kinds of customized requirements, enterprises maintain relatively stable cooperation with others in the community but establish transient collaborations with their customers. The enterprises could be geographically agglomerated or not (actually all are distributed in the Internet). The enterprise organization mode has turned into a networked and visualized one. As to the enterprise network, much research has been done on the network structure, functional analysis, and network dynamics, etc. The scientific discipline of the collaboration network (including virtual organizations, virtual enterprises and dynamic supply chains) are studied. Little work has been devoted to the distributed enterprises organizing from the perspective of layered network topology, inherent connections of enterprises, especially the SMSEs.

A bi-level social manufacturing-oriented enterprise relationship network (SMERN) is proposed to facilitate the organizing of SMSEs [4]. The generation rule and formal description of the SMERN are defined based on ontology and relation algebra. Furthermore, an enterprise classification model together with an improved clustering algorithm is proposed to aggregate enterprises into MCs for initial recommendation. The topological and physical characteristics of SMERN are discussed to discover the features of enterprise organizing and enterprise collaborations.

4.1.2 Definition and Identification of Social Business Relationship

Definition 4.1 Social Interactions in social manufacturing (SocialM) paradigm are the processes by which prosumers [5] act and react to each other, and usually characterized by requirements, preferences, situations, experiences, or feedbacks (e.g., a safer, more effective, and more comfortable one). It is clearly the prerequisite and foundation for establishing and maintaining prosumer relationships.

Definition 4.2 Social Business Relationship in SocialM paradigm is the interactive collaboration reality among prosumers resulted from the matchmaking between manufacturing demands and capabilities/supplies, and it includes various service (e.g., designing, machining, and product) outsourcing (i.e., individual-to-individual or individual-to-multiple specified prosumers) and crowdsourcing (i.e., individual-to-group prosumers), as shown in Fig. 4.1.

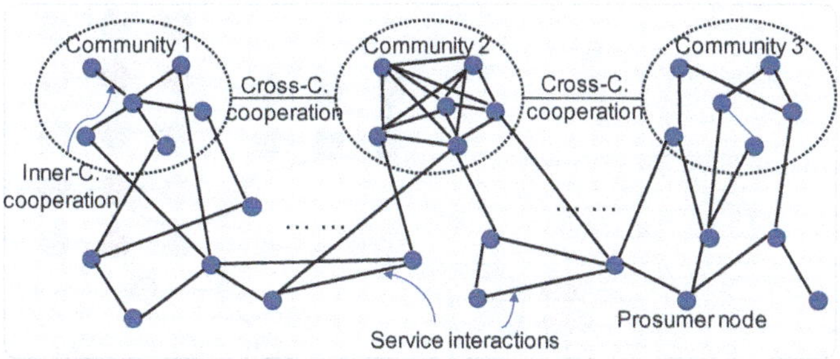

Fig. 4.1 Social business relationship in SocialM paradigm

4.1.3 Methods to Enable Social Manufacturing Through Social Business Relationship

Social business relationship is embodied by product-order-driven service interactions, which is the foundation to enable social manufacturing through communities and order-driven static organizational shapes. SocialM has special resource organization logic, so the SMRs clustering method and SMEs selection method should be reconsidered.

Different from the conventional order-driven static organizational shapes, community is a major way to enable social manufacturing through social business relationship. The community is evolved in the self-adaptive progress of prosumer relationships and finally achieve a self-organized eco-system [6]. Social context is formed from both the medium and outcome of social interactions and manufacturing operations. By using proper analysis techniques, it is a source of community and participants' growth, as well as a starting point for various manufacturing knowledge such as how to enhance product performance or productivity based on findings from the context data [7, 8].

4.2 Establishing of Communities and Organizational Shapes

4.2.1 Clarify the Resources Community of Social Manufacturing

SocoalM is an SMRs-based aggregation manufacturing mode which clusters resources into resource communities by similarity, selects and combines proper resources to form the SMSEs community for satisfying customer requirements [9–11]. The community is a dynamic unity of interrelated prosumers who are held together by the common interest or goal of making an individualized product to meet a certain functional requirement or performance experience [12].

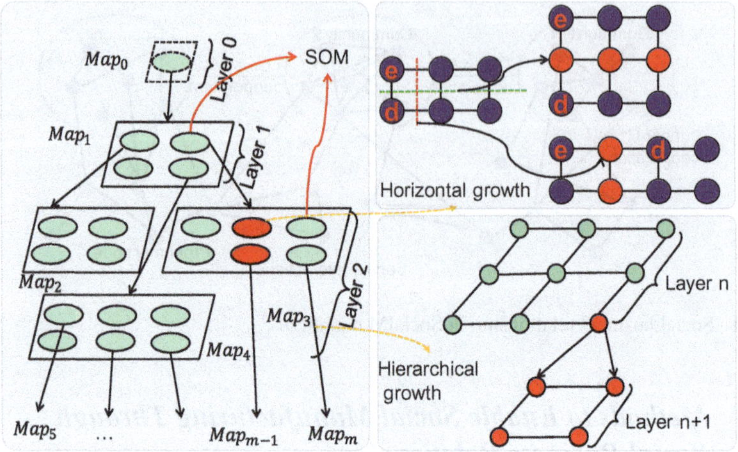

Fig. 4.2 The GHSOM with horizontal and hierarchical growth

4.2.2 The GHSOM Method for Resources Clustering

To establish the SMRs communities under the circumstance of SocialM, the growing hierarchical self-organizing map (GHSOM) is used to cluster the SMRs into communities according to their capabilities.

GHSOM is developed based on the method of self-organizing map (SOM), which is a type of artificial neural network trained through an unsupervised learning to produce a low-dimensional (typically two-dimensional) discretized representation of the input space of the training samples [13]. The architecture of SOM has two layers: input layer and output layer. Input layer consists of a set of input vectors $X = (X_1, X_2, \ldots X_n)^T$, which contains features value of the SMRs. And $X_n = (MF_n, MP_n, SA_n, SQ_n)$ represents the n-th SMR's details on machining function, machining performance service activity and service quality. Output layer consists of neurons and each neuron represents a cluster. A competitive learning mechanism is introduced in SOM to find the best matching unit (BMU) which the input vector belongs to.

Under the context of SocialM, the number of clusters cannot be determined in advance and a hierarchical structure is needed for identifying different discrimination among SMRs. To overcome the SOM limitations and the constraints, the growing hierarchical self-organizing map, which can dynamically fit its multilayered architecture according to the structure of the data, is used to tackle the clustering problem [14]. GHSOM consists of several layers, called as "*level*", and each "*level*" has one or more "*Maps*" with 2-D rectangular that can be arranged and visualized as a quad-tree-like structure, and each "*Map*" has even number of SOM (Fig. 4.2 shows).

Due to its dynamic nature, GHSOM is constantly growing. Figure 4.2 shows the two different growing ways of GHSOM [15]: (1) Horizontal growth is growing in a

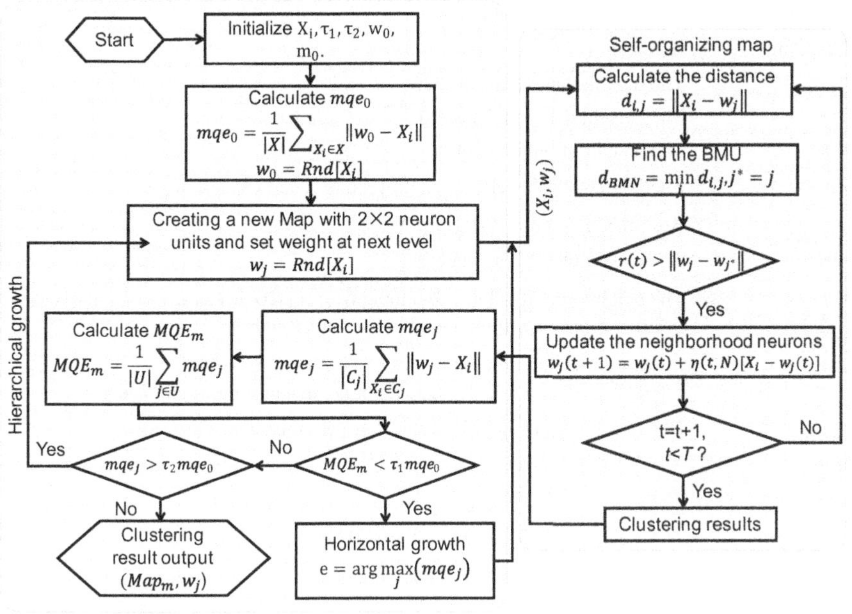

Fig. 4.3 Flowchart of GHSOM

"*Map*" unit which is enlarged by inserting a row or a column to subdivide SMRs' clusters, and (2) the clusters are checked, and if the SMRs mapped to one neuron are highly different than a presupposed threshold, another 2×2 neuron unit will be added in the next level below the parent neuron (hierarchical growth).

The flowchart of GHSOM is shown in Fig. 4.3 and the corresponding notations are detailed in Table 4.1. The two growths are controlled by two parameters, and the Map unit stops to grow after a certain point. For the horizontal growth, as long as $MQE_m < \tau_1 \cdot mqe_0$ exists, the training of current "*Map*" is continued. If the stopping criterion is not met, the unit *e* with largest mqe_j is selected and then the most dissimilar unit *d* is computed in the same "*Map*" with *e*. Between the neuron *e* and *d*, a row or column neurons are inserted into this "*Map*". The weights of new neurons are the average weight value of *e* and its neighborhood. After a horizontal growth reaches the stop point, the neurons in the "*Map*" are checked, if a neuron still needs a more detailed representation, this neuron extends to next level with 2×2 neurons unit for hierarchical growth. If $mqe_m < \tau_2 \cdot mqe_0$, the hierarchical growth of the current "*Map*" stops. The weights of new neurons are determined by the weight of their parent neuron. After horizontal growth or hierarchical growth, the SOM algorithm runs again for refreshing the results of clustering.

Table 4.1 Notations in GHSOM algorithm

Notation	Remarks
T	The maximal iterations of SOM
t	The current iterations of SOM
w_j	The weight of output neuron j
d_{ij}	Euclidian distance between input vectors X_i and weight w_j
d_{BMN}	Best matching unit (BMU) for X_i
w_{j*}	The weight of winning neuron
$r(t)$	The neighborhood radius of winning neuron, r(t) $= C_1(1 - t/T)$, C_1 is a positive constant with the numbers of output neurons
$\eta(t, N)$	Learning speed in training process, $\eta(t, N) = \eta(t)e^{-N}$, $t \uparrow \Rightarrow \eta \downarrow, N \uparrow \Rightarrow \eta \downarrow$, N is the topology distance between j and j^*
τ_1, τ_2	Horizontal growth and hierarchical growth threshold, respectively
mqe_j	The mean quantization error of neuron j, mqe_0 represents the layer 0's
C_j	The subset of the samples for which unit j is the BUM
MQE_m	The mean quantization error of mth "*Map*"
$\|X\|, \|U\|$	The number of samples and the number of X belongs to uth "*Map*"

4.3 Modeling of Product Order-Driven Service Interactions

Under the service-dominant logic, prosumers provide different kinds of competitive production services to the core enterprises, build outsourcing production orders with them, and further form an ecological production cluster within which the intertwined collaborations and interactions are operated. As SPs are in large amount, service interactions between core enterprise and SPs become vital and need to be well managed. Service interaction derives from the concept of service, and is defined as "*the direct interactions between service providers and customers to provide customers with timely and relevant information to enable them to make informed decisions, complete their work easily, and co-create added value*". The modeling and design method for service interaction is always empirical and lacks a unified graphical way to solve it.

With the aid of new information and communication technology (ICT), inter-enterprise interactions become all-around, efficient and online-offline integrated. Except for the inter-enterprise offline interactions as usual (e.g., material/finished parts transition), through online public platforms and interfaces of enterprise information systems (EIS), core enterprise and SPs could achieve real-time interaction and communication information such as collaborative design, progress monitoring, quality feedback and so on. Thus, online-offline integrated service interactions among core enterprise and SPs should be comprehensively considered.

4.3.1 Product Orders Under Service Interactions

To build the service interaction models, some definitions should be given first, e.g., service interaction, interaction content, etc.

Definition 4.3 Service interactions are defined as the collaboration or communication during a period of time where core enterprise directly interacts with its SPs. They can be divided into two kinds, i.e., online service interactions and offline service interactions.

Definition 4.4 Service interaction contents are defined as the information and data that core enterprise and SPs exchange, e.g., outsourcing order progress, service plans, production quality, etc.

Definition 4.5 Interaction events are defined as the activities triggered by core enterprise or SPs at certain time points to deal with the service interaction contents. We formalize the interaction events as follows:

$$E_i ::= \{Type, Tr, t, L, Dt, Info\} \tag{4.1}$$

where E_i is the i-th event occurred between core enterprise and SP; $Type$ is the online or offline interactions; Tr represents who triggers the event (core enterprise, SP or both); L is the event occurrence location; Dt is the state duration time after the event occurred; $Info$ gives the other information of the event.

Definition 4.6 Interaction event connectors are defined as the connection relationships between two events, including three kinds: ordinal connector ("+"), parallel connector ("||"), alternative connector ("⊕"). E1 + E2 indicates that event E1 and E2 occur in a time sequence; E1||E2 indicates event E1and E2 occur consequently; E1 ⊕ E2 indicates that there could be only one event occurs among event E1 and E2. The priority of the three connectors is defined as: "+">"||">"⊕"

Definition 4.7 Interaction event flow is generated by connecting events with different connectors. It provides inspections to the overall outsourcing service interaction processes. It can be formalized as:

$$EF = \{E_1, E_2, \ldots, E_n\} \bowtie \mathbf{M}_{n\times n} \tag{4.2}$$

where $\mathbf{M}_{n\times n}$ is the event connection matrix, made up of the three kinds of connectors, $\bowtie$ is the natural connector in relational algebra.

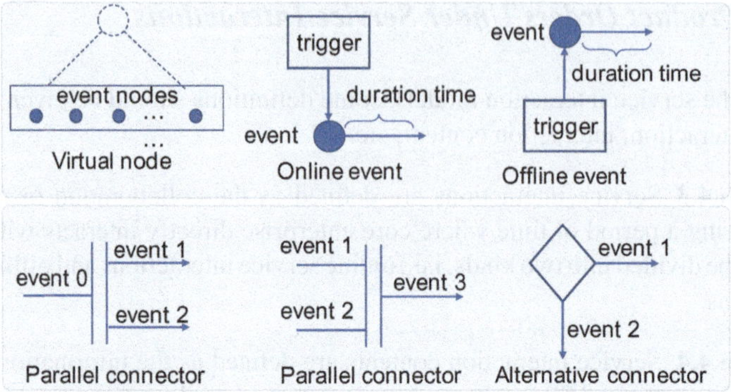

Fig. 4.4 Graphical units for the service interaction

4.3.2 Order Decomposition and Prosumers Selection

To efficiently execute the outsourcing plans, the production orders around a certain product need to be decomposed and composed first. The decomposition method is based on the product Bill of Material (BOM) and its parts grouping methods (which solves the customization/personalization problems of different customer orders).

We just simply describe the SP selection processes. The detailed method and tools can be referred in our previous work [16, 17]. First, core enterprise releases its order requirements at the online platform, and then SPs can apply for it based on its current manufacturing capability and production capacity. Second, core enterprise evaluates the SPs with two aspects: (1) Estimate their capabilities and capacities using online evaluation APP tools; and (2) Investigate the SPs offline. Note that for the stable SPs, they always have been investigated before, thus only the temporary SPs need to be investigated in this step. Third, optimal SPs are selected, core enterprise collaborates and interacts with them to efficiently accomplish the outsourced production tasks. Multiple outsourced orders correspond to multiple service interaction flows, these inter-enterprise service interactions are complex and intertwined, thus unified graphical modeling and analyzing method should be developed for it.

4.3.3 Graphical Modeling Method

Based on the above definitions and preparation work, a graphical modeling method for inter-enterprise service interactions is proposed, and a new graphical modeling method is extended into inter-enterprise service interaction situation.

(1) *Graphical units*: graphical units are firstly formalized as shown in Fig. 4.4 to build a service interaction flow (Fig. 4.5). The units are classified into 6 kinds: 2 event units (online event and offline event), 3 connector units and 1 virtual node unit.

For online/offline event unit, it describes who triggers the event, and after the event occurs, the state sustains a duration time. To reduce the complexity, some

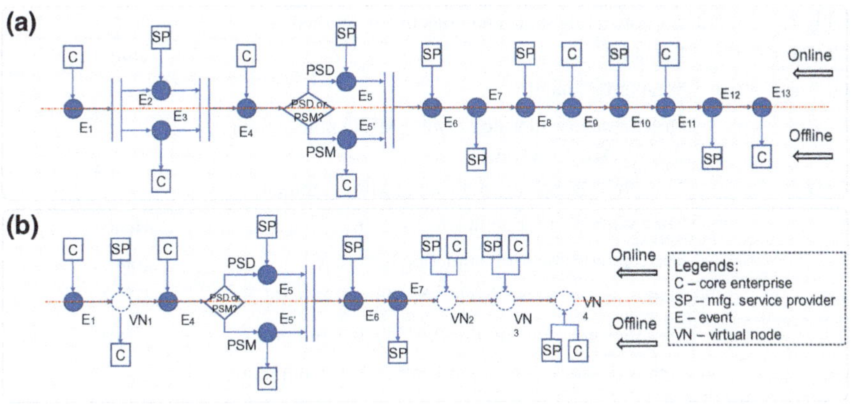

Fig. 4.5 Service interaction event flow

inter-connected event units and connector units can be abstracted as a virtual node unit, which is helpful to analyze the interaction granularity [18, 19] and simplify the interaction processes. The virtual node unit can be formalized as:

$$VN_i = \{E_1, E_2, \ldots, E_m\} \bowtie \mathbf{M}_{m \times m} \quad m < n, \tag{4.3}$$

(2) *Interaction event flow modeling*: by connecting the above graphical units, the real service interaction process flow can be modeled as an interaction event flow. It provides better instructions for core enterprise and SPs. Take a part manufacturing order as an example. The core enterprise outsources the task to an SP and reaches a production order with it

4.4 Order-Driven Social Manufacturing Network

According to customer requirements on capability, suitable SMRs communities are selected. Since customer requirements on performance have strong relationship with SMEs' production cost and scheduling, the SMRs should be mapped into SMEs communities with different order allocation strategies. Order allocation is not a novel problem in engineering field and a lot of intelligent algorithms are applied to tackle it. Since customer requirements on performance are multifold, it is a multi-objective optimization problem for order allocation. Swarm intelligence algorithm is one of the most effective ways to find the optimum results for multi-objective optimization, such as genetic algorithm (GA), particle swarm optimization (PSO) and so on [20]. But most allocation optimizations easily trap in local optimum, a modified multi-objective bird swarm algorithm (MOBSA) is used to find the most suitable SMEs community.

Table 4.2 Notations of operating logic and objective functions

Notation	Remarks	Unit
O	Total order quantity	
λ_n^m	Order quantity allocated to *n*th SME in *m*th community. m ∈ [1, M], n ∈ [1, N_m]	
SME_n^m	The *n*th SME belong to the *m*th SMR community	
C	The totally cost for completing the order	RMB
T	The longest time for completing the order	min
pc_n^m	Unit processing cost of *n*th SME in *m*th community	RMB
ic_n^m	Unit inventory cost of *n*th SME in *m*th community	RMB
pt_n^m	Unit processing cost of *n*th SME in *m*th community	min
mr_n^m	Number of equipment devoted of for *n*th SME in *m*th community	
lc	The logistics cost for completing the order	RMB
lt	The logistics time for completing the order	day
α	Unit processing cost elastic coefficient with λ_n^m	
β	Unit inventory cost elastic coefficient with λ_n^m	
γ	Unit processing time elastic coefficient with λ_n^m	
pc^L, pc^U	Upper and lower limit of unit processing cost	
mr^U, mr^L	Upper and lower limit of devoted equipment	

Based on the results of SMRs clustering, the SMRs communities which can satisfy the capability requirements of customers are selected. According to the definition of SMRs communities, the SMEs in SMRs community are qualified. With order allocation strategies on SMEs, different SMEs communities are established. There are more than one SMEs communities can meet the performance requirements of customers. The production cost and delivery time are chosen to represent the performance of SMEs community. A modified multi-objective bird swarm algorithm is proposed to select one optimal SMEs community. This multi-objective order allocation problem has two objective functions: cost function and time function, corresponding notations are detailed in Table 4.2.

$$\min_{\lambda_n, mr_n^i} C_n = \lambda_n \cdot \left(\sum_{i=1}^{K} pc_n^i \cdot \lambda_n^{-\alpha} \cdot \left(mr_n^i\right)^{\beta} + lc_n \right) + \frac{ic_n}{2} \cdot \sum_{i=1}^{K} pt_n^i \cdot \lambda_n / mr_n^i, \quad (4.4)$$

$$\min_{\lambda_n, mr_n^i} T_n = \left| T_{per} - \sum_{i=1}^{K} \frac{\lambda_n}{mr_n^i} \cdot pt_n^i - lt_n \right|, \quad (4.5)$$

subject to $\lambda^U \leq \lambda_n \leq \lambda^L; \lambda_1 + \lambda_2 + \ldots + \lambda_n = O; mr_n^{iL} \leq mr_n^i \leq mr_n^{iU};$

Cost function of an SMEs community consists of production cost, logistics cost and inventory cost. Production cost can increase with the rise of unit production cost pc and quantity. Inventory cost is determined by processing time $\sum_{i=1}^{K} pt_n^i \cdot \lambda_n / mr_n^i$ and unit inventory cost ic. Since the unit process cost can decrease with the rise in λ_n and increase with increasing mr_i^n, α and β elastic coefficient are used to adjust the cost.

Time function consists of production time and logistics time. The object of time function equals to preferred production interval minus actual service time and logistics time. The preferred production interval is predefined according to customer requirements. Here, delivery time represents the difference between preferred production interval and actual production time. Since SocialM is a multi-provider manufacturing mode, the total cost function and time function can be calculated by adding every SME's values. $C = \sum_{n=1}^{N} C_n, T = \sum_{n=1}^{N} T_n$, N is the quantity of SMEs in the community.

4.4.1 The MOBSA for Order Allocation to Social Manufacturing Communities

Bird swarm algorithm (BSA) is proposed by Meng in 2016 [21], which is based on the swarm intelligence extracted from the social behaviors and social interactions in bird swarms. It mimics the foraging behavior, vigilance behavior and flight behavior of birds. Foraging behavior means that each bird searches for food according to its previous experience and the swarms' experience. This activity aims at searching for feasible solutions and then identifying the dominate solutions. Vigilance behavior means that birds try to move to the center of the swarm for foraging, and they would inevitably compete with each other. To avoid this phenomenon, some birds would not directly move towards the center of the swarm and keep vigilance (avoid trapping in local optimum). Flight behavior means that birds may fly to another site by the frequency FQ. When arriving at a new site, some birds acting as producers can search for food patches, while others acting as scroungers follow the producers.

BSA is applied to find the global optimal solution of single objective optimization, but SMEs community selection has two objectives. We modify BSA with the global best g_j and the nondominated solutions filter mechanism to form a multi-objective bird swarm algorithm (MOBSA). The flowchart of MOBSA is shown in Fig. 4.6 and the corresponding notations are detailed in Table 4.3. Each bird is composed of $N \cdot (1 + K)$ dimension variables: $\lambda_1, mr_1^1, mr_1^2, \ldots, mr_1^K, \lambda_2, mr_2^1, \ldots, mr_2^2, mr_2^K, \ldots, \lambda_n, mr_n^1, mr_n^2, \ldots, mr_n^K$.

Unlike the single-objective optimization problem, MOBSA for multi-objective optimization needs a procedure to select the global best positions during iterations. We can draw lessons from multi-objective particle swarm optimization on how to

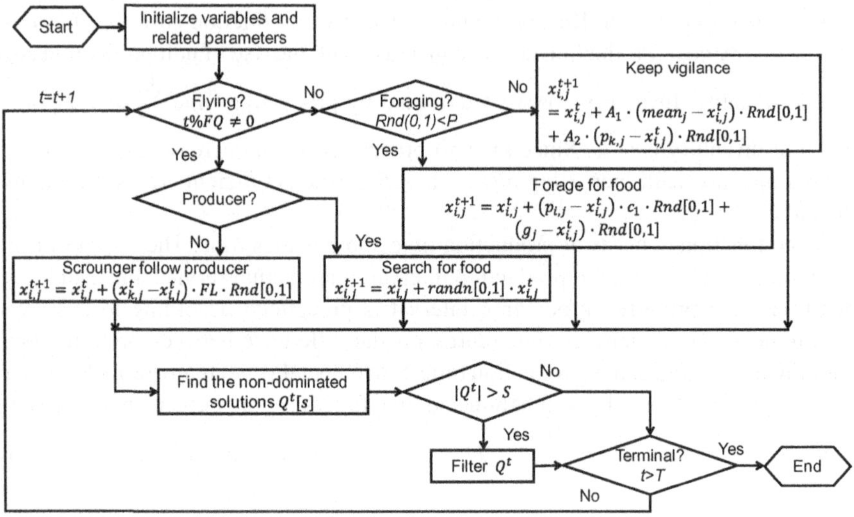

Fig. 4.6 Flowchart of MOBSA

Table 4.3 Notations in MOBSA algorithm

Notation	Remarks
P	The number of candidate birds in MOBSA
T	The predefined max iteration number of MOBSA
FQ	Unit interval of a bird flies to another place $PF \in (0, 1)$
PF	The probability of foraging food
c_1, c_2	Cognitive and social accelerated coefficients, respectively
FL	Coefficient means the scrounger would follow the producer $FL \in [0, 2]$
$Q^t[]$	A set stores the potentially nondominated solutions
S	The predefined max size of nondominated solutions.
x_i^t	Strategy of ith bird in tth iteration
$p_{i,j}$	The best previously visited position of ith bird for jth dimension
g_j	The best previously visited position shared by the swarm for jth dimension
t	Current iteration of MOBSA
A_1	$A_1 = a_1 \cdot exp\left(-\frac{pFit_i}{sumFit+\varepsilon} \cdot N\right)$
A_2	$A_2 = a_2 \cdot exp\left(\left(\frac{pFit_i - pFit_k}{\lvert pFit_k - pFit_i \rvert + \varepsilon}\right) \cdot \frac{N \cdot pFit_k}{sumFit+\varepsilon}\right)$

select global best g_j from nondominated solutions. The simplest method is to randomly select a point from $Q^t[s]$. The crowding distances method is proposed to calculate distance using binary tournament, and the least crowed solution with the highest distance is regarded as global best [22]. Another useful selection mechanism is called Sigma method that calculates the sigma value of each particle, and then calculates the distance between two particles to find minimum distance particle as the global best.

We have proposed a hybrid selection mechanism which combines crowding distances method and Sigma method. Firstly, all nondominated solutions are sorted in descending order according to crowding distance and only top 10% are selected for the candidates. Then, the sigma method is adapted to find the global best among the candidates. In order to filter $Q'[]$, the crowding distances method is applied once again.

The crowding distance and σ equations are shown as follows:

$$I[i]_{distance} = I[i]_{distance} + \frac{(I[i+1]\cdot m - I[i-1]\cdot m)}{\left(f_m^{max} - f_m^{min}\right)}, i\ iterate\ from\ 2\ to\ P-1 \quad (4.6)$$

$$\sigma = \frac{(K_2 f_1)^2 - (K_1 f_2)^2}{(K_2 f_1)^2 + (K_1 f_2)^2} \quad (4.7)$$

4.4.2 An Example of Competitive Order Allocation Inside a Community

An example of competitive order allocation inside a community from a National High-Tech Industrial Development Zone of China is presented. The main business of the firm is gravure press which operates at high speed, carries a layer of ink to a doctor blade disposed at a relatively low angle to the cylinder surface. Almost 75% components of gravure press are outsourced. According to the records of the firm, around 120 SMEs has participated in the manufacture of the gravure press. We select the cone head of the gravure press as an example, since it is a typical and individual part which demands high machining precision and is largely used in gravure press. It has six operations from workblank to the finished product and can be divided into three types: turning, milling and drilling. Customer releases 1200 jobs to the Web-based SocialM platform. Based on calculation, the platform selects the most suitable SMEs community to complete the order.

In the first place, we have chosen 106 types of machine tools in 52 SMEs from the SocialM platform and 43 types of machine tools from internet for training GHSOM. The input parameters for the algorithm are $\tau_1 = 0.6$, $\tau_2 = 0.001$. These SMRs can be clustered into 57 SMRs communities. Based on the training results, all the SMRs are clustered into these communities. The turning-lathe communities are chosen to represent the clustering result as Fig. 4.7 shows. There are 15 SMEs with 54 types of turning-lathes which can be clustered into 20 communities by 4 layers. The more similar these SMRs are, the deeper layer can distinguish them.

According to the technological requirements of cone head, three turning-lathe communities have the machining capability (red cycles in Fig. 4.8). The three com-

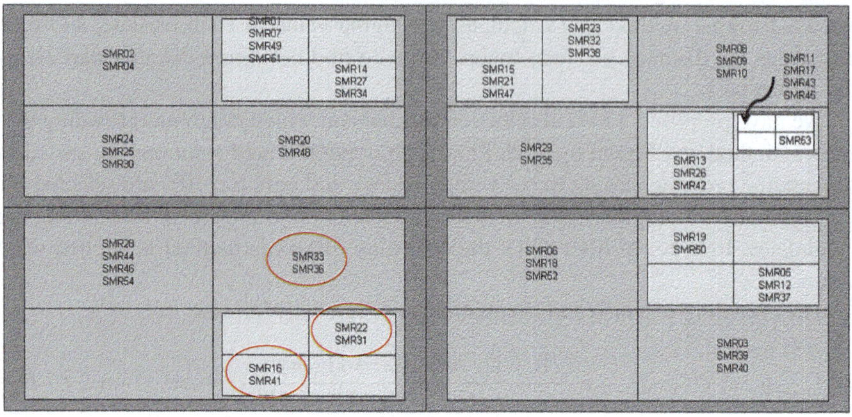

Fig. 4.7 The SMRs communities for turning-lathe

Fig. 4.8 Nondominated solutions without filter mechanism

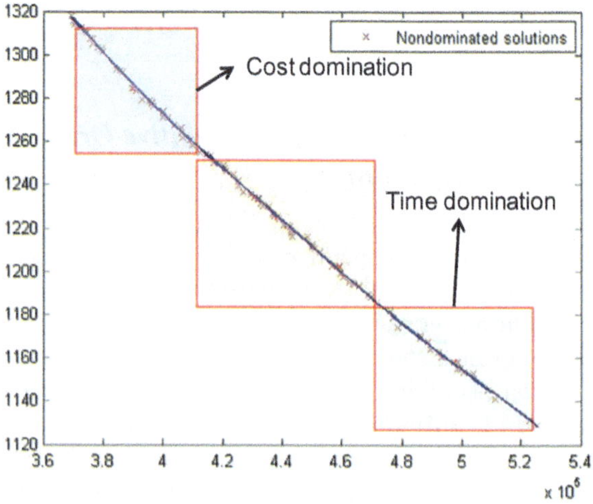

munities can be seen as one community that shows the high adaptability of GHSOM. SMR22 and SMR41 belong to the same SME, so there are five SMEs for turning processes. By the same process, milling SMRs communities and drilling SMRs communities are selected. The intersection of the three communities is the SMEs that can satisfy the capability requirements. In this case, there are three SMEs in the intersection. Then the performance requirements should be considered for order allocating and selecting of the best SMEs community. The input parameters of MOBSA are shown in the Table 4.4. The nondominated solutions without filter mechanism are shown in Fig. 4.8 that can be divided into three parts: cost domination, time domination and equilibrium part. By applying filter mechanism, the best solution is shown in Table 4.4. The production cost and delivery time are 4.32×10^5RMB and 1.22×10^3min, respectively (Table 4.5).

Table 4.4 The input parameters of MOBSA

	P	T	FQ	c_1, c_2	a_1, a_2	M	α	β	O	T_{per}
SME_1	100	200	15	1	1.5	5	−0.014	0.1	1200	14
SME_2							−0.018	0.13		
SME_3							−0.013	0.12		

Table 4.5 The result of the demonstrative case

	λ_n	mr_n^1	mr_n^2	mr_n^3	mr_n^4	mr_n^5	mr_n^6	C (RMB)	T^a (min)
SME_1	374	5	3	3	5	3	3	432,540	122.5
SME_2	337	4	3	3	2	3	3		
SME_3	289	4	1	2	3	2	2		

[a]Assuming that every SME works 10 h per day

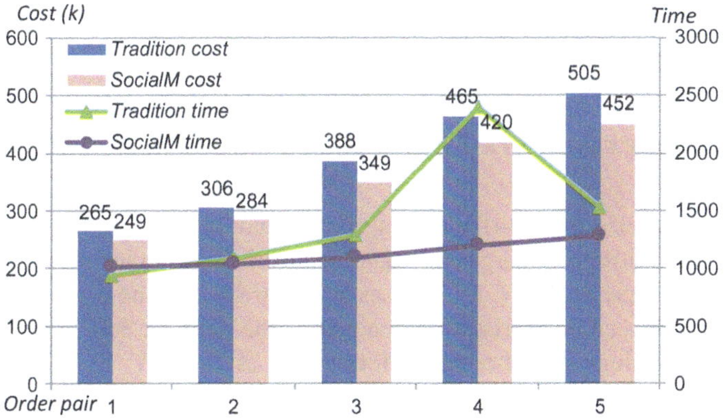

Fig. 4.9 Comparison between traditional manufacturing and SocialM

As a new manufacturing mode, SocialM allocates the order to multiple SMEs. The professional printing firm in the case study usually outsourced orders to previous manufacturer. For fast responding to the market and customers, this printing firm optimized the outsourcing strategies by applying SocialM. Five pairs of orders were selected to compare SocialM with traditional manufacturing on production cost and delivery time. Each pair of order has same quantity within the year between 2012 and 2016. The comparison result is shown in Fig. 4.9. The histogram represents the comparison on production cost and line graph represents the comparison on delivery time.

By analyzing the Fig. 4.9 and algorithm results, we can draw some conclusions as follows: For production cost, traditional manufacturing and SocialM have little difference when the quantity of order is small. Since a single enterprise can complete the order in specific time, then deliver to the customer. If the quantity of order is more than 1000, the SocialM production cost is far lower than traditional manufacturing.

For delivery time, SocialM doesn't show the absolute advantage. The delivery time of SocialM and traditional manufacturing are almost the same. However, SocialM is more stable than traditional manufacturing. In the fourth order pair, traditional manufacturing takes a really long time to finish the order. We look up the document of that order, the manufacturer cannot deliver on time since due to insufficient supply of raw materials. SocialM shares the risks to each SME that can ensure the delivery time.

According to the algorithm results, the rules can be found when the quantity of order is less than 1000, two SMEs participate in the SMEs community; When the quantity of order is between 1000 and 1700, three SMEs participate in the SMEs community; When the quantity of order is more than 1700, four or more SMEs participate in the SMEs community.

4.5 Modeling and Analyzing of Social Manufacturing Network

4.5.1 Complex Network Theory-Based Network Modeling

In order to organize the distributed enterprises and analyze the characteristics of enterprise relationships under social manufacturing, the complex network theory is applied to build the SMERN. Enterprises are abstracted as the network nodes, the relationship between enterprises are denoted as the network edges. On this basis, the network topology model could be formulated as $G = (V, E, W)$, where $V = (v_1, v_2, \ldots, v_N)$ represents the enterprise node set, $E = (e_1, e_2, \ldots, e_M)$ represents the relationship set between pairs of enterprises, and $W = (w_1, w_2, \ldots, w_M)$ is the holistic weight of edges. The adjacency matrix is denoted as $A = \left(A_{ij}\right)_{N \times N}$, where A_{ij} is proposed as a vector and denoted as $A_{ij} = \left[S_{ij}, F_{ij}\right]^T$, S_{ij} and F_{ij} stand for the enterprise similarity and enterprise collaboration relationships between v_i and $v_j$vj, respectively.

4.5.2 Characteristic Analysis

After building the network model, some characteristics of SMERN should be discussed to evaluate the organizing result of enterprises and the network performance. Each complex network presents specific topological features which characterize its connectivity and highly influence the dynamics of processes executed on the network. They are measured by some indices such as node degree, degree distribution, average distance, clustering coefficient, centrality, vulnerability, etc. These indices could reveal some important features and phenomena of the existing network. We build a series of characteristics indices from view of both network topological structure and its physical meanings to measure the SMERN.

4.5.3 Topological Characteristics

Definition 4.8 The strength of enterprise node. The strength of enterprise v_i belongs to [0,1] and depicts the strength v_i links with other enterprises, so we define the strength of enterprise as follows:

$$D_i = \sum_{j}^{j \leq n} S_{ij} \quad \forall i, j, S_{ij} \in [0, 1] \tag{4.8}$$

where S_{ij} represents the similarity value between enterprise v_i and v_j. Larger D_i indicates more links with other enterprises.

Definition 4.9 Enterprise node betweenness. Enterprise node betweenness is one of the standard measures of node centrality, and usually represents the importance of an enterprise in the whole network. The betweenness of enterprise v_i is denoted as b_i, which can be defined as follows:

$$b_i = \sum_{m,n} \frac{g_{mn}(i)}{g_{mn}} m \neq n, m, n \neq i \tag{4.9}$$

where A_{ij} is the adjacency matrix, g_{mn} is the number of the shortest paths from enterprise v_m to v_n; $g_{mn}(i)$ represents the number of the shortest paths connecting v_m and v_n and passing through enterprise v_i.

Definition 4.10 Clustering coefficient. Clustering coefficient c_i describes the degree to which enterprises tend to cluster together in into a network. The larger c_i of an enterprise indicates more enterprises tend to connect with it. c_i is defined as follows:

$$c_i = \frac{N_i}{(1/2)k_i(k_i - 1)} = \frac{\sum_{j,m} A_{ij} A_{jm} A_{mi}}{k_i(k_i - 1)} \tag{4.10}$$

where k_i is the number of adjacent enterprises connecting with enterprise v_i. $(1/2)k_i(k_i - 1)$ is the maximum number of possible edges between v_i and its adjacent enterprises. N_i is the actual number of edges connecting enterprise v_i and its adjacent enterprises. Obviously, $0 \leq c_i \leq 1$. Based on that, the average clustering coefficient of network is defined as the average value of $\boldsymbol{c_i}$ of all enterprises in the network and is described as follows:

$$\boldsymbol{C} = \frac{1}{\boldsymbol{N}} \sum_{i=1}^{\boldsymbol{N}} \boldsymbol{c_i} \tag{4.11}$$

To better understand the special physical meanings, several indicators are proposed, which is helpful for analyzing and optimizing the network performance.

Definition 4.11 The extending range (ER). Extending range of enterprise v_i is written as ER_i, which describes the business extensibility of enterprise v_i in the whole network.

$$ER_i = \frac{100}{N-1} \sum_{m \in N, n \neq m} \frac{1}{g_{mn}(i)} \tag{4.12}$$

It should be noted that $g_{mn}(i)$ may be ∞ if there is no directed edge connecting v_m to v_n.

Definition 4.12 The inter-MC correlation coefficient (IC^2IC³). The inter-MC correlation coefficient of enterprises is defined to find versatile enterprises which could undertake multiple OMS tasks. IC^2 is described as follows:

$$IC_i^2 = \frac{D_i}{\frac{1}{2}\sum_k^{k \leq N} D_k} \tag{4.13}$$

where $\frac{1}{2}\sum_k^{k \leq N} D_k$ denotes total strength of the MC, N is the number of enterprises in the MC. There is a minimum intra-MC correlation coefficient IC_{min}^2, For each enterprise in the MC, if $IC_i^2 > IC_{min}^2$, the enterprise is a versatile and the bigger the IC_i^2 is IC_i^3, the higher extent its production diversification on OMS can be.

Definition 4.13 The collaboration importance (CI). In the e-production chain, enterprises share and spread production information, and collaborate for their common goals and interests. We define CI to describe the collaboration importance when subjected to the removal of enterprises (i.e. cancel the certain collaboration).

$$CI_i = \frac{100}{N(N-1)} \sum\nolimits_{m,n \in G'', m \neq n} \frac{1}{g_{mn}(i)}, \forall i \in G' \tag{4.14}$$

where $g_{mn}(i)$ is the shortest path from enterprise m to n after the removal of the node i and all its edges, $G' = (V', E', W')$ represents the second level network. The removal of important collaboration would cause the collapse of the network's main branches.

Definition 4.14 Collaboration stability (CS). The collaboration stability is defined to depict the stability and orderliness of enterprise collaborations, which is derived from the concept of entropy in thermodynamics discipline. The $CS_i$$\mathrm{CS}_\mathrm{i}$ is formulated as follows:

$$CS_i = -\sigma \cdot \sum_i P(i) \cdot \log(P(i)) \tag{4.15}$$

where $P(i)$ is the distribution of enterprise node strength, σ is an index to normalize the CS_i. Note that the collaboration stability changes with time dynamically because of the add-in and removal of some certain collaborations with core product manufacturers.

4.5.4 An Example

We take 151 SMSEs specialized at different types of OMS tasks to form the SMERN and analyze its characteristics. For simplicity of gathering data, these enterprises are mostly from the *Weinan* National High-Tech Industrial Development Zone and *Xi'an* Economic & Technological Development Zone of China, which play the most important role in China's northwest region industrial revitalization. Firstly, the formal description of enterprises is implemented and all the SMSEs registered their SMRs and production capabilities in the social manufacturing platform. Then they aggregate to join in different MCs. For example, *Dadong* Machinery Company, *Zhengqi* Printing Machinery Company, *Qinya* Printing Machinery Company, etc., form a horizontal MC to undertake the guide roller machining tasks. Thus, the first level network of SMERN is formed and the characteristics are analyzed. After that, according to the manufacturing tasks, proper enterprises are selected to reach collaborations through matchmaking. Around a certain product's manufacturing tasks, a vertical MC is built and enterprises in this vertical MC collaborate with each other to satisfy the production control of the certain product. For example, around the *FR300* printing machinery's manufacturing tasks, *Shaanxi Beiren* Printing Machinery Company form a vertical MC with other 29 SMSEs, each of them undertakes a certain kind of manufacturing task. The relationship matrix of enterprises is set and the second level network of SMERN is formed. Note that the second level network composes of all the vertical MCs formed in the social manufacturing platform. The characteristic analysis of the second level network could indicate the key enterprises, collaboration importance and stability, etc.

To form the first level network, the enterprise similarity should be calculated and then the clustering algorithm is implemented. Take *Juying* Machinery Company and *Sailong* Machinery Company which are specialized at shell class part machining as an example. The data for similarity calculation is listed in Table 4.6. There are 3 kinds of SMRs and the similarities between them are calculated based on the SMR ontology similarity calculation in Sect. 4.2.2. $S_{SMR}(\text{Juying, Sailong}) = 0.78$, $S_{PC}(\text{Juying, Sailong}) = 0.665$. Given that $\omega = 0.5$, then the comprehensive similarity between Juying and Sailong is $S(\text{Juying, Sailong}) = 0.5 \times 0.78 + (1 - 0.5) \times 0.665 = 0.72$.

Accordingly, the similarities between other enterprises are calculated in the same way. Based on the enterprise similarity, the adjacency matrix is built and the clustering algorithm is implemented. Based on that, the first level network with 11 kinds of MCs is established in Fig. 4.10. The task types that the MC could undertake and its scale are listed in Table 4.7. Note that the above results are recommended by the platform

Table 4.6 The data for similarity calculation

Items		Juying	Sailong
SMRs	CNC planer type milling machine	XK2408C, 3 sets	XK2730, 2 sets
	Gantry machining center	TK621, 2 sets	TK621, 2 sets
	CNC boring and milling machine	TK6213, 3 sets	TK6916, 4 sets
Production capability	max_A	3 pcs	5 pcs
	min_C	800 RMB	1100 RMB

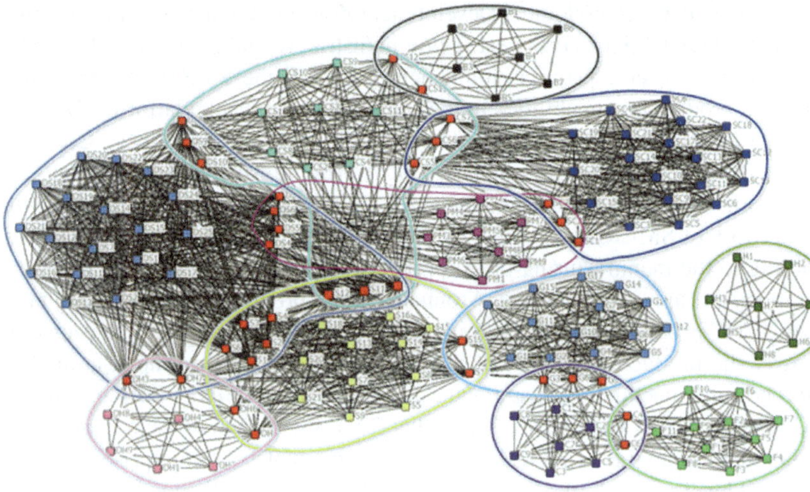

Fig. 4.10 The first level network (with MCs)

as the default MC structure. In reality, SMSEs could dynamically join in or quit from a MC autonomously during the operation phase of the platform based on the recommended results. Here, we omit the self-organization process for simplicity.

The results show that the MC specialized at disc-sleeve type components machining (DS) is the largest one in the network. We can infer from the above that the demand of this kind of OMS is huge, while the threshold to get into this industry is relatively low, but in the meantime, the profits may be low and the competition may be fierce. The minor scale MCs are deep hole processing, blade processing and heat treatment processing, which means the competition in these MCs may not be fierce or the threshold to get into this industry may be relative high. The distribution of task types also reflects the industry structure of the two industrial parks.

After forming the network, the clustering coefficient, node strength, node betweenness, intra-MC correlation coefficient and extending range are calculated.

Table 4.7 The task type and the number of versatile enterprises in each MC

Task type	MC scale	Versatile
Deep hole processing (DH)	10	4
precision mold processing (PM)	19	1
Complex shaped parts processing (CS)	19	11
Blade processing (B)	9	2
Shell class part machining (SC)	25	6
Shaft machining (S)	25	12
Disc-sleeve type components machining (DS)	35	17
Gear processing (G)	22	5
Casting processing (C)	12	5
Forging processing (F)	13	2
Heat treatment processing (H)	8	0

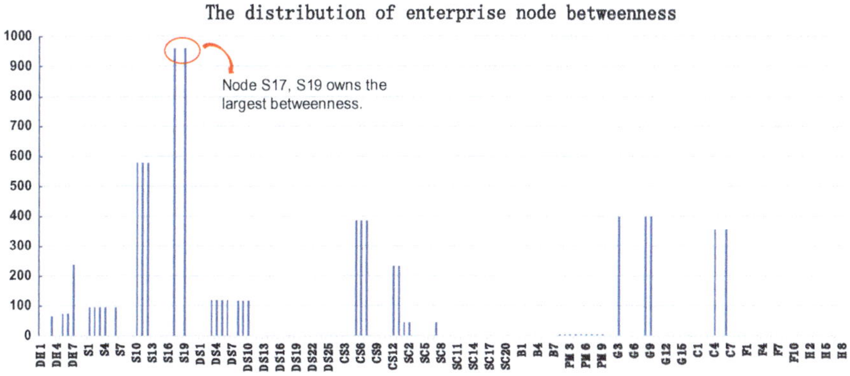

Fig. 4.11 The distribution of node betweenness

The average clustering coefficient of network is 0.713, which means the network could be easily clustered. Enterprises S10-S12 have the biggest strength $D = 31.2$, and their links with other enterprises are denser. That is easy to understand, because enterprises S10-S12 belong to three large MCs. Enterprises H1-H8 have the smallest degree $D = 4.6$, which means they are relatively independent. The result of node betweenness (as shown in Fig. 4.11) shows that enterprise S17, S19 has the biggest node betweenness $b = 964.1$, which means the shortest paths through these two enterprises are the most, and these two enterprises play a more important role in the network. The extending range (ER) of enterprises depicts the extendibility of enterprise business scope, the enterprise S10-S12 have the smallest $ER = 1.33$ and the enterprise F1-F11 has the biggest $ER = 4.85$ except for the standalone H1-H8, which indicates that they could easily expand their business to other task types.

As for the versatile enterprises, we take the MC specialized at shaft machining for example. In Fig. 4.12, the IC^2 of enterprise nodes are calculated. Except for

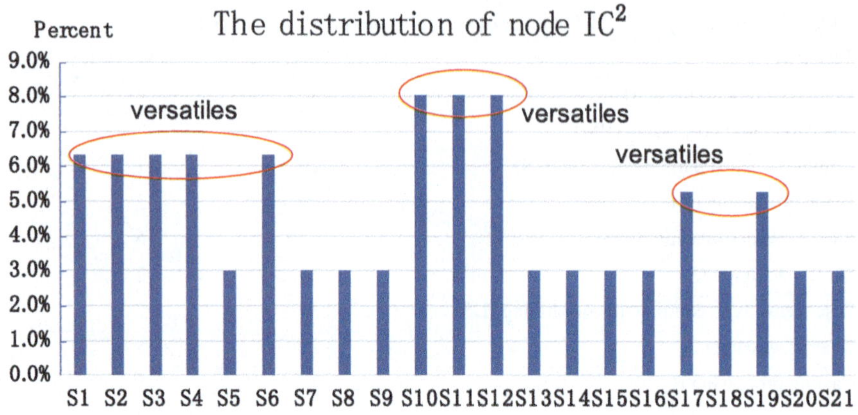

Fig. 4.12 The distribution of node IC^2

the minimum IC^2IC^3, the enterprises with bigger IC^2IC^3 are all versatiles and the bigger the IC^3IC^2 is, the more kinds of OMS tasks the enterprise can take. Here, the enterprises S10-S12 have the biggest IC^2 value, they can provide 3 kinds of OMS. The number of versatile enterprises of each MC is also listed in Table 4.7.

In this section, we build the second level network based on the OMS tasks from 4 different core product manufacturers, whose products are *FR300* printing machine, *FX90L2* textile motor, *ZD160-3* bulldozer and *ZE150E* excavator, respectively. There are 100 SMSEs from different MCs undertaking some of the OMS tasks.

Firstly, 4 core product manufacturers negotiate with different MCs according to the task type they need to outsource. Note that the MCs could neglect the tasks if the payment or negotiation do not meet with their expectation because they manage their SMRs by themselves. Here, we take the *Shaanxi Beiren* Printing Machinery Company as an example. The OMS tasks are from its FR300 printing machine. There are 7 kinds of OMS tasks, 24 SMSEs from 7 MCs are selected, forming an e-production chain (No.1 in Fig. 4.13). There are not only inter-MC collaborations between core enterprise and SMSEs, but also intra-MC collaborations among SMSEs. Their outsourcing relationships data were get from the social manufacturing platform. Table 4.8 gives partial relationships data between these enterprises. If relationship $F_{ij} = 1$, enterprise j undertakes 100% of an OMS task from enterprise i and reaches an agreement, if $0 < F_{ij} < 1$, enterprise j undertakes F_{ij} percent of an OMS task from enterprise i and reaches an agreement, while $F_{ij} = 0$ means enterprise i and j do not have an agreement.

Based on the initialization of collaborations around the 4 core product manufacturers, the second level network is formed. Note that the formed network is a weighted undirected network. As shown in Fig. 4.13, the red nodes are the core product manufacturers, and others in different colors represent enterprises from different MCs. Around the core product manufacturers, 4 e-production chains are established where enterprises interact with each other through social media [23], sharing the production

Table 4.8 Partial relationships data between enterprises

	C1	S1	S2	S3	S4	S5	S6	S7	S8	S9	S10	…
C1	1	0	1	0.26	0	0.76	0	0.33	1	1	0.36	
S1	0	1	0	0	1	1	0	1	1	0	0	
S2	1	0	1	0	0.51	1	0.61	0	0	0.53	0.87	
S3	0.26	0	0	1	0	0.32	0.51	0.89	0.43	0.46	0	…
S4	0	1	0.51	0	1	0	1	0	1	1	0	
S5	0.76	1	1	0.32	0	1	0	1	1	0	1	
S6	0	0	0.61	0.51	1	0	1	0	0	1	0	
S7	0.33	1	0	0.89	0	1	0	1	1	0	1	…
S8	1	1	0	0.43	1	1	0	1	1	0	1	
S9	1	0	0.53	0.46	1	0	1	0	0	1	0	
S10	0.36	0	0.87	0	0	1	0	1	1	0	1	
S10	0.36	0	0.87	0	0	1	0	1	1	0	1	…
…		…			…			…			…	

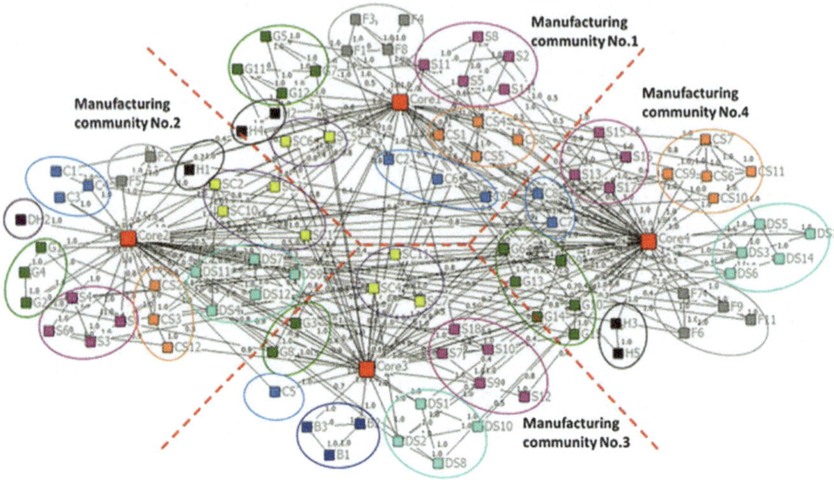

Fig. 4.13 Forming the ECN and manufacturing communities

progress and material flow information to finish the production tasks under proper makespan and cost. Based on that, core product manufacturers could adjust the production chain planning dynamically and make beneficial decisions to react to the dynamic market proactively.

Some characteristics of the second level network are discussed in this section. The distribution of enterprise node strength is shown in Fig. 4.14, which has not displayed a power law distribution. Therefore, the second level network does not have the scale-free property. However, the network has both a smaller shortest path (2.307) and a higher clustering coefficient (0.36). Therefore, the network has the small-world property. The collaboration importance CI in the network is calculated. Here, we take the No.1 e-production chain (Beiren-centered) for example. The CI value after the removal of nodes is depicted in Fig. 4.15. From the curve, we could easily find that when deleting the collaborations with enterprise SC5 (KESAI Mechanical & Electrical Equipment Company), the network has the smallest value $CI = 31.04$, thus CS5 is the most important enterprise in this e-production chain. The collaboration stability (CS) of enterprises indicates the service loyalty and service quality of OMS tasks. The enterprise Core4 (Weinan Aoma Machinery Company) has the smallest $CS = 0.35$, i.e. the collaborations with Core4 is relatively stable. On the other side, enterprise Core4 is highly loyal to its cooperators and gives high evaluation to their service quality.

Note that the second level network is dynamic and self-organized. The existing collaborations between enterprises may end and dissolve, and potential enterprises may have new collaborations.

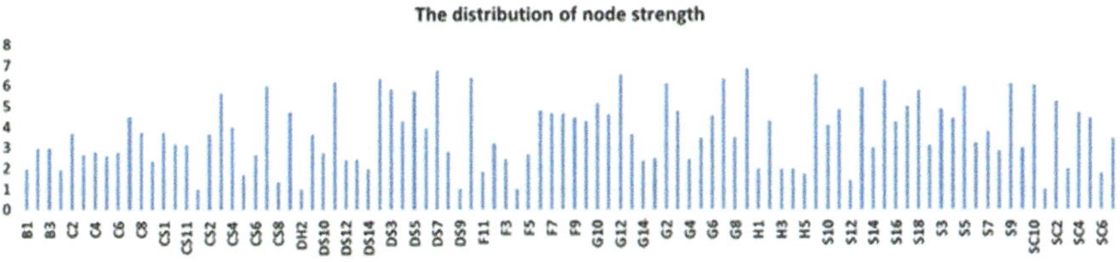

Fig. 4.14 The distribution of enterprise node strength

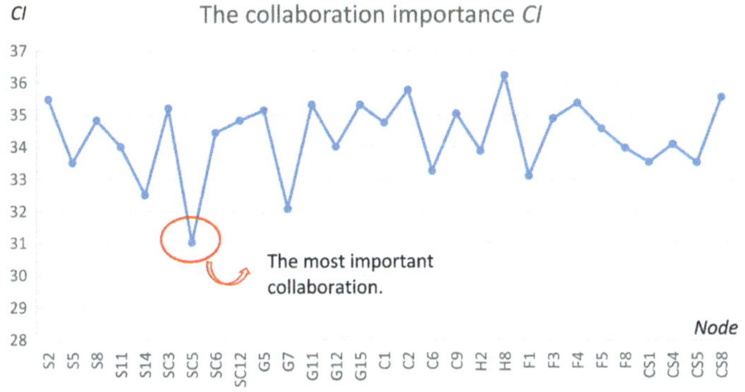

Fig. 4.15 The *CI* value of the Beiren-centered e-production chain

4.6 Concluding Remarks

In this chapter, with respect to massive enterprise relationships and social business relationships distributed in diverse social media, the manufacturing communities (MCs) were proposed to integrate and virtualize aggregate SMSEs by recommendation and self-organization for providing manufacturing services and promoting intelligent business as well as all-around order management. A bi-level social manufacturing-oriented enterprise relationship network (SMERN) was proposed as a theoretical model to facilitate the organizing of SMSEs. The generation rule and formal description of the SMERN were defined based on ontology and relation algebra. Furthermore, an enterprise classification model together with an improved clustering algorithm was proposed to aggregate enterprises into MCs for initial recommendation. The topological and physical characteristics of SMERN were discussed to discover the features of enterprise organizing and enterprise collaborations.

References

1. Errasti A, Rudberg M, Corti D (2015) Managing international operations: configuration of production network for SMEs. Prod Plann Control 26(9SI):691–692
2. Heinrichs H (2013) Sharing economy: a potential new pathway to sustainability. GAIA-Ecol Perspect Sci Soc 22(4):228
3. Vargoa SL, Luschb RF (2008) From goods to service(s): divergences and convergences of logics. Ind Mark Manage 37(3):254–259
4. Ding K, Jiang PY, Leng JW, Cao W (2016) Modeling and analyzing of an enterprise relationship network in the context of social manufacturing. Proc Inst Mech Eng Part B: J Eng Manuf 230(4):752–769
5. Ritzer G, Dean P, Jurgenson N (2012) The coming of age of the prosumer. Am Behav Sci 56(4):379–398
6. Shen H, Cheng X, Cai K, Hu M (2009) Detect overlapping and hierarchical community structure in networks. Phys A-Stat Mech Its Appl 388(8):1706–1712
7. Leng JW, Jiang PY (2016) A deep learning approach for relationship extraction from interaction context in social manufacturing paradigm. Knowl-Based Syst 100:188–199
8. Leng JW, Jiang PY (2017) Mining and matching relationships from interaction contexts in a social manufacturing paradigm. IEEE Trans Syst Man Cybern: Syst 47(2):276–288
9. Jiang PY, Leng JW, Ding K, Gu P, Koren Y (2016) Social manufacturing as a sustainable paradigm for mass individualization. Proc Inst Mech Eng Part B: J Eng Manuf 230(10):1961–1968
10. Jiang PY, Leng JW, Ding K (2016) Social manufacturing: a survey of the state-of-the-art and future challenges. IEEE International Conference on Service Operations and Logistics, and Informatics, Beijing, China, 10–12 July 2016
11. Jiang PY, Ding K, Leng JW (2016) Towards a cyber-physical-social-connected and service-oriented manufacturing paradigm: social manufacturing. Manuf Lett 7(2016):15–21
12. Jiang PY, Leng JW (2017) The configuration of social manufacturing a social intelligence way toward service-oriented manufacturing. Int J Manuf Res 12(1):4–19
13. Kohonen T (2013) Essentials of the self-organizing map. Neural Network 37:52–65
14. Dittenbach M, Merkl D, Rauber A (2000) The growing hierarchical self-organizing map. IEEE-INNS-ENNS Int. Joint Conference on Neural Networks, Como, Italy, 27 July 2000
15. Chattopadhyay M, Dan PK, Mazumdar S (2014) Comparison of visualization of optimal clustering using self-organizing map and growing hierarchical self-organizing map in cellular manufacturing system. Appl Soft Comput 22:528–543
16. Leng JW, Jiang PY, Ding K (2014) Implementing of a three-phase integrated decision support model for parts machining outsourcing. Int J Prod Res 52(12):3614–3636
17. Leng JW, Jiang PY, Zheng M (2017) Outsourcer—supplier coordination for parts machining outsourcing under social manufacturing. Proc Inst Mech Eng Part B: J Eng Manuf 231(6):1078–1090
18. Leng JW, Chen Q, Mao N, Jiang PY (2018) Combining granular computing technique with deep learning for service planning under social manufacturing contexts. Knowl-Based Syst 143:295–306
19. Leng JW, Jiang PY (2017) Granular computing based development of service process reference models in social manufacturing contexts. Concur Eng Res Appl 25(2):95–107
20. Ting C, Wu K, Chou H (2014) Particle swarm optimization algorithm for the berth allocation problem. Expert Syst Appl 41(4):1543–1550
21. Meng X, Gao XZ, Lu L, Liu Y, Zhang H (2016) A new bio-inspired optimisation algorithm: Bird Swarm Algorithm. J Exp Theor Artif Intell 28(4):673–687
22. Deb K, Pratap A, Agarwal S, Meyarivan T (2002) A fast and elitist multiobjective genetic algorithm: NSGA-II. IEEE Trans Evol Comput 6(2):182–197
23. Andreadis G (2015) A collaborative framework for social media aware manufacturing. Manuf Lett 3:14–17

Chapter 5
Open Product Design for Social Manufacturing

Maolin Yang and Pingyu Jiang

5.1 Designer Roles Changing in Social Manufacturing

As with other manufacturing paradigms, product design for social manufacturing is one of the most important starting points [1]. It is clear that such a design methodology is a kind of service-driven mode, which is quite different from traditional one in many aspects. In order to gain a better understanding of this novel design mode, we first discuss the designer role changing in the design practice of social manufacturing.

5.1.1 Product Design Model Changing in Social Manufacturing

Under the context of social manufacturing, we may observe the following changes, all of which are the reasons why designer role has been changing [2].

Socialization in production. Micro-and-small-scale enterprises and even individual entrepreneurs are spring up and providing various manufacturing services to the entire society. At the same time, such enterprises are now able to use manufacturing resources from both inside and outside to fulfill their manufacturing tasks.

Driven force of product innovation. Emerging socialnomics has fundamentally changed how enterprises interact with consumers. Community based open product design practice has emerged as important ways for product innovation. For instance, many crowdsourcing products and platforms (e.g. *Rally Fighter* automobile from *Local Motors*) are focusing on co-creative value development by encouraging their fans or consumers to participate in various activities related to a product lifecycle, and some new born companies (e.g. *Xiaomi* mobile phone) succeed even without any production lines.

P. Jiang, *Social Manufacturing: Fundamentals and Applications*, Springer Series in Advanced Manufacturing, https://doi.org/10.1007/978-3-319-72986-2_5

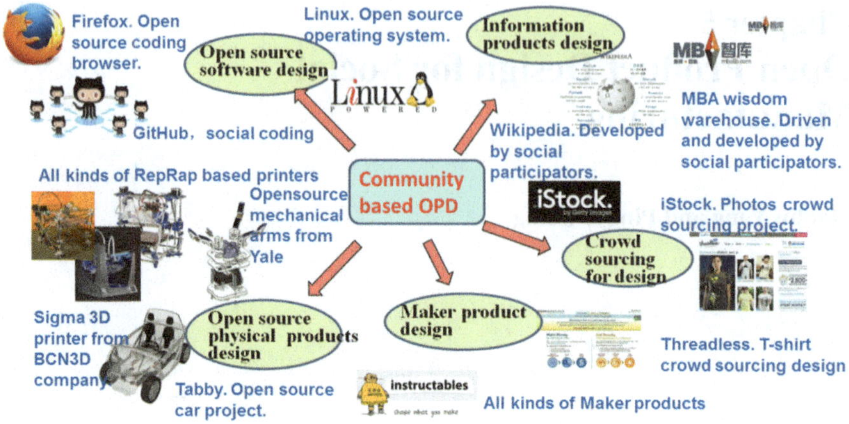

Fig. 5.1 Booming community based open product design approaches

Virus-like information propagation. Under the context of social manufacturing, social-media–like tool sets and platforms enable peer to peer decentralized interactions. In this way, their "*virus-like*" information propagation mechanisms make the demands of consumers and the capabilities of providers far more visible than ever before.

All these transitions are breaking down the technique barrier that once separated product consumers from product designers [3]. Traditional product consumers are now capable and motivated to share their product reviews and innovative ideas online with the help of booming social media and information technologies [4]. In this way, the consumers are actually participating in the product development process by providing their innovative ideas. Such a trend results in significant changes of consumer's behaviors of purchasing, using, and reviewing products [5]. Hobbyists, part-timers, and dabblers now have chances to be more involved in the design process of their own interested products. Under this background, the boundaries between product consumers and product designers are becoming increasingly blurry, as shown in Fig. 5.1. These non-professional participators are defined as social designers (SDs), and the product design process carried out by SDs as a kind of open product design (OPD).

5.1.2 Designer Role Changing in Typical Open Product Design Approaches

In OPD activities, SDs may initiate projects of their own or join the others'. Because of the overwhelming diversity of SDs' design ability, all kinds of OPD projects can be developed by these SDs. For software and information products, examples such as

Linux, *Apache*, *Wikipedia*, and many other projects on *GitHub* exhibit the innovation capability of SDs. For physical products, 3D printers, convenient houses, and even cars are designed by SDs. All these product design approaches are different from traditional product development processes. In these product development approaches, product innovation and development processes are not the privilege of professional designers anymore, and SDs can exchange and share ideas [6] in virtual communities and develop their own innovative ideas into highly personalized products. In this section, three typical product design approaches that embody the designer role changing characteristics of OPD mode are introduced, which are crowdsourcing design, open source design, and maker movement.

Crowdsourcing design. Crowdsourcing (CS) is an activity that launches a specific topic or challenge to the outside crowd in the form of open calls. In this way, new ideas from social wisdom can be used to help the internal R&D department of the enterprise for innovation tasks [7, 8]. Because of its capability for mobilizing the creative power of endless social wisdom, CS has drawn much attention from both business and academic fields [9]. All these researches indicate that in CS practices, SDs are contributing to the innovation design practice and that's why CS project succeed [10].

Open-source design. Open source (OS) software can be dated back to 1984 when *Free Software Foundation* and their popular GNU project were developed [11]. The word "*Free*" here doesn't mean "*at no cost*" but "*everyone has the freedom to join the product development process and share the results*". From then on, many famous software products (e.g. *Apache* and *Linux*) have proved that this novel design approach can produce software with high quality and reliability [12, 13]. Open source software is becoming a viable development approach not only for enthusiasts and amateurs but also for professional software development companies [14]. Relatively, OS physical products development approach is not as convenient as OS software, but it also shows great values and potentials. Recently, physical products such as 3D printers, homemade products, robotic arms, and even convenient houses and cars are being designed in OS approach.

The most noteworthy characteristic of OS design is that it can be unrestrainedly participated by designers from all over the society [14]. If there are many participators for an OS product, the product can be developed very fast with high quality and less mistakes or bugs. And the mistakes or problems during the product development process can be detected and solved easily. But there are also drawbacks for an OS product. For example, if there aren't enough participators, the design process would be very slow, and the development results are not predictable. Furthermore, the *Matthew* effect is very common in OS projects that if the project doesn't attract many participators at the beginning, there isn't much chance that it will have many participators later [13].

Maker movement. Fast technology development is offering amateur designers with access to low-price and convenient manufacturing equipment. And they are now able to organize their own DC in different forms, such as *Makerspace*, *Hackerspace*, *TechShop*, and *FabLab* [15]. Collectively, we describe them as makers. However, each of them has its own characteristics and focus. Traditional *Hackerspaces* mainly focus on electronics and programming project, while *Makerspace* [16] represents a wider vision of public access for professional and amateur designers, and even school kids can make their own interested product from scratch. *TechShop* is earlier than the term "*Makerspace*", it offers public with access to craft areas with supporting equipment for profit. *FabLab* [17], inspired by an MIT course, is equipped with a core set of tools (e.g. basic electronic equipment, cutters, CNC machines) that allow amateurs to fabricate their own interested stuffs. All these approaches have one same characteristic that they are all openly participated by non-professional designers to develop their own fascinating ideas into practical products [18].

5.1.3 Comparison Between Traditional Product Design and Open Product Design

The most noteworthy characteristic of OPD mode is that it integrates tremendous socialized designers, who actively participate into an interactive product design process, and organize these socialized designers into design communities (DC) with customer-centric markets [19]. And this brings great changes in the entire product design process. Generally speaking, key designers design the core parts and the architecture of a product, and then define the interfaces (including mechanical, electrical, and information/software) for potential modules, which may be produced by other prosumers. Then the SDs involved in the project finish their final products by adding individualized modules and combinations. The detailed differences are listed in Table 5.1 [20].

All these differences bring OPD mode with many advantages, including higher level of innovation [10, 21], faster innovative product development process [22], deeper customer involvement [22–24], and lower development cost [24]. With all these advantages, micro-and-small-scale enterprises, even self-organized communities can provide high quality design services in the customer-centric markets [20]. However, open and self-organization characteristics of OPD mode also impart itself with disadvantages, namely unreliable design process caused by unreliable design community (DC). Many negative events (e.g. core designers quitting, designers losing original design capacity, excessive inter-community competition, insufficient information exchanging, opponent disturbances) threaten the performance of DC. And it's worth mention that new intellectual property protection mechanisms should also be considered in the future to make sure that the collaborative open product design process can be carried out orderly.

Table 5.1 Comparison between traditional product design and open product design

	Traditional	OPD
Organization	Hierarchical	Network structure
Project initiation	Initiated after market analysis	Initiated by participants
Task assignment	Tasks assigned to participants	Participants self-select tasks
Participants motive	Economic driven	Non-economic driven
Proximity	Primarily local ties	Local and non-local ties
Available resources	Within the enterprise	Socialized resources
Cost	High	Relatively low
Systems	Systematically designed	Evolve over time
Decision-making	Hierarchical	Decentralized
User involvement	Low during the development	Users are the developers
Trust insurance	High	Relatively low
After sales	Guaranteed	Unguaranteed
Design process	controllable	uncontrollable
Design trend	Predictable	unpredictable
Design result	Guaranteed	Unguaranteed
Product complexity	High	Relative low
Product modularity	Depends	Relative high
Core secrets involved	Positive	Negative
IP protection	Positive	Negative
Product up-gradation	Premeditated	spontaneously
Production innovation	Depends	Relatively high
Product customization	Depends	Relatively high

5.2 Architecture of Crowdsourcing-Driven Open Product Design

Crowdsourcing design is one of the most typical and well developed open product design approaches. Hence, in this section, crowdsourcing design approach is used as an example to provide a general understanding of OPD mode.

5.2.1 Conceptual Architecture

Before further introduction to crowdsourcing design, there are a few concepts that should be demonstrated first.

Crowdsourcing. Crowdsourcing is a distributed problem-solving pattern. Problems or tasks are openly distributed to the crowd via Internet for solutions. There are mainly four kinds of crowdsourcing patterns, namely crowd wisdom, crowd creation, crowd selection, and crowd funding. Crowd wisdom indicates the activities that utilize the knowledge of social crowds to predict future events and solve problems (e.g. *Quora.com*). Crowd creation indicates the activities in which social crowds voluntarily use their creativity power to generate news, amusement, and all kinds of creative stuff (e.g. *Facebook*). Crowd selection indicates the activities that use mass social crowds to filter and select information (e.g. *Amazon*). Crowd funding indicates the activity that social crowds devote their money for a common goal (e.g. *Kickstarter*).

Socialized manufacturing resource. In this chapter, socialized manufacturing resource specifically indicates socialized design resource that is mainly enabled by SDs who are willingly and capable to fulfill certain product design tasks. And these SDs are interested in participating in the crowdsourcing tasks for self-satisfaction and financial remuneration.

Design community. After the establishment of crowdsourcing tasks, SDs, including professional designers and amateurs, gather together to form dynamic virtual open designer communities which are self-organized, self-driven, and centreless-self-control. Due to its open, social, and self-driven, and self-organization characteristics, a design community usually possesses huge amount of design resources and evolves constantly over time, but is also unreliable when performing collaborative design tasks.

The service architecture of crowdsourcing-driven OPD mode is made up of three layers, which are physical layer, technical support layer and application layer.

Physical layer, indicating the socialized manufacturing resources including all the equipment, mainly represents the resource flow from customer requirement (CR) to product design process. As shown in Fig. 5.2, under the context of social manufacturing, there are mainly two kinds of socialized manufacturing resources in the physical layer of product crowdsourcing system, namely the CR source provided by product consumers and the design source provided by designers. It has to be mentioned that, the consumers and designers may overlap in the context of social manufacturing.

Technical support layer connects the physical layer and the application layer, and provides essential supporting techniques for the well function of all the activities in the system. During the matching from CR to product design process, deep learning methods can be utilized to classify the CR into specific categories. In this way, the CR resources hidden in the resource pool can be excavated and used for decision support.

Application layer embodies all the functions and applications that the crowdsourcing system provides to the users, including CR release, design resource formalization, CR excavation, crowdsourcing design order initiation, design decision supports, etc.

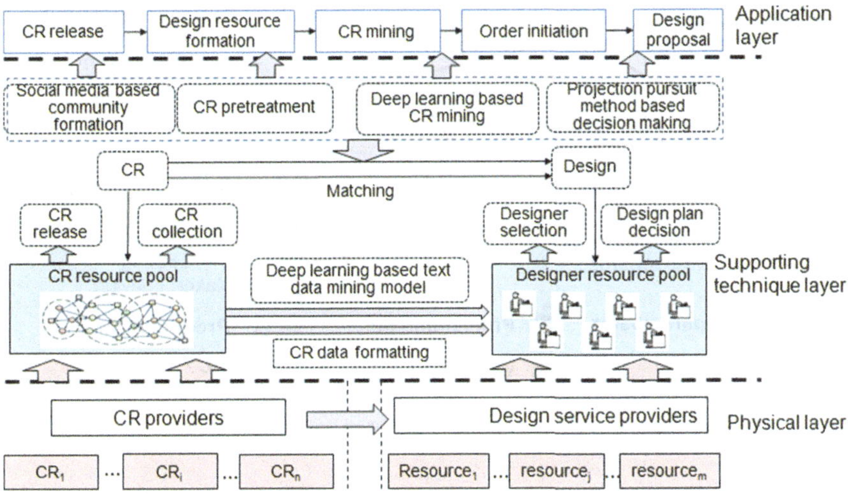

Fig. 5.2 Service architecture of crowdsourcing-driven open product design

5.2.2 General Workflow

According to the service architecture in the last section, it can be seen that the general workflow of crowdsourcing driven open product design activities starts from project initiation and design community establishment, and ends up with product launching and DC disbanding, as shown in Fig. 5.3. In step 1, a project related to a potential product is initiated from innovation ideas. In step 2, the initiator starts the project on a social-media-like platform. In step 3, design community is generated targeting the product bill of materials (BOM). In step 4, the design community dynamically adjusts and grows due to the design ability CR of the product and this in turn leads to the iteration of the product. In step 5, the product designed by the design community flow into the market. And in step 6 the designers get rewards. If the design community disbands, designers will leave the community as shown in step 7, and maybe join another community as shown in step 8, and it's all depend on the designer's own decision. In all these steps, 1, 2, 3, 4, 5, 6 represents the product design process and 2, 3, 4, 7, 8 represents the growing and decaying process of design community.

5.2.3 Key Enabled Technologies

Based on the general workflow, three most important key enabled technologies are identified, which are analyzing of customer requirements, shaping and analyzing for design community, and evaluating of design results.

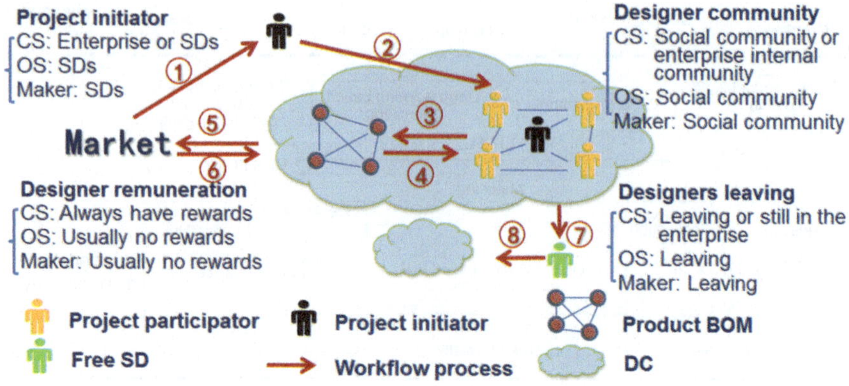

Fig. 5.3 General workflow of crowdsourcing-driven open product design process

The first key enabled technology deals with analyzing customer requirements. As with other design methods, open product design activities are also carried out based on the customer requirements (CRs). However, the CR acquisition under the environment of open product design is hugely different from traditional mode in many ways. Hence, specific techniques should be developed for solving the problems related to the CR acquirement analysis.

The second key enabled technology deals with shaping and analyzing for design community. Designers in an open product design environment are from different fields and with different background. Not all the designers have enough design abilities for the design tasks and they don't even know each other before they join in the same DC. Therefore, shaping and analyzing for DC is the second key enabled technology.

The third key enabled technology deals with evaluating design results. The open and free characteristics of open product design bring it with great unreliability. Hence, design results evaluation has to be considered during OPD practice.

All these three key enabled technologies are focused on the most fundamental problems in OPD practices and will be detailed discussed in the next three chapters.

5.3 Customer Requirements Analysis in Crowdsourcing-Driven Open Product Design

5.3.1 Customer Requirements Acquisition

Under the context of social manufacturing, interactions and communications among socialized manufacturing resource owners are becoming more and more convenient and efficient with the help of Internet technologies and social media technologies.

Huge amounts of data which contain CR information, production capacity information, production intention, etc. are generated during these interactions and communications. If we can extract all the product design requirements from the data and utilize them for decision making, it would help us to produce deep customized product much more easily.

However, the data generated in social media is usually multi-dimensional, heterogeneous and complex. All these data are disorderly spread on the Internet and this makes it difficult to extract. Furthermore, the data have to be processed into computer-sensible structure before further analysis. In another hand, the data contain too much information decision makers have to thoroughly analyze so as to identify the customer production requirements and filter out the redundant information before they can precisely acquire what they need.

There are many ways for CR acquisition, such as investigation and survey, sales record analysis, Internet data crawling, etc. However, due to the fast development of social media and Internet technology, Internet data crawling will be the mainstream approach for CR acquisition.

However, Internet data crawling can only get rough data that cannot be utilized directly for CR analysis. Hence, pretreatment should be performed before further analysis. In our research, mainly two kinds of pretreatments are applied to turn the raw data into computer-sensible format, namely Chinese text segmentation and text vector space representation.

During Chinese text segmentation, all the Chinese words are separated with the help of dictionary and statistical rules based on the word segmentation algorithm. In this way, the words are organized into phrases to extract customer requirements. The steps of Chinese text segmentation are listed below:

Step 1: Separate the text into Chinese character strings.

Step 2: Use bidirectional maximal matching algorithm to compare the Chinese character strings with dictionary entries.

Step 3: Find out if there are any discrepancies during the matching process.

Step 4: For the character strings with discrepancies, keep separating them with the help of statistical rules until all the strings match with the corresponding entries in the dictionary.

Step 5: Finish.

The Chinese text segmentation separates the contiguous Chinese words into phrases, but the phrases are still difficult to be understood by computers. Hence, unstructured phrases have to be transformed into structured and computer-sensible data for further analysis. Hence, vector space model is utilized to represent the text data in the form of binary data. In the model, text is considered as a combination of independent phrases, which are represented as $d = ((t_1, w_1), (t_2, w_2), \ldots, (t_n, w_n))$ where t_n is the nth feature item and w_n its weight.

Suppose that $D = \{p_1, p_2, \ldots, p_n\}$ represents all the phrases in the text, and each phrase is represented in the form of vector, for example $v_i = \{q_1, q_2, \ldots, q_n\}$, where q has only two values, namely 0 and 1. When the value equals 0, the vector doesn't match the phrase; and when the value equals 1, the vector matches the phrase. And

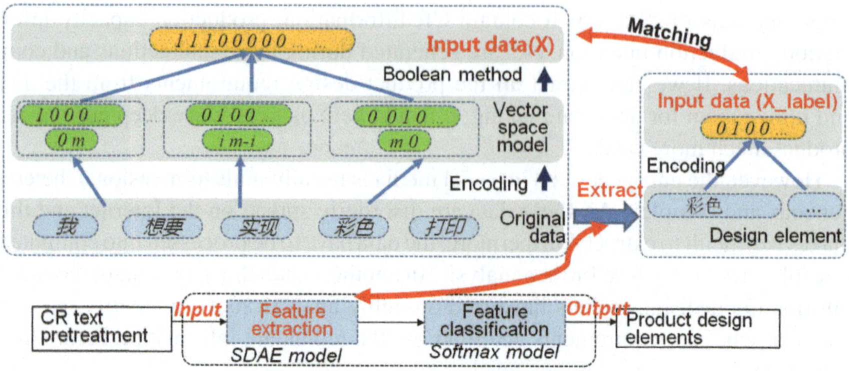

Fig. 5.4 Original CR text data pretreatment and matching

then, the text can be represented by adding all the vectors together, as shown in Fig. 5.4. In this way, the unstructured Chinese character strings are transformed into structured binary data which are readable in computer program.

5.3.2 A Deep Learning Approach for Customer Requirements Analysis

In the previous subsection, raw data that contain CRs are acquired and pretreated into structured and computer-sensible binary data. And in this section, specific customer interests and propensity are derived in the form of text features with the help of text data mining. Text data mining is an important application of deep learning method. It extracts text features by separating them into different groups and thus derives information that contains CRs and match the CRs to product design elements.

The deep learning method in this section is mainly separated into two steps. In the first step, the pretreated data are utilized to derive features. And in the second step, deep learning model is trained based on the data that labeled with the features derived in the first step. And then, the well trained deep learning model can be used to derive features from new data. The entire process is illustrated in Fig. 5.4.

In order to derive features more effectively and efficiently, feature learning method is introduced. The most important part in feature learning is the extraction of feature word. However, there are a lot of redundant data and noisy data in the raw dataset. This results in high error rate and low recognition rate and hence hinders the learning rate. In our research, stacked de-noise auto encoder (SDAE) method [25] is utilized before feature classification. With the help of SDAE method, low level features (words) can be integrated and organized into high level concepts, as shown in Fig. 5.5. And then, with the help of SoftMax regression model, the features are classified into different groups. And thus, the design elements hidden in the text data are extracted.

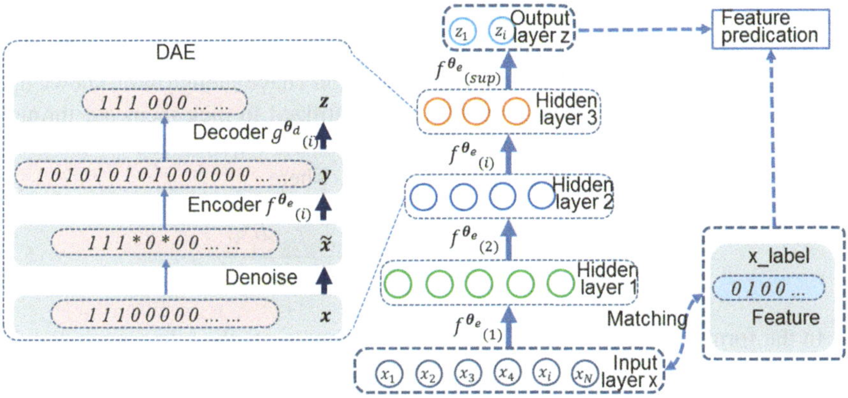

Fig. 5.5 SDAE based CR text feature extraction

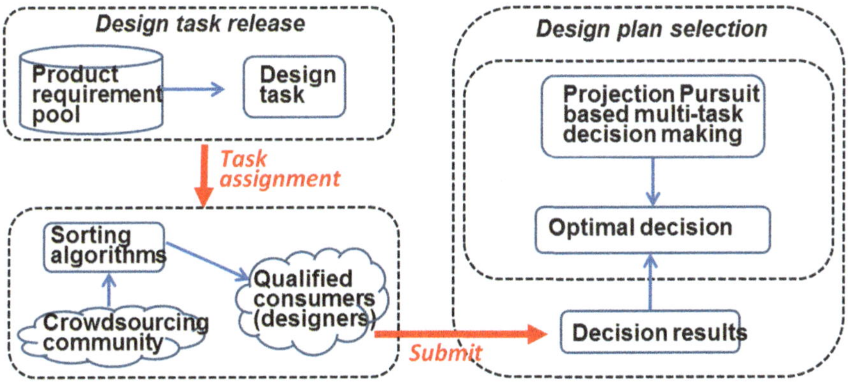

Fig. 5.6 Runtime logic of CR-driven crowdsourcing design and decision making process

5.3.3 *Product Design Decision Support Based on Customer Requirements Analysis*

It's worth mention that open product design approach should take full use of CRs so as to provide the products that meet the requirements of customers. As demonstrated in Fig. 5.6, a comprehensive approach that selects qualified consumers and makes product design decisions according to these selected consumers is proposed in this section.

Nowadays, more and more customers are willing to provide their ideas and considerations about the products they are interested in. However, the customers are from different background and have different cognitive level, and some of them are malicious participators who try to disturb the well function of customers' interaction process. Hence, the customers who do not have the basic knowledge of the certain product and seemingly trying to disturb the design process have to be picked out, and only the requirements of qualified customers should be acquired and considered during the decision-making process.

For the first kind of unqualified customers who don't have enough basic knowledge of the product, feature distance method can be utilized to pick them out through comparing the requirement of each customer with the requirements of the other customers, and calculating their random tractions, where

$$RandomSpam = \frac{\sum_{j\in J_w}\sum_{i\in J_{j,\overline{w}}} dis_{ij}}{\sum_{j\in J_w}\left|J_{j,\overline{w}}\right|} \tag{5.1}$$

In the formula, w represents the designer, J_w represents the relevance judgment set of consumer w, $J_{j,\bar{w}}$ represents the relevance judgement set of other consumers, dis_{ij} represents the distances between the requirement of w and other customers. If dis_{ij} equals 0, then the requirements of the two compared objects are the same, and if dis_{ij} equals 1, they have different requirements. And the threshold value can be set as 0.7 to pick out unqualified customers.

For the second kind of unqualified customers who are suspected trying to disturb the well function of product design process, sequence discrepancy can be used to pick them out.

$$UniformSpam_w = \frac{\sum_{s\in S}|s|(f_{s},{}_{Jw}-1)\left\{\sum_{j\in J_{s,w}}\sum_{i\in J_{j,\overline{w}}}(disagree_{ij})^2\right\}}{\sum_{j\in J_w}\left|J_{j,\overline{w}}\right|} \tag{5.2}$$

In the formula, S represent all the possible sequences of classification, $disagree_{ij}$ represents the amount of disagree between the relevance judgement provided by customer w ($J_{s,w}$) and other customers (in sequence S) on the same question j, $f_{s},{}_{Jw}$ represents the frequency that design task s marked by customer w appeared in J_w. And the threshold value can be set as 1.6 to pick out malicious customers.

Assume that the decision-making problem is F, and there are n design plans f_i ($i=1, 2\ldots n$), and there are p indicators for each plan, namely C_i ($i=1, 2\ldots p$), then we have an $n\times p$ matrix that contains all the plans, which is defined as $A_{n\times p}$. And there are m qualified customers providing their marks for the plans, then we have an $m\times n\times p$ three-dimensional matrix, $R_{m\times n\times p}$. The multi-target decision making problem in open product design is to turn $R_{m\times n\times p}$, which represents the choice of the qualified customers, into a one-dimension sequence for comparative analysis. Hence, the dimension of $R_{m\times n\times p}$ has to be reduced.

In order to reduce the dimension of $R_{m\times n\times p}$, the *Projection Pursuit* method [26] can be utilized. *Projection Pursuit* method casts high-dimension data to low-dimension space for data processing. And the method has good disturb resistance, accuracy, and robustness. In this way, the data structures and features of the original high-dimension data can be revealed easily and thus makes it convenient to analyze the data.

5.4 Shaping and Analyzing of Design Community

5.4.1 Formalization of Design Community

With the help of social media technologies, all kinds of media, websites, resources, and customers gather on certain platforms. And then, on these platforms, design communities made up of product designers and consumers are formed targeting product design activities in centerless-self-control mechanism.

One of the most significant characteristics of DC is self-organization. It indicates the phenomenon that the DC can spontaneously transform from simple and rough state to a complex and delicate state under an intrinsic motive driven force. Along with the popularization of network technology, self-organized virtual networks emerge. It is a virtual community made up of people with the common interests and values and is physically decomposed from the original DC. In the virtual communities, people with the same goals can share ideas, information, resources, and connections conveniently and effectively. With the same driven force, people in the resource pool gather into virtual communities. For example, designers with the same design abilities can gather into a design ability community, designers focusing the same certain product can gather into the same product community, as shown in Fig. 5.7.

5.4.2 Design Community Classification and Analysis

Different from the hierarchical organization structure of traditional manufacturing mode, the designers' virtual communities related to a DC are organized in a network structure. Depending on how to control a DC, two main types of design community

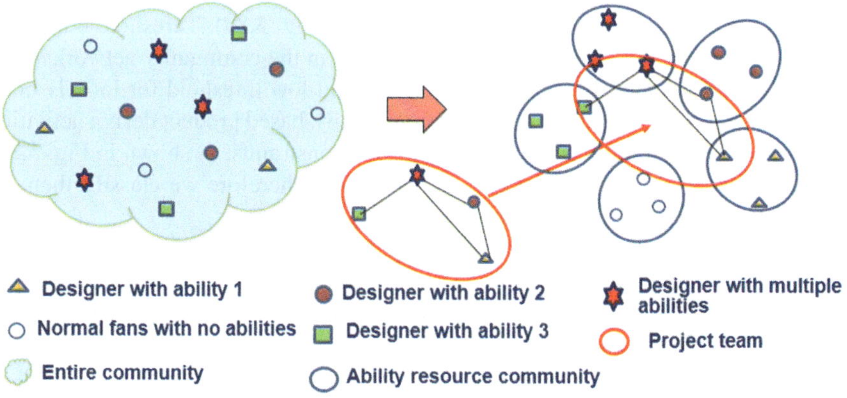

Fig. 5.7 Formation of designer communities

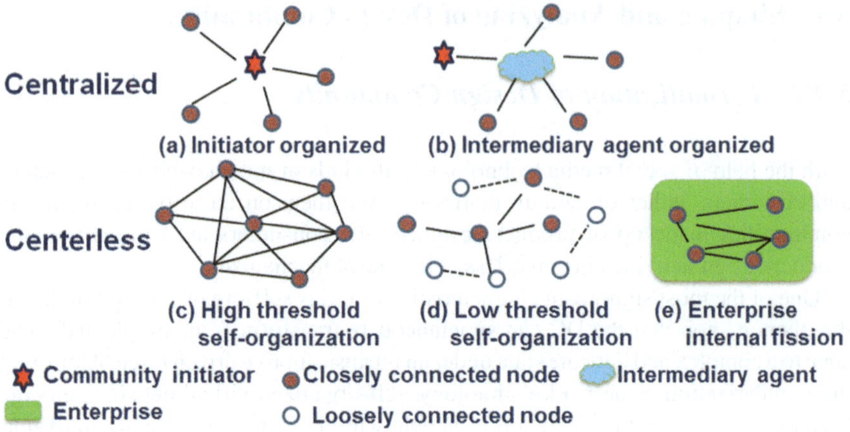

Fig. 5.8 Classification of design community network structures

are identified from designers' networked interaction behaviors. They are centralized-control-structure-driven DC and centerless-self-control-structure-driven DC.

Centralized-control-structure-driven DC defines such a situation when a DC is initiated by initiators who represent certain individual or organization and the initiators are involved in all the design processes. In this situation, DC works in centralized manner, and the initiators are the "*center*" of the DC. There are two cases in centralized-control-structure-driven DC, which are that initiators organize the community themselves (Fig. 5.8a) and initiators organize the community via an intermediary agent (Fig. 5.8b).

Centerless-self-control-structure-driven DC works in the different way. In this case, no obvious "*center*" can be found in it. The DC is usually self-organized around self-initiated product design projects. Due to different threshold conditions for new designer participation, the DC basically develops into either a high threshold community (Fig. 5.8c) or a low threshold community (Fig. 5.8d). Threshold condition influences the connection strength between nodes in the community networks, i.e., high threshold for closely connected networks and low threshold for loosely connected networks. It's worth mention that community-based product design activities also happen in traditional enterprises with some constraints, as shown in Fig. 5.8e. These communities usually don't have a "*center*". Therefore we classify them as centerless-self-control-structure-driven DC.

All the community structure mentioned above have different network structures due to different organization mechanisms, and thereby embody different resilience characteristics. The characteristics and existing examples of these communities are elaborated in Table 5.2.

Table 5.2 Characteristics and examples of different DC structure forms

		Centralized-control structure DC		Centerless-self-control structure DC		
		Initiator organized	Intermediary agent organized	Self-organized		Enterprise internal fission
				High threshold	Low threshold	
Network structure characteristics		loosely connected member nodes	loosely connected member nodes	Highly connected; Relatively stable members	Loosely connected; Highly dynamic	Relatively stable members;
Resilience characteristics	Trust & Credit level	Low	Medium	Medium	Low	High
	Interaction level	Low	Low	High	Low	High
	Operation level	Low	Medium	Low	Low	Medium
	Connection and bounding	Low	Low	Medium	Low	High
	Participation motivation	compensation	compensation	Self-satisfaction	Self-satisfaction	Enterprise requirement
	Compensation	Financial	Financial	Nonfinancial	Nonfinancial	Financial
Innovation capability		Medium	Medium	High	High	Low
Existing examples		*Threadless*	*iStockphoto*	*Vechicle Forge*	*RepRap*	*Haier*

5.4.3 Designer Interaction Relationship Analysis

There are mainly two kinds of interactions between designers, which are participating design activities concerning the same product design tasks and exchanging a variety of information [27]. First, we build an unweighted bipartite two-mode network $G_1 = \{B \times D_2, E_2\}$, where B represents product BOM nodes set and D_2 the corresponding designers set, and E_2 a set of edges that each edge in E_2 represents a designer's participation on a product BOM component (as shown in Fig. 5.9a where $B = \{b_1, b_2, b_3\}$ and $D_2 = \{d_1, d_2, \ldots, d_6\}$). And then, G_1 is transformed to one-mode designer interaction networks $G_2 = \{D_3, E_3\}$ by removing all the nodes in B, and each edge in E_3 connects two designers who participate in the same BOM node design or have any kind of information exchanging, as shown in Fig. 5.9 (b). The joint strength of edges in E_3 is defined as $s_{\mathrm{ij}} = \alpha \times n + \beta \times m$, where n and m indicate that designer i and designer j participate in n BOM nodes design and have m pairs of information exchanging. For example, in G_2, $s_{34} = 2\alpha + \beta \times m$ because they participate in the design task of two same parts. It has to be mentioned that d_7, which is not in D_2, is added to D_3 because d_7 is connected to d_6 through information exchanging with a joint strength $s_{67} = \beta \times m$. The models represent the designer resource distribution and designer interaction relationship of certain DC and can be used for further analysis.

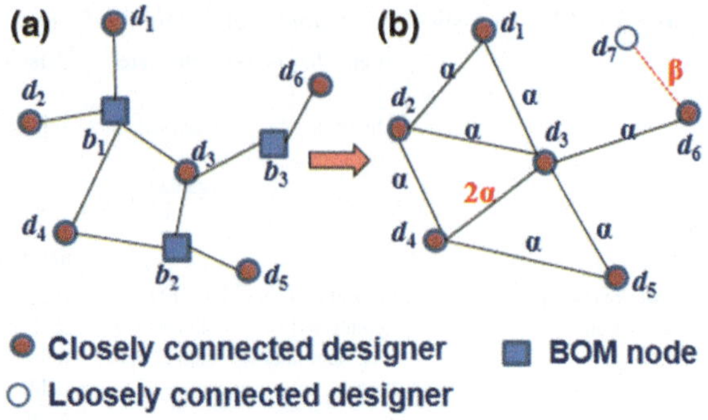

Fig. 5.9 Product BOM-driven two-mode designer networks and one-mode designer networks

5.5 Evaluation of Crowdsourcing-Driven Open Product Design

In this section, approaches from different entry points are introduced to provide guidelines for OPD evaluation.

5.5.1 Design Community Resilience Evaluation

The most noteworthy advantage of virtual community based OPD mode is that a well-organized DC can introduce socialized design resources in its product design process and connect it with customer-centric markets. However, open and self-organization characteristics also bring disadvantages, namely unreliable design process caused by an unreliable DC. Many negative events (e.g. core designers quitting, designers losing original design capacity, excessive inter-community competition, poor information exchanging mechanism, opponent disturbances) threaten the orderly design process of DC. In this section, an integrated resilience assessment approach from different perspectives is proposed to deepen our understanding to DC, because assessing DC resilience is the first step to build a resilient and reliable community.

During research review, we find that resilience issue of DC in OPD mode has its unique characteristics and requires specialized analysis. However, hardly any efforts have been paid on DC resilience and this leaves it with no specified theoretical support and thus limits its further development and applications. Hence, we start our study from DC analysis and classification, followed by resilience manifestation analysis and resilience concept boundary definition. And then we established an integrated resilience assessment approach with quantitative indicators.

In order to analyze DC resilience, three resilience manifestations are defined in advance.

The term "***static stability***" is defined as the ability to function well during peaceful time and fulfill certain product design tasks.

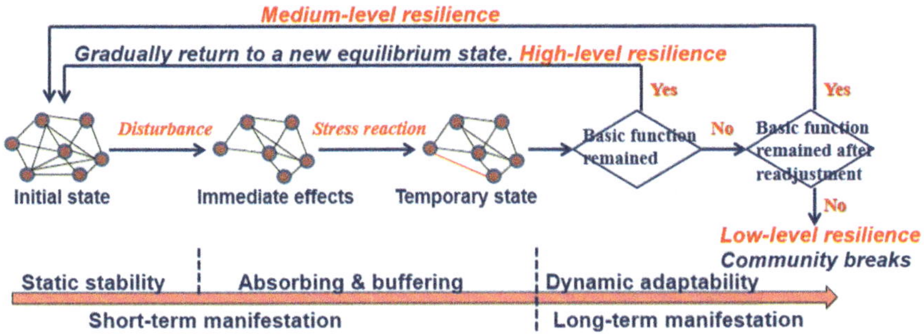

Fig. 5.10 Dynamic rebounding process of disturbed DC

The term "***absorbing and buffering***" is defined as the ability to absorb small disturbances by quickly replacing the damaged nodes or ties and maintain the original design function without much network structure changing.

Unlike absorbing and buffering ability that absorb small disturbances without much network structure changing, the term "***dynamic adaptability***" refers to the ability to active sleeping nodes [28] to replace the damaged nodes, generate new ties, and of course change the network structure to a new equilibrium state.

Dynamic rebounding process of resilient DC during crisis is illustrated in Fig. 5.10, in which different resilience levels lead to different rebounding performance.

And then, different assessment methods are developed to analyze the three resilience manifestations respectively. The three assessment perspectives are discussed below.

Design capacity perspective in DC: The organization process of a DC is driven by product bill of material (BOM), and all the designers are attached to one or many product components in the BOM. In our DC design capacity model, we assume that every component in the BOM is related to a certain kind of design ability which is mastered by one or more designers in the DC. Hence, the model is abstracted into a 2-mode network, $C = \{A \times D_1, E_1\}$, where C defines the design capacity of the DC, $A = \{a_1, a_2, \ldots, a_S\}$ defines all the design abilities mastered by all the designers, $D_1 = \{d_1, d_2, \ldots, d_N\}$ represents all the designers in the DC, and E_1 contains a set of edges that each edge in E_1 connects a node in A to a node in D_1 indicating that the a type of ability is mastered by a designer, as shown in Fig. 5.11 where $A = \{a_1, a_2, a_3\}$ and $D_1 = \{d_1, d_2, d_3\}$. For each ability a_i, there are totally K_i designers can master it, $K_i = (0, 1, 2, 3, \ldots)$. In this model, only the ability to accomplish the design task is considered, hence as long as a designer masters the ability required by a product component, the weight of the edge linking them scores the same, and it doesn't matter whether or not the designer do participate in the design of that product component. Based on the model, two indicators specifically for DC design capacity assessment are developed which are diversity and redundancy.

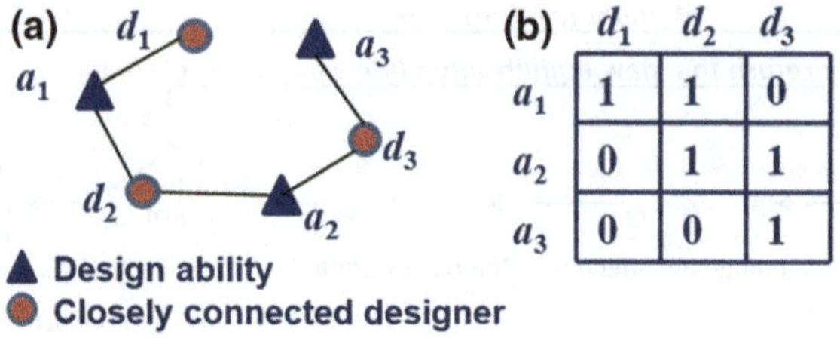

Fig. 5.11 Two-mode DC design capacity model

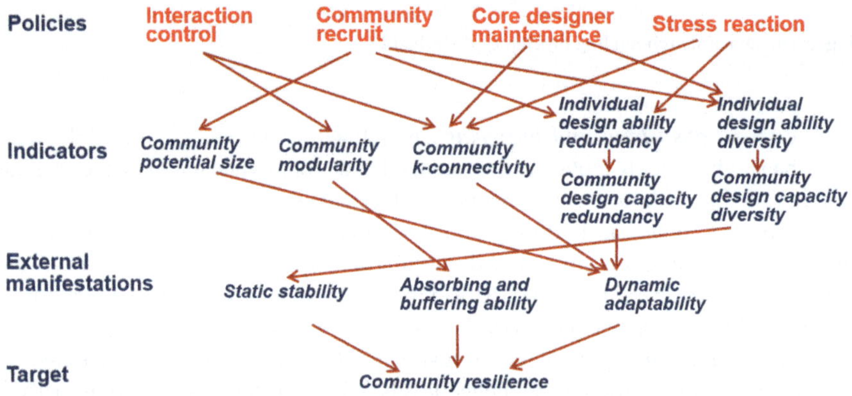

Fig. 5.12 Causal map analysis for DC resilience influencing aspects

Interaction relationships in DC: Optimal DC interaction relationships influence DC from many aspects, such as information exchanging, design cooperation, disturbances dispersion, designers' intimacy, community reorganization possibility, etc. Based on the designer interaction relationship model established before (Sect. 5.4.3), many original indexes in complex network theory (CNT) can be utilized to evaluate DC resilience. However, not all of them are suitable for the physical characteristics in DC. Hence, specific indexes should be developed based on the basic CNT indexes.

Leading policy perspective: Different from traditional enterprise-based product design mode, community based OPD mode doesn't have strong and compulsive controlling policies. But a serious of practical leading policies can help guarantying DC resilience. Therefore, four important key points for leading policies are identified in the form of causal map, as shown in Fig. 5.12. The four points are the main influence factors for DC resilience, which are interaction control, community recruit, core designer maintenance, and stress reaction.

5.5.2 *Project Feasibility Evaluation*

Many OPD project are not successful due to many reasons. Hence, a refined *Bayesian* causal map (BCM) approach is utilized to derive and express experts experience for OPD project feasibility analysis. The approach is mainly divided into two phases, which are BCM structure detection and BCM parameters derivation. Detailed steps are described as follow.

Step1: ***Mind-map-based variable pool initiation***. It is difficult for experts to fully express their experience during time limited interviews. Hence, the basic clues of OPD project feasibility factors have to be shown to the experts so that they can express their judgments and beliefs based on the clues, as shown in Fig. 5.13.

Step 2: ***Variable exploration and raw causal value identification***. In this section, a semi-structured interview [29] for OPD project feasibility is designed based on the mind maps to elicit the causal statements from domain experts. In order to ensure a balanced interview result, interviewees in our research includes one expert with CS experience, one expert with OS project experience, and one successful maker. The interview results are recorded in the form of causal statements [30, 31] and then transformed into causal maps (CMs) which contain the newly acquired experts' judgement.

Step 3: ***Multi-expert experience merging and raw CM construction***. The CMs from different experts have different focuses. In order to get a comprehensive result, all the CMs are integrated in one single CM, as shown in Fig. 5.14. In the integrated CM, concepts with similar meanings are considered as the same, and all the causal judgments from interviewees are included.

Step 4: ***CM structure modification***. The CM acquired in the last step is modified into BCM based on the following rules [30, 31]: direct and indirect relationship analysis, conditional independencies analysis, eliminating circular relations, causal concepts integration. After all these modifications, the final structure of OPD project feasibility BCM is established.

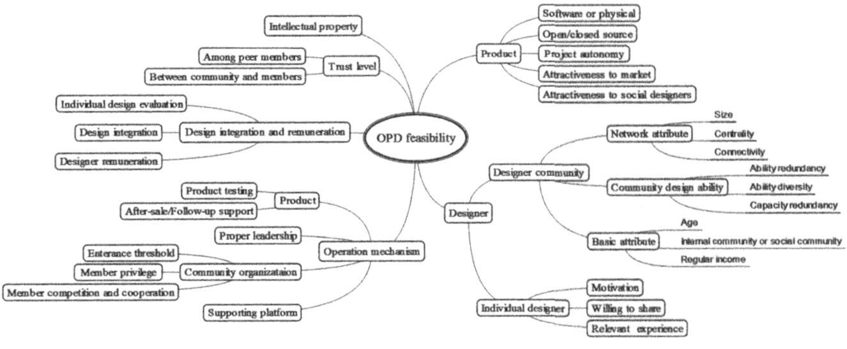

Fig. 5.13 Influencing aspects for OPD project feasibility

Fig. 5.14 Integrated CM for OPD project feasibility analysis

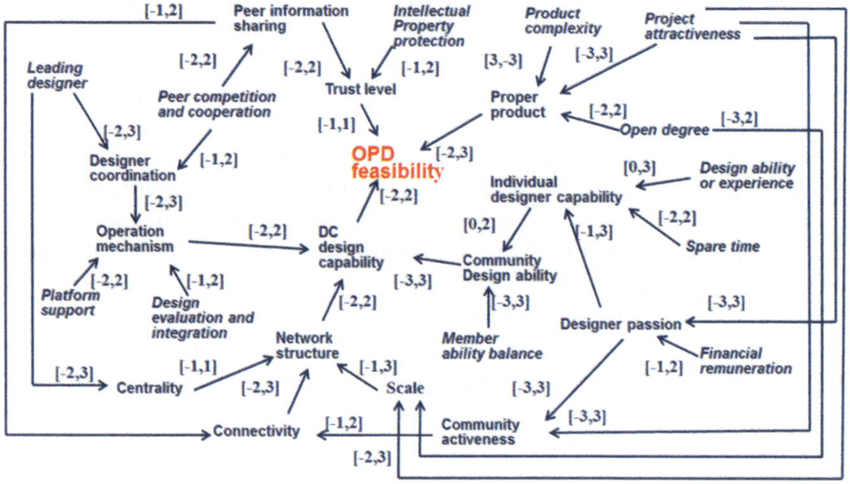

Fig. 5.15 OPD project feasibility analyses BCM

Step 5: ***Enriched causal value detection***. The BCM constructed so far doesn't consider variable uncertainty and thus cannot be used for probabilistic inference. Hence, probabilistic parameters of each variable will be determined for quantitative probabilistic inference [32]. In order to make precise inference, causal values are defined as $[V_1, V_2]$ for variables with different level of states. The V_1 represents the influence degree of parent variable on the child variable when the parent is at its low level (negative state). And V_2 represents the influence degree of parent variable on the child variable when the parent is at its high level (positive state). All the values are numerical and can be positive, negative or null, as shown in Fig. 5.15.

Step 6: ***From causal values to CPTs***. Causal value and conditional probability table (CPT) both represent the influence of parent variables on child variables. However, CPT is much more difficult to acquire because it's very time consuming (if

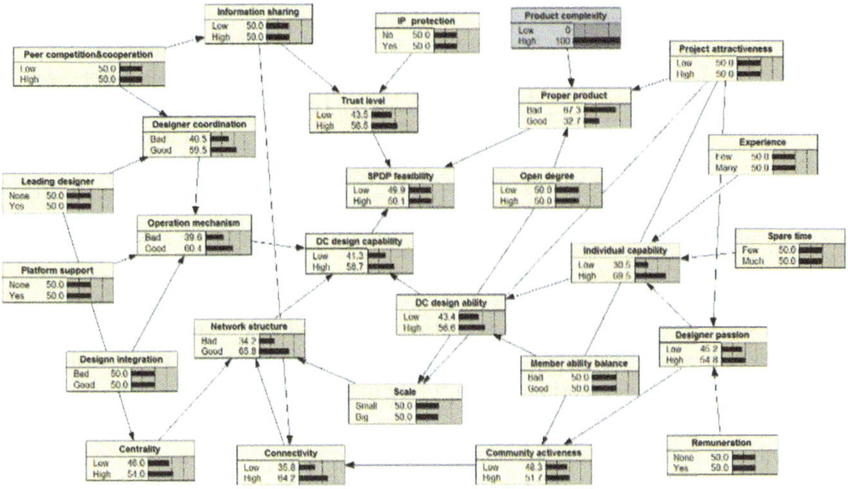

Fig. 5.16 Influence of product complexity on OPD project feasibility

possible) for experts to determine the probabilities of all the possible child variable combinations. Hence, CPT should be derived from causal values [32].

Step 7: ***BCM completion and validation***. After the CPTs are determined, the final BCM can be constructed with the help of *Bayesian* network software. In our research, the BCM is implemented using *Netica,* a commercial *Bayesian* network software. The resulting BCM has 27 nodes and 34 links, as shown in Fig. 5.16.

The main reason for building the OPD feasibility BCM model is to deepen our understanding of OPD and provides decision aids for OPD practices. In our research, the model is utilized to perform sensitivity analysis, top-down analysis, and bottom-up analysis to explore OPD mode.

5.5.3 Design Result Evaluation

In order to perform effective evaluation for the design results of OPD projects, specific evaluation criteria are identified from both user perspective and designer perspective, as shown in Fig. 5.17. And the most distinctive criteria from designer perspective are demonstrated below.

Degree of openness: This criterion indicates whether the product is of open-source or not. If the product is of open-source, it will have more chances to be noticed, used, and developed by more SDs.

Modularity: Open product design process is carried out by self-organized designers. Hence, it's difficult to organize the design process when the product is over complex. If the product has high modularity, it would be easier for the designers to develop the product collaboratively.

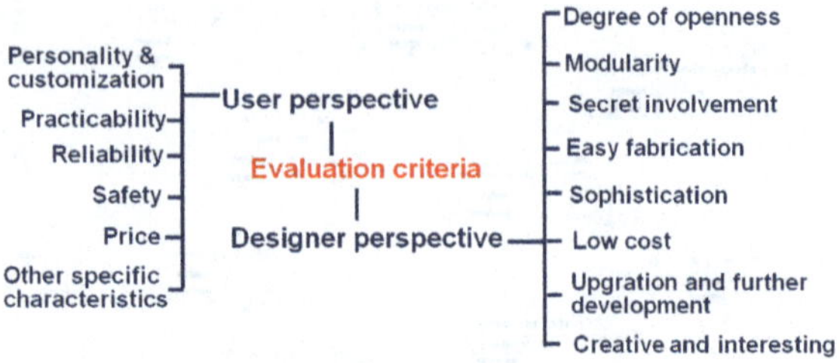

Fig. 5.17 Evaluation criteria for open product design result

Interesting: As we know, the designers in open product design pattern are mainly driven by interests and not profits. Hence, the product has to be very creative and interesting to draw the attention of SDs. Only in this way, the product can be continuously developed and upgraded.

Based on these objectives, full-scale, and comprehensive evaluation criteria, many evaluation algorithms, e.g. analytic hierarchy process, fuzzy mathematics, gray theory, projection pursuit, can be utilized to provide a specific and quantitative evaluation for open product design results.

5.6 Concluding Remarks

Open product design, featured by its tremendous capability to take advantage of socialized resources for product implementation, is the corresponding design mode of social manufacturing paradigm. Following the story line of its emerging background, operation characteristics, and its design project runs, we can draw the following conclusions.

- OPD mode emerges because technology development has given product consumers the desire and ability to participate in the design process of their interested product.
- One of the most distinguishing characteristic between OPD mode and the traditional one is that OPD project is carried out by a huge number of socialized designers in the form of open and self-organized community.
- To successfully run an OPD project, participants must consider the specificity of the enabled technologies, because these technologies are totally different from traditional ones when they are used in the product design process related to the OPD project.

All in all, OPD is an operable and effective design approach, but further researches are still required to explore all the potentials.

References

1. Jiang PY, Leng JW, Ding K, Gu PH (2016) Social manufacturing as a sustainable paradigm for mass individualization. Proc Inst Mech Eng Part B: J Eng Manuf 230(10):1961–1968
2. Cao W, Jiang PY (2012) Cloud machining community for social manufacturing. Appl Mech Mater 61–64(2012):220–223
3. Jiang PY, Ding K, Leng JW (2016) Towards a cyber-physical-social-connected and service-oriented manufacturing paradigm: social manufacturing. Manuf Lett 7(2016):15–21
4. Howe J (2006) The rise of crowdsourcing. WIRED 14(2006):14–24
5. Liu A, Lu C (2016) A crowdsourcing design framework for concept generation. CIRP Ann Manuf Technol 65(1):177–180
6. Stacey M (2006) The FAB LAB network: a global platform for digital invention, education and entrepreneurship. Innovations Technol Governance Globalization 9(1):221–238
7. Simula H, Ahola T (2014) A network perspective on idea and innovation crowdsourcing in industrial firms. Ind Mark Manage 43(3):400–408
8. Geiger D, Seedorf S, Schulze T, Nickerson R (2011) Managing the crowd: towards a taxonomy of crowdsourcing processes. Am Conf Inf Syst 2011(1):1–11
9. Brabham D (2008) Crowdsourcing as a model for problem solving: an introduction and cases. Int J Res New Media Technol 41(1):75–90
10. Schee B (2009) Crowdsourcing: why the power of the crowd is driving the future of business. Am J Health-Syst Pharm 67(4):1565–1566
11. N.N. (2017) The GNU project, Wikipedia, available at: https://en.wikipedia.org/wiki/GNU_Project, http://www.gnu.org/philosophy/fr
12. Konstroffer M, Weis T, Braun S (2000) Open-source in electronic commerce—a comparative analysis. Int Conf Electron Commer Web Technol 4(5):266–279
13. Merilinna J, Matinlassi M (2006) State of the art and practice of opensource component integration. Euromicro Conf Softw Eng Adv Appl 10(2006):170–177
14. Petrinja E, Nambakam R, Sillitti A (2009) Introducing the opensource maturity model. ICSE Workshop Emerg Trends Free/Libre/Open Source Softw Res Dev 9(2009):37–41
15. N.N. (2017) Is it a Hackerspace, Makerspace, TechShop, or FabLab. Maker Magazine, available at: http://makezine.com/2013/05/22/the-difference-between-hackerspaces-makerspaces-techshops-and-fablabs/
16. Mikhak B, Lyon C, Gorton T, Gershenfeld N et al (2002) Fab Lab: an alternate model of ICT for development. ResearchGate, available at: https://www.researchgate.net/publication/228600651_Fab_Lab_An_alternate_model_of_ICT_for_development
17. Troxler P (2010) Commons-based peer-production of physical goods: is there room for a hybrid innovation ecology. The 3rd Free Culture Research Conference. Berlin, Germany, 8–9 Oct 2010
18. Song G, Chen K, Zhang N, Tang Q, Zhu H (2008) Innovation mode of Fab Lab and its enlightenment. Sci Manage Res 26(2008):1–4
19. Kohler T, Matzler K, Füller J (2009) Avatar-based innovation: using virtual worlds for real-world innovation. Technovation 29(6–7):395–407
20. Panchal H, Le Q (2018) Product development by self-organized virtual communities. ResearchGate, available at https://www.researchgate.net/publication/245442686_Product_Development_by_Self-Organized_Virtual_Communities
21. NN (2018) The wisdom of crowds. Wikipedia, available at https://en.wikipedia.org/wiki/The_Wisdom_of_Crowds
22. Majchrzak A, Malhotra A (2013) Towards an information systems perspective and research agenda on crowdsourcing for innovation. J Strateg Inf Syst 22(4):257–268
23. Chesbrough HW (2003) Open innovation: the new imperative for creating and profiting from technology. ResearchGate, available at: https://www.researchgate.net/publication/235700923_Open_Innovation_The_New_Imperative_for_Creating_and_Profiting_From_Technology
24. Miller DT (2008) Rediscovering social innovation. ResearchGate, available at: https://www.researchgate.net/publication/242511521

25. Vincent P, Larochelle H, Lajoie I, Bengio Y (2010) Stacked denoising autoencoders: learning useful representations in a deep network with a local denoising criterion. J Mach Learn Res 11(2010):3371–3384
26. Huber PJ (1985) Projection pursuit. Ann Stat 13(2):435–475
27. Keller AI (2018) For inspiration only; designer interaction with informal collections of visual material. ResearchGate, available at: https://www.researchgate.net/publication/27343368ForInspirationOnlyDesignerinteractionwithinformalcollectionsofvisualmaterial
28. Janssen M, Bodin R, Anderies J, Elmqvist T (2006) Toward a network perspective of the study of resilience in social-ecological systems. Ecol Soc 11(1):1599–1604
29. Baumbusch J (2010) Semi-structured interviewing in practice-close research. J Spec Pediatr Nurs 15(3):255–258
30. Nadkarni S, Shenoy P (2004) A causal mapping approach to constructing Bayesian networks. Decis Support Syst 38(2):259–281
31. Nadkarni S, Shenoy P (2001) A Bayesian network approach to making inferences in causal maps. Eur J Oper Res 128(3):479–498
32. Sedki K, Beaufort L (2012) Cognitive maps and Bayesian networks for knowledge representation and reasoning. IEEE Int Conf Tools Artif Intell 13(2012):1035–1040

Chapter 6
RFID, Social Sensors and Extended Cyber-Physical System

Chao Liu and Pingyu Jiang

6.1 Roles of RFID, Social Sensors and Existing Cyber-Physical System in Social Manufacturing

6.1.1 Role of RFID in Social Manufacturing

Radio Frequency Identification (RFID) is one of the most promising technologies with big potentials in a broad variety of business processes, such as material flow monitoring [1, 2], quality assurance [3], production scheduling and decision making [4, 5], object locating [6], supply chain and inventory management [7, 8], and access control [9]. Abstractly from its application spectrum perspective, the fundamental principle of RFID applications in manufacturing industry is to track and monitor the dynamic changes of position, state and other attributes of physical objects attached with RFID tags, to make sure that right objects perform right tasks under right states at right time and in right positions during a process flow [10]. Once RFID-tagged objects enter or leave the coverage space of RFID signals, they can be automatically detected and identified through wireless communication with RFID readers/antennas deployed onto gateways. Within this process, production disturbances or exceptions can be captured, and further handled and fed back timely by RFID readers/antennas for dynamic control and decision-making. Under the context of social manufacturing (SM), RFID can help to provide visibility and controllability of production and logistics processes at both intra-enterprise level and inter-enterprise level.

In the intra-enterprise level, RFID-enabled monitoring and tracking applications can perfectly collect real-time process-related data (e.g., timestamp, object ID, operators, states and positions) via collaborating with other existing sensor networks. Then, on the basis of Auto-ID computing methods coupled with RFID tags and backend databases, these data can be fused into valuable manufacturing information for different production scenarios. For example, with the aid of RFID data collection and analysis, real-time occupation rate of machines can be calculated according to

P. Jiang, *Social Manufacturing: Fundamentals and Applications*, Springer Series in Advanced Manufacturing, https://doi.org/10.1007/978-3-319-72986-2_6

the arrival time and leaving time of workpieces. From the aspect of RFID application scenarios and modelling, Jiang and Cao proposed an early formalized method related to a systematic RFID-driven graphical formalized deduction model (rfid-GDM) for monitoring the time-sensitive state and position changes of Work-In-Progress (WIP) material flows in a job shop [1]. It provides guidelines for shop floor operators to identify where to deploy RFID devices and how to use them for collecting real-time on-site data. On the basis of previous work, Ding et al. developed this method further to an enhanced rfid-GDM to describe the RFID-enabled process flow, which is helpful for physical object tracking [10]. Four kinds of RFID application scenarios were summarized and depicted by the enhanced state blocks. The tracking scheme based on the enhanced rfid-GDM can provide a more explicit way to build the RFID-enabled process flow. From the aspect of RFID-enabled production control in a shop floor, RFID technology realizes decentralized production control that is adapted to *Industry 4.0*, production disturbance detection and rescheduling, and order completion time prediction, etc. For example, Wang and Jiang leveraged the nature of RFID tags in data storage capability to develop a hybrid-data-on-tag-enabled decentralized manufacturing control system [11]. In this solution, the workpiece-related data such as basic machining information and its index were stored in the RFID tags. The detailed information such as engineering drawings and production plans was stored in the backend server. Besides, the real-time shop floor load conditions can be obtained by capturing the real-time RFID data such as type and waiting list information of all WIPs in the in-buffer and out-buffer of machines, and the real-time machining time of all WIPs [12]. Based on that, the order completion time can be predicted.

In the inter-enterprise level, community members such as core producers, micro-and-small-sized enterprises, factories, shop floors, logistics service providers, public warehouse providers, and even individuals (e.g., makers, leading users) need to collaborate with each other to fulfill the outsourcing/crowdsourcing tasks. The mass collaboration between community members requires social business interactions and information sharing in an efficient manner. Real-time RFID data driven by production collaboration can be utilized to deal with the real-time data capturing, production and transportation monitoring, and timely production and transportation decision-making. Currently, many studies have been devoted to the real-time goods flow tracking in logistics and supply chains. For example, Lee et al. integrated RFID technology with artificial intelligence to design and develop a flexible logistics information system that was featured by the fast responsiveness in dealing with customer requirements through the integration of various value chain activities [13]. Ding and Jiang proposed a RFID-enabled social manufacturing system to realize the real-time monitoring and dispatching of inter-enterprise production and transportation tasks [14]. By deploying RFID devices in each enterprise's shop floors and transportation vehicles, real-time production and transportation data can be collected and processed for inter-enterprise production processes management. Based on the real-time RFID data processing, dynamic dispatching decisions and manufacturing tasks for inter-enterprise production and transportation can be made even if the unexpected disturbances exist.

In summary, with the aid of RFID technology, machining process monitoring and tracking, and other related RFID applications like production scheduling, order completion time prediction, and decentralized production control in the intra-enterprise level can be achieved. In the inter-enterprise level, core producers, socialized manufacturing resources (SMRs) providers and other stakeholders can leverage RFID technology to collaborate with each other to deal with inter-enterprise production management and decision optimization under the context of SM.

6.1.2 Role of Social Sensors in Social Manufacturing

The term "*social sensors*" is defined as a kind of hardware-software-integrated interactive media [15] that not only facilitate the social business interactions among social entities (humans and machines) but also enable SMRs to be interconnected.

In the inter-enterprise level, cooperative enterprises collect production data manually or extract from enterprise information systems (enterprise resource planning system, manufacturing execution system, etc.) and then send them to the partners by emails, phones, or other social media such as *WeChat*, *WhatsApp* and *Skype*. In this way, the production data sharing is not applicable to time-varying customer requirements and unexpected production disturbance. Furthermore, the high cost and poor efficiency in communication, and the information barriers between different cooperative enterprises' information systems will result in rigid, unsmooth, and unreliable production collaboration. On the other hand, the distributive SMRs are interconnected to form a manufacturing community by the social business relationships that are built through manufacturing outsourcing or crowdsourcing orders. SMRs in the community need to be queried, positioned, and utilized in an efficient way so as to improve the resource utilization. Thus, a kind of media called social sensors is utilized to facilitate inter-enterprise production interactions and coordination, and equip the SMRs with access to the Internet. Social sensors change the way that SMRs stay in touch with their upstream or downstream partners as well as customers. From the viewpoint of customers, they use social sensors to communicate with each other and share their interests and requirements. Besides, whether specialists or non-specialists, customers can use social sensors to interact with SMRs about their product orders, such as checking product information, expressing opinions and requirements, and monitoring order states. From the viewpoint of SMR providers, they use social sensors to interact with their suppliers or customers and transmit production order information. Besides, the distributive SMRs can freely join the socialized manufacturing network and are self-organized into different manufacturing communities to meet different requirements.

In the intra-enterprise level, the role of workers in smart factory has shifted from repetitive manual laborers to decision makers. This requires that these workers must obtain the right information at the right time and place (anywhere and anytime). The proposed social sensors can connect humans, machines, sensors and robots to build a shop floor-level production network that is able to execute the production tasks.

During a production process, social sensors can record and handle the interaction data generated via human-to-human (H2H), human-to-machine (H2M), machine-to-human (M2H), and machine-to-machine (M2M) interactions. Therefore, enterprises use social sensors to link with their staffs and machines in factories, transmit commands to them and receive production feedback from them.

In summary, with the aid of social sensors, SMRs interconnect with each other to form self-organized manufacturing communities and SM network, which facilitates the production interactions and collaboration among different stakeholders in the inter-enterprise level. In the intra-enterprise level, different factory objects such as machines, operators and software components are interconnected via social sensors, which enable the transparency and information flow of a production process.

6.1.3 Role of Existing Cyber-Physical System in Social Manufacturing

Industry 4.0 is featured by the horizontal and vertical integration of networked manufacturing systems through value networks and end-to-end digital interconnections [16]. It is enabled mainly by cyber-physical system (CPS) technologies. CPS is a system composed of collaborating computational entities which are intensively connected with their surrounding physical world and on-going processes, and provide and use data-accessing and data-processing services available on Internet [17]. Under the context of SM, CPS is an enabling technology to realize wide interconnection and management of physical elements (e.g., machines, tools, sensors, actuators and controllers) in the physical space and computational elements (e.g., software, applications, functions and information systems) in the cyber space.

In the inter-enterprise level, CPS enables enterprise's machines to be separately accessible for the external customers and suppliers. It means they know which machines are responsible for their product orders and what are current states of machines [18]. Thus, CPS-enabled production is decentralized, self-organized, self-coordinated, and rapid-responsive by means of using cyber technologies in the physical space [17], which is suitable for inter-enterprise production collaboration.

In the intra-enterprise level, all the involved production resources (e.g., machines and their auxiliaries, industrial robots, workpieces and workers) are equipped with various sensors (e.g., acceleration, noise and energy) to improve the adaptability, flexibility and transparency [19]. The integration of CPS with production enables machines to equip the capabilities of self-awareness, self-maintenance and intelligent adaptive control on one hand [20], and provide factories with real-time production planning, control and monitoring, near-zero downtime and intelligent decision-making on the other hand [21].

In summary, CPS follows the close-loop of "*sensing-computation-communication-control-feedback*". It contains physical elements and cyber elements. The interconnection and communication among these elements determine the feasibility and efficiency in different industrial applications. Products are manufactured with the help of various CPS nodes among which there exist seamless interoperations and plug-and-play configurations. By clouding the CPS nodes from SMRs under certain authority and privacy, the cyber-physical interconnection is built and the production information sharing can be realized.

6.2 RFID Used for Tracking and Tracing of Production and Supply States

6.2.1 Pickup Modes of RFID Tags in a Limited Detecting Space

(1) ***RFID-related concept clarification***

To build the basis for RFID applications, RFID-related concepts are clarified as follows.

The term "***RFID-tagged object***" is defined as a physical object attached with an RFID tag for automatic identification. The binding of RFID tag and physical object enables the RFID-tagged object to be exclusively identified. Object attributes such as tag code, position and state can be collected through the wireless communication between RFID tags and readers/antennas.

The term "***RFID workstation***" is defined as a place where RFID readers and antennas are installed and relevant processes are monitored. It can be fixed (e.g., machining workstations of machines, quality inspection workbenches) or movable (e.g., vehicles, forklifts).

The term "***RFID detection space***" is defined as the physical coverage space of radio frequency signals where RFID reader/antenna could exclusively identify the RFID-tagged object via unique tag code. In accordance with RFID workstations, the detection space can also be fixed or movable.

The term "***RFID tag event***" refers to the reading of RFID reader/antenna when RFID-tagged objects move in, move out, or pass through the detection space. The tag events are comprised of four kinds of events, i.e., start event, finish event, pass-through event, and random event. Start event and finish event make up an event pair indicating the start and finish of an operation. Pass-through event represents that the RFID-tagged objects pass through a gateway or fixed position. Random event is usually an unexpected event that happens occasionally. All the events are transient events and have no duration time.

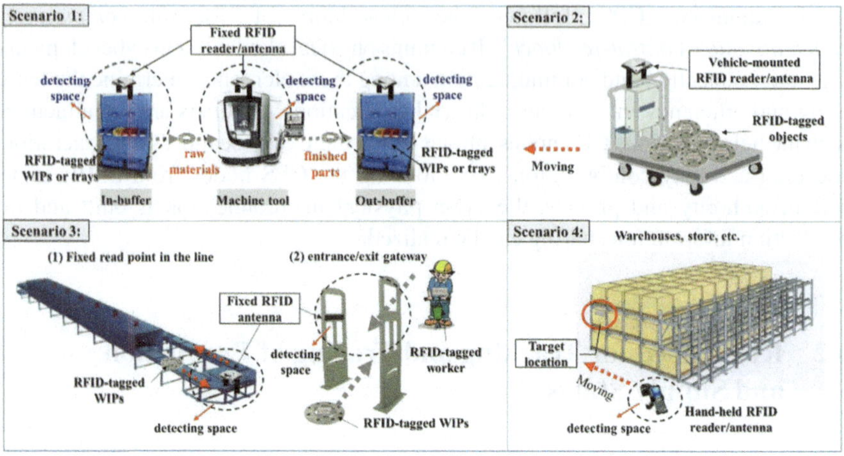

Fig. 6.1 Four kinds of RFID application scenarios in manufacturing field [10]

The term "***RFID tag state***" refers to the state of RFID tag when it enters, stays in or leaves the detection space. The tag sates are comprised of three states, i.e., tag-moving-in, tag-staying-in, and tag-moving-out.

(2) ***Four kinds of RFID application scenarios***

According to RFID-related concept clarification and RFID application cases in manufacturing, logistics and supply chain, four kinds of application scenarios have been summarized [1, 10]. Figure 6.1 shows the examples applied in the manufacturing field.

Scenario 1 demonstrates the case related to "*Fixed-Reader/Antenna-Based Fixed Detection Space Control Mode*". Here, fixed RFID reader/antenna monitors the RFID-tagged objects' entering into and moving out of a fixed detection space.

Scenario 2 shows the case concerned with "*Fixed-Reader/Antenna-Based Movable Detection Space Control Mode*". Here, fixed RFID reader/antenna, which is usually mounted on a vehicle or forklift, monitors the RFID-tagged objects' entering into and moving out of a moveable detection space.

Scenario 3 illustrates the case related to "*Fixed-Reader/Antenna-Based Gateway Access Control Mode*". Here, fixed RFID reader/antenna monitors the RFID-tagged objects' passing through a gateway or other fixed points.

Scenario 4 just describes the case concerning "*Movable-Reader/Antenna-Based Random Detection Space Control Mode*". Here, moveable RFID reader/antenna, which is usually in the form of hand-held RFID reader/antenna, monitors the random events executed on objects.

(3) ***Enhanced state blocks***

An enhanced state block (ESB) model is built to describe the RFID-tagged object such as process, operator, triggered event, and start/finish time. It is defined as a

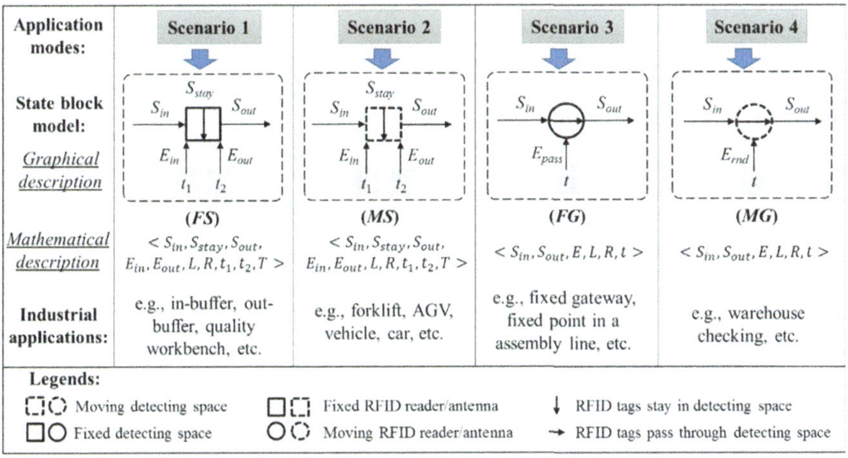

Fig. 6.2 Four kinds of enhanced state blocks [10]

Table 6.1 Notations and meanings in the formal description of enhanced state blocks

Notations	Meanings
SB	The enhanced state block
FS, MS, FG, and MG	Four kinds of RFID application scenarios, respectively
E_{in} and E_{out}	The start event and finish event of a process, respectively E_{in} triggers the changes of object's state from S_{in} to S_{stay} E_{out} triggers the changes of object's state from S_{stay} to S_{out}
E	The pass-through event E_{pass} or random event E_{rnd} E triggers the changes of object's state from E_{in} to E_{out}
S_{in}, S_{stay}, and S_{out}	Three states of RFID tags: tag-moving-in, tag-staying-in, and tag-moving-out, respectively
L	RFID workstation
Fr and To	The location where objects come from and where objects finally go
R	The relevant operator
t_1, t_2 and t	The triggering time of E_{in}, E_{out}, and E, respectively The duration time of tag-staying-in state in state block FS and MS equals to $T = t_2 - t_1$
SB_SD, SB_ED, and SB_RD	The state domain, event domain, and RFID-related domain of the enhanced state blocks

graphical unit that illustrates the state, triggering event and its triggered time, position, and other attributes of RFID-tagged object [10]. As shown in Fig. 6.2, four kinds of ESBs are generated based on the four kinds of RFID application scenarios. The ESB is formulated below, and the corresponding notations and meanings are illustrated in Table 6.1.

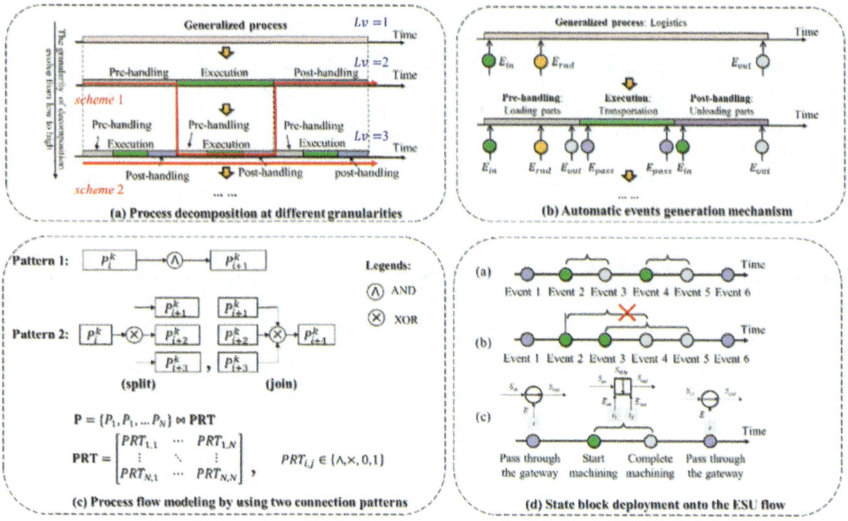

Fig. 6.3 Modeling framework of RFID-based event-state-position graphical deduction model

$$SB = \{FS, MS, FG, MG\} \tag{6.1}$$

$$\begin{cases} SB\big(S_{in}, S_{stay}, S_{out}, E_{in}, E_{out}, L, Fr, To, R, t_1, t_2\big), SB \in \{FS, MS\} \\ SB(S_{in}, S_{out}, E, L, Fr, To, R, t), SB \in \{FG, MG\} \end{cases} \tag{6.2}$$

$$S_{stay} = \emptyset, \; t_2 = t_1, \; \forall SB = FG, MG \tag{6.3}$$

$$\begin{cases} SB\_SD = \prod_{S_{in}, S_{stay}, S_{out}}(SB) \\ SB\_ED = \prod_{E_{in}, E_{out}, t_1, t_2}(SB) \\ SB\_RD = \prod_{L, Fr, To, R}(SB) \end{cases} \tag{6.4}$$

6.2.2 A RFID-Based Event-State-Position Graphical Deduction Model

The modeling framework of RFID-based event-state-position graphical deduction model includes four steps, i.e., multi-granularity process decomposition, automatic events generation, RFID-enabled process flow modelling, and the final rfid-GDM construction. Figure 6.3 just demonstrates the logic flow of the framework.

(1) ***Multi-granularity process decomposition***

As shown in Fig. 6.3a, each process is decomposed into three sub-processes, i.e., pre-handling, execution, and post-handling based on fractal theory. Each sub-process can be further decomposed into three fine-grained sub-processes. The decomposition granularity of each process is identified according to the specific monitoring and tracking requirements. Let the set F represents the decomposition results, and the basic subset of F contains three sub-processes and the length of each basic subset is 1/3. Thus, it is a special example of uniform Cantor Set, $F = \bigcup_{Lv=1}^{\infty} P^{Lv}$, where P^{Lv} is the set of sub-processes at the Lv-th decomposition granularity.

By linking the required sub-processes, a process decomposition solution can be achieved, which is formalized as

$$P_i = \left\{P_{i,1}, P_{i,2}, \ldots, P_{i,A_i}\right\} \tag{6.5}$$

where P_i indicates the i-th process, $P_{i,j}$ represents the j-th sub-process of P_i, A_i represents the number of sub-processes ($A_i \leq 3^{Lvm}$), and Lvm is the maximum decomposition granularity, $Lvm \in \mathbf{R}^+$.

(2) ***Automatic events generation***

As shown in Table 6.1, the state, position, and other attributes of RFID-tagged objects are changed via the triggering of four kinds of events, i.e., start event E_{in}, finish event E_{out}, pass-through event E_{pass}, and random event E_{rnd}. Figure 6.3b illustrates the automatically generated events of a process at two granularities. The event pair E_{in} and E_{out} is used to monitor the process, the pass-through event E_{pass} is used to monitor the object's entering and moving out the gateways, and the random event E_{rnd} is manually added into the process if a random event happens on the object at a certain time (for example, an object-checking event E_{rnd} is inserted into the loading-parts sub-process for random control).

After the automatic events generation, events of each process are sequentially connected to form an event sequence unit (ESU), which can be formalized as

$$ESU_i = \left\{E_{i,1}, E_{i,2}, \ldots E_{i,n_i}\right\} \tag{6.6}$$

where ESU_i indicates the i-th ESU, $E_{i,j}$ is the j-th event of ESU_i, n_i is the number of events and $n_i = 2A_i + N(E_{rnd})$, where $N(E_{rnd})$ is the number of manually-inserted random events.

(3) ***RFID-enabled process flow modelling***

Two kinds of logical connectors AND ($\wedge$) and XOR ($\times$) are applied to connect processes/sub-processes or ESUs to form process/sub-process flow or ESU flow. Correspondingly, two connection patterns are summarized, as shown in Fig. 6.3c.

Pattern 1: Sequence connection (AND). In this pattern, process P_{i+1} is executed after P_i sequentially.

Pattern 2: Exclusive connection (XOR). In this pattern, process P_{i+1} is an optional choice of the target process P_i, and if P_{i+1} cancels or cannot satisfy the requirements, P_{i+2} will take the place of P_{i+1} and become the direct subsequent process of P_i. Note that XOR connector may have one incoming and multiple outgoing directions (split) or multiple incoming and one outgoing direction (join).

By using two connection patterns, the RFID-enabled process flow can be formalized as

$$\mathbf{P} = \{P_1, P_2, \ldots, P_N\} \bowtie \mathbf{PRT} \tag{6.7}$$

$$\mathbf{PRT} = \begin{bmatrix} PRT_{1,1} & \cdots & PRT_{1,N} \\ \vdots & \ddots & \vdots \\ PRT_{N,1} & \cdots & PRT_{N,N} \end{bmatrix} \tag{6.8}$$

where **PRT** is the connector matrix, $PRT_{i,j}$ indicates the connector between P_i and P_j, $PRT_{i,j} \in \{\wedge, \times, 0, 1\}$. If $i = j$, $PRT_{i,j} = 1$, and if there is no connection between P_i and P_j, $PRT_{i,j} = 0$.

It should be pointed out that when the processes are decomposed into fine-grained sub-processes, there are still two connection patterns to connect sub-processes. Moreover, there is an one-to-one mapping relationship between the process and the ESU, so an ESU flow can be formed by substituting P_i with ESU_i.

(4) ***The final rfid-GDM construction***

After modelling the RFID-enabled process flow, ESBs are deployed onto the ESU flow depending on the event types and event occurrence places, as shown in Fig. 6.3d. The deployment of ESBs can establish relationships between process flow and RFID devices configuration. Considering that two event pairs cannot be executed concurrently at two RFID workstations for the same RFID-tagged object, the overlap of the event pairs is not allowed. Therefore, a stack structure algorithm (first in and last out) is applied to deploy ESBs automatically onto the ESU flow. The pseudo code of the algorithm is shown in Fig. 6.4a.

Based on the algorithm, a simple example is given in Fig. 6.4b. A batch of RFID-tagged workpieces are transported from the warehouse to the shop floor (pass through the entrance-gateway of the shop floor), and then either Machine 1 or Machine 2 is selected to process the workpieces. After that, the workpieces are transported to the quality inspection workbench to check their machining quality. Finally, they are transported from the shop floor to the warehouse (pass through the exit-gateway of the shop floor). By connecting the ESBs logically, the final rfid-GDM is automatically built, as shown in Fig. 6.4b. The description of rfid-GDM can be formalized as

$$\mathbf{GDM} = \{SB_1, SB_2, \ldots, SB_i, \ldots, SB_M\} \bowtie \mathbf{BRT} \tag{6.9}$$

where **GDM** represents the *rfid*-GDM, SB_i stands for the i-th ESBs, and **BRT** is the matrix of their relationships. M is the amount of ESBs derived from the model. It should be pointed out that the final rfid-GDM is also a guideline to deploy RFID

```
Stack<Integer> stack = new Stack<Integer>();  // define a new stack
for (int i=0; i<N; i++)
    for(int j = 0; j < n; j++)
        if (E_{i,j}.E_T = E_in) stack.push (j);        // E_{i,j} is a start (entrance) event
        else if (E_{i,j}.E_T = E_out)                  // E_{i,j} is an end (exit) event
            stack.pop ();
            if (E_{i,j}.LT = fixed)                    // fixed event occurrence place
                deploy a state block FS between E_{i,j-1} and E_{i,j};
            else if (E_{i,j}.LT = moving)              // moving event occurrence place
                deploy a state block MS between E_{i,j-1} and E_{i,j} ;
        else if (E_{i,j}.E_T = E_pass)                 // E_{i,j} is a pass-through event
            deploy a state block FG onto E_{i,j};
        else                                           // E_{i,j} is a random event
            deploy a state block MG onto E_{i,j};
    end for
end for
```

(a) Pseudo code of the stack structure algorithm

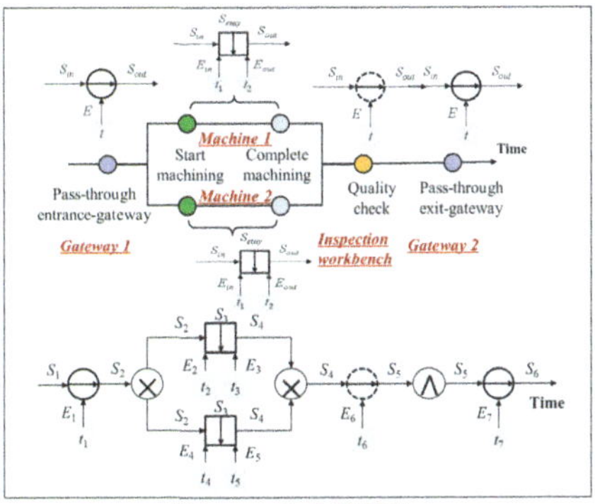

(b) An example of rfid-GDM

Fig. 6.4 Enhanced state blocks deployment onto event sequence units

devices onto any process flows since the ESBs in rfid-GDM contain the information of RFID workstations and RFID devices.

6.2.3 *RFID-Based Graphical Deduction Model (GDM) for Tracking and Monitoring Machining-Process Material Flows*

(1) *Machining-process material flows tracking and monitoring based on rfid-GDM*

Generally, the logic flow of parts machining in a shop floor includes four steps. Firstly, raw materials are checked and prepared in the warehouse. Secondly, they are transported through RFID gateways to the machines' workstation waiting for to be machined. All the parts in the machining workstation are successively in the states of "*in-buffer*", "*machining*", and "*out-buffer*". Thirdly, after the machining processes are finished, the parts are transported to the quality inspection workbench for quality checking. Finally, the qualified parts are transported to the warehouse and the unqualified parts are transported to the waste part area.

Based on the description of parts machining in a shop floor and the proposed rfid-GDM model, the tracking and monitoring implementation for machining-process material flows is illustrated in Fig. 6.5 and consists of the following four steps.

Step 1: Suppose there are seven up-level processes, namely, one raw material check process, two transportation processes, two machining processes, one quality inspection process, and one inventory check process. The transportation processes are divided into three sub-processes, including loading parts, transportation execution, and unloading parts. The machining processes are also divided into three

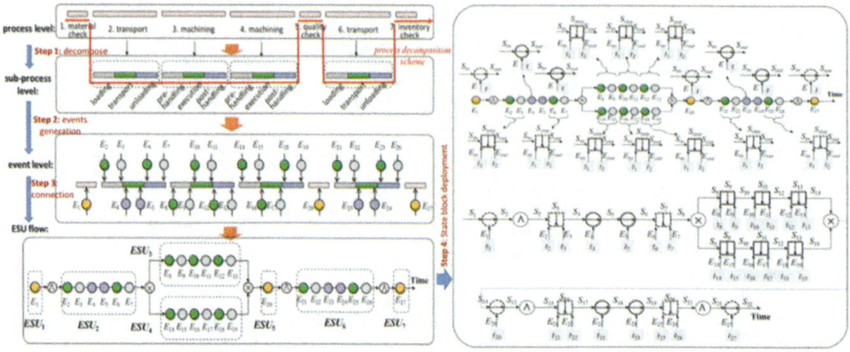

Fig. 6.5 Tracking and monitoring model of machining-process material flows based on rfid-GDM

sub-processes, including pre-handling in in-buffer, machining execution, and post-handling in out-buffer.

Step 2: Depending on the automatic events generation mechanism, different events of each process and its sub-processes are extracted. For the up-level processes (raw-material-check, quality-check, and inventory-check processes), the random events are extracted. For the low-level sub-processes (loading-goods, uploading-goods, pre-handling, machining-execution, and post-handling sub-processes), the start events and finish events are extracted as event pairs. For the low-level sub-process (transportation-execution sub-process), two pass-through-gateway events are extracted. In correspondence with the seven up-level process and the extracted events, seven ESUs and the relevant twenty-seven events are generated. Note that the attributes of processes, sub-processes and events are formalized.

Step 3: Two kinds of connectors are utilized to connect the up-level processes and sub-processes to form a process flow. The connectors are also used to connect the ESUs to generate an ESU flow.

Step 4: Four kinds of ESBs are automatically deployed onto the ESU flow, and the final rfid-GDM is built by connecting the ESBs logically.

(2) ***RFID devices configuration based on the tracking and monitoring model of machining-process material flows***

The constructed rfid-GDM deals with the problem that which kind of RFID device should be deployed in certain places. Four kinds of RFID devices configuration modes are presented in Table 6.2 in accordance with the four kinds of ESBs. Each RFID workstation is equipped with RFID readers and antennas in a certain configuration mode. The RFID devices configuration mode can be defined as

$$Rc_i(id, name, CM, Pos, WS, MA, MR, RF) \tag{6.10}$$

where id and $name$ are the configuration id and the configuration name, CM represents the configuration mode, Pos denotes the position where RFID devices are

Table 6.2 Four kinds of RFID devices configuration modes

Configuration mode	Enhanced state block	Reader type	Reader number	Antenna type	Antenna number
FS	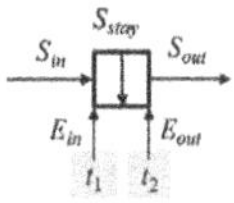	Fixed	1, or shared	Fixed	1
MS	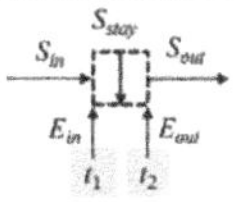	Vehicle-mounted	1	Vehicle-mounted	1
FG	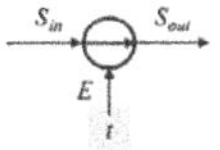	Fixed	1, or shared	Fixed	1
MG	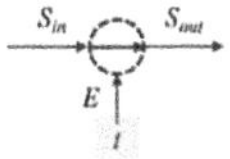	Hand-held	1	Integrated	–

placed, WS is the working state, MA and MR represent the identification accuracy and range of RFID devices, RF is the shared radio frequency. The solution space of the RFID devices configuration for an RFID-enabled process flow is described as

$$\mathbf{Rc} = Rc_1 \times Rc_2 \times \ldots \times Rc_M \tag{6.11}$$

The RFID configuration process can be easily carried out based on four kinds of RFID devices configuration modes. Firstly, parts bundled with RFID tags are uniquely identified and data-indexed. Two configuration schemes are adopted, that is, "*one-to-one tagging scheme*" which suits for the high-end or one-of-a-kind physical objects needing to be tracked separately and "*many-to-one tagging scheme*" which suits for the low-end or batch-type physical objects tracked in batches. Secondly, the fixed reader and antenna are deployed in the fixed RFID workstations (such as gateways, in-buffers and out-buffers of the machines, and inspection workbench) which form the fixed detection spaces to monitor the entrance and exit of RFID-tagged parts. Vehicle-mounted reader and antenna are installed on the forklift to form a movable detection space. Hand-held readers and inspection workbenches are equipped in the warehouse for inventory check and quality inspection.

(3) ***Tracking and monitoring scheme based on rfid-GDM***

Based on the rfid-GDM, the object tracking and monitoring scheme is achieved which can be formalized as

$$TTS = T \bowtie TT \bowtie \mathbf{P} \bowtie \mathbf{GDM} \bowtie \mathbf{Rc} \tag{6.12}$$

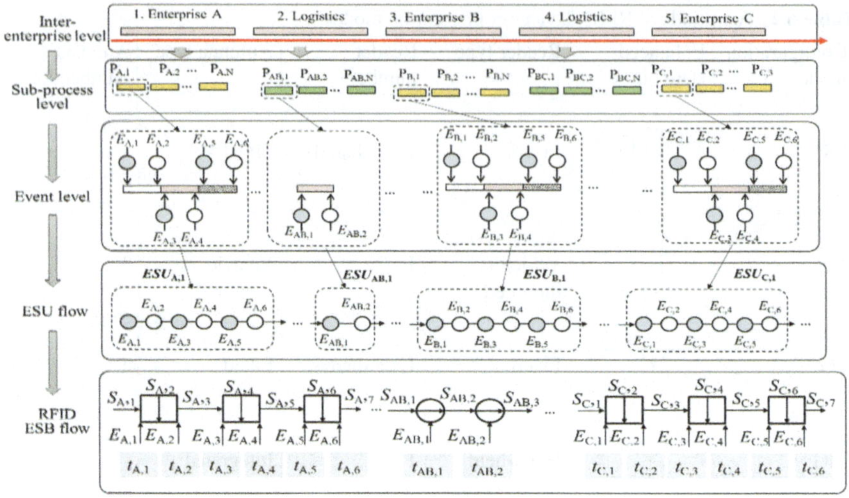

Fig. 6.6 Manufacturing supply chain tracking and monitoring model integrating intra-enterprise production and inter-enterprise logistics

where TTS denotes the tracking scheme, T is the set of RFID-tagged objects, TT is the set of RFID-tagged delivery units, **P** is the process flow, **GDM** is the rfid-GDM, **Rc** is the RFID device configuration solution.

Based on the TTS, RFID-tagged parts are tracked in their process flows. At the same time, the comprehensive state, position, and other attributes of parts are monitored. Furthermore, Auto-ID computing technology is utilized to transform the collected process-related real-time data into semantic information.

6.2.4 RFID-Based Graphical Deduction Model for Tracking and Monitoring Manufacturing Supply Chain

RFID applications related to the inter-enterprise-level collaboration aim to track and monitor both the production and logistics. Suppose that three enterprises (enterprise A, B, and C) collaborate with each other when machining a batch of outsourcing parts. Based on the proposed rfid-GDM, a tracking and monitoring model for manufacturing supply chain that integrates the intra-enterprise-level production with the inter-enterprise-level logistics is illustrated in Fig. 6.6.

Proper decomposition granularity of each process is selected according to the tracking requirements. There are two kinds of top-level processes, namely, intra-enterprise-level production and inter-enterprise-level logistics. The intra-enterprise-level production process can be divided into several sub-processes, such as raw material check process, transportation process, machining process, quality inspection process, and inventory check process, which is the same with part machining in a shop floor mentioned in Sect. 6.2.3. The inter-enterprise-level logistics process deals with the parts delivery from one enterprise to another. It can be divided into stock-out process, delivery process, and stock-in process, from which start event, finish event, random event, and pass-through-gateway event are extracted. After that, the events are generated and connected, and the ESBs are deployed onto the ESU, through which the final rfid-GDM for tracking and monitoring inter-enterprise-level manufacturing supply chain is built.

6.3 Social Sensors for Interactions among Humans and Machines

6.3.1 Social Sensor Clarification and Operational Logics

(1) ***Definition and components***

Social sensors as a kind of the hardware-software-integrated interactive medium are very useful for humans to realize the social communication from three dimensions, including H2H, H2M/M2H, and M2M interactions during order production process under the context of SM [15, 18]. They capture input data from social interactions, and then merge them into meaningful interaction results as output data through embedded algorithms and methodologies, and finally transfer the processed data via networks and receive feedback from the other side. A social sensor can be viewed as an integration of both physical sensors that are its hardware and non-physical data processor which is its software applications. The hardware includes fixed or wearable physical sensors, embedded devices, mobile devices, etc. The physical sensors aim to capture input data (e.g., production environment states, working conditions of machines, and human interactive data such as text, voice and gesture) that are transferred to the embedded devices or mobile devices for further processing. The software embedded into the embedded devices or mobile devices have several functions, which include building communication interfaces for H2H, H2M/M2H, and M2M interactions, crawling the production states from other social sensors via the embedded crawler for global production cooperation, aggregating the collected data and translating them into engineered information for production decision making.

The "*social*" features of social sensors are summarized into two aspects, including that social sensors are geographically distributive around the world rather than a certain manufacturing factory and act as the social media for interconnection and sharing among humans and machines under the age of mobile Internet and social network. They provide ubiquitous sensing and processing services related to Internet-based connecting and communicating behaviors, environment, capabilities, commands, and states for H2H, H2M/M2H, and M2M interactions. With the help of social sensors, requirement data from customers, production data from enterprises, and industrial

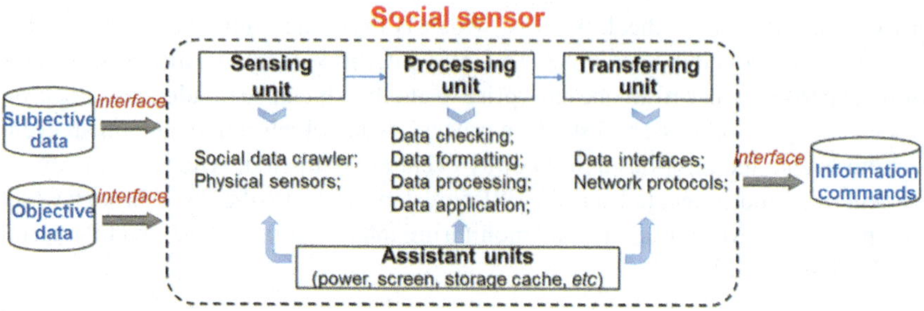

Fig. 6.7 Components of social sensors [18]

field data from machines can be captured and shared. Therefore, social sensors facilitate the interaction relationships and enable the high efficient production coordination.

According to the definition of social sensors, the components of a social sensor consist of sensing unit, processing unit, transferring unit, and assistant unit, as shown in Fig. 6.7. Its sensing unit proactively captures objective data from physical sensors and other social sensors, and perceives subjective human interaction data from human "*natural*" language such as text, voice and gesture. Its processing unit is responsible for checking the error data or duplicated data, and then normalizing the multi-source heterogeneous data to a machine-readable and easy-transfer format (e.g., JSON, XML). Besides, a storage cache from the assistant unit is used to cope with intermittent connectivity. Thus, the social sensor can synchronize data whenever network connectivity is available. Its transferring unit transfers the formatted data from the current social sensor to other social sensors or to the cloud database for sharing or further processing via the certain interfaces and protocols such as HTTP, TCP/IP, or Web Socket. Its assistant unit provides the assistant services, e.g., power, screen, and storage cache. Note that the screen can be regarded as human-machine interface (HMI) which displays production information and provides man-machine interface for commands inputting.

(2) ***Classification and operational logic***

Social sensors can be divided further into three types, i.e., H2H social sensors, H2M/M2H social sensors, and M2M social sensors.

H2H social sensors deal with enabling H2H social interactions and collecting business social data during order productions, such as interactions between customers and enterprises, interactions among enterprises, and interactions among enterprise's employees. Thus, H2H social sensors are applied in both the inter-enterprise level and the intra-enterprise level. Smart mobile terminals such as smartphones, tablets, and wearable devices installed with customized Apps can be viewed as a concrete instantiation of H2H social sensors. The enabling technologies for H2H-social-sensors-based interactions deal with social computing, big data analysis, and so on. Taking the production interactions between customers and enterprises as an example, the

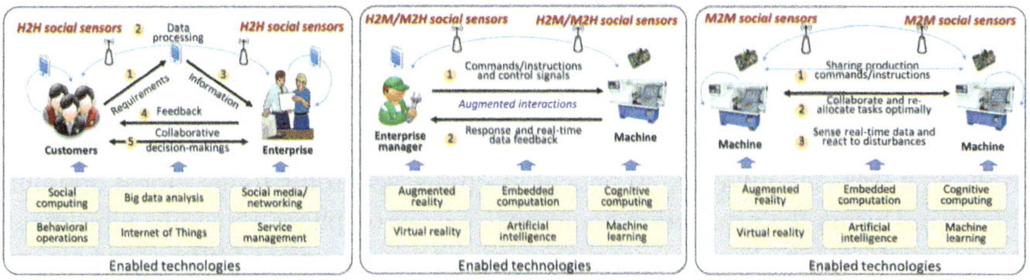

(a) Operational logic of H2H social sensors (b) Operational logic of H2M/M2H social sensors (c) Operational logic of M2M social sensors

Fig. 6.8 Operational logics of three kinds of social sensors

operational logic of H2H social sensors is illustrated in Fig. 6.8a. Firstly, customers interact with enterprises via H2H social sensors in the form of smartphones to express their feelings, intentions and requirements. Then, both the objective data and the subjective data, e.g., longitude and latitude, camera images, are fused into meaningful information and transferred to the H2H social sensors of enterprises. Finally, H2H social sensors of enterprises collect and handle these data and feed back to customers. With the aid of H2H social sensors, both enterprises and customers collaboratively predict the market trends and make production decisions.

H2M/M2H social sensors deal with enabling H2M/M2H interactions and sensing physical data during order productions. By using H2M/M2H social sensors, humans can efficiently interact with the machines, including assigning production tasks, ordering commands and receiving feedbacks. HMIs or chat bots are the concrete instantiations of H2M/M2H social sensors. The enabling technologies for H2M/M2H-social-sensors-based interactions include chat-bot-based autonomous information processors, virtual/augmented reality, embedded intelligence, cognitive computing, and artificial intelligence. Taking the chat bot application in the production stage of a product life cycle as an example, the chat bots inside H2M/M2H social sensors provide an intuitive interface for the direct text/voice/gesture-based H2M/M2H interactions. The operational logic of H2M/M2H social sensors can be described in Fig. 6.8b. Firstly, operators issue the production commands in the form of text, voice or gesture. Then, the chat bots inside the H2M/M2H social sensors perceive the input social context data and translate them into machine-readable format for the further execution. Finally, H2M/M2H social sensors at the machine end collect real-time operating data of machines and merge these data into operating state information to give suggestions for operators to make production decisions.

M2M social sensors deal with enabling M2M interactions and sensing physical data during order productions. Through M2M social sensors, machines can socially communicate with each other to react autonomously to unexpected events, make collaborative production decisions, and upload the execution results to the management center. The embedded device, which integrates various physical sensors and installs communication protocols and data processing algorithms, can be viewed as a concrete instantiation of M2M social sensors. The operational logic of M2M social sensors is described in Fig. 6.8c. Firstly, each M2M social sensor senses real-time data from its bundled machine, such as working state, current task, ambient temperature/humidity, and so on. At the same time, the crawler embedded in the M2M social sensor capture the production information of other M2M social sensors via IP address. After interactions among M2M social sensors have been settled, the production tasks are allocated to optical machines according to the global optimum objective. During the production commands execution phase, dynamic adjustments may be made if disturbances occur. Thus, coordination among M2M social sensors is periodically made according to real-time data.

6.3.2 Social Sensor Implementation

(1) ***Formal description of social sensor by using SocialSensorML***

Most physical sensors are geographically distributive and difficult to be integrated into a unified sensor platform that aims to realize global discovery, access, processing and sharing. The nature of SM such as distribution and socialization makes it clear that multi-source and heterogeneous social sensors need to be virtualized and formalized for discoverability, interconnection, and data interchange, sharing and handling. Thus, the formal description of a social sensor is formulated as

$$SocialSensor = \{URI, Type, PhySenList, SoftModList, Owner, State, SecuCons, Info\} \tag{6.13}$$

where URI is the unique address of the social sensor, $Type$ represents the category of the social sensor including one of three types of social sensors, $PhySenList$ is the set of physical sensors integrated into the social sensor, $SoftModList$ is the set of software modules installed into the social sensor, $Owner$ is the individual or enterprise who owns the social sensor, $State$ indicates the running states of the social sensor including normal and abnormal, $SecuCons$ describes the data security and authority constraints, and $Info$ contains the information of capabilities, function components, etc. It must be pointed out that each social sensor is configured with a URI (usually in the form of IP address) on Internet, which enables them accessible to others under certain authority. All the social sensors form a global social sensor network for interconnection, sharing, and interoperability. Besides, different social sensor network can aggregate into a social community autonomously for further seamless interactions and authorized sharing.

The eXtensible Markup Language (XML) schema can be used to formally describe sensors including location, capabilities, interfaces, protocols and other attributes. Based on the XML, Sensor Model Language (SensorML) Standard has been proposed. It provides an information model and encoding methods that enable discovering and tasking of Web-resident sensors and sensor observation exploitation [22]. SensorML can be used to describe a wide range of sensors, including both dynamic and stationary platforms as well as in situ and remote sensors. However, the social interactions between sensors cannot be described appropriately by using SensorML due to the lack of related functions and mechanisms. Furthermore, SensorML mainly deals with objective data collecting and handling, and rarely handles the subjective data from social interactions. Thus, based on the principles of SensorML, we proposed a formal description language called SocialSensorML to describe the metadata and functions of social sensors. SocialSensorML focuses on the function model of social sensors, not just the hardware description model. The more important concepts in SocialSensorML include "*SocialSensor*", "*Process*", "*AggregateProcess*", and "*ProcessMethod*". Other relevant definitions can be found in reference [22].

Here, "*Process*" is defined as an operation that there are one or more inputs and outputs generated based on a set of parameters, relevant metadata, and methodologies for discovery and human assistance. "*Process*" described in SocialSensorML is discoverable and executable. Within SocialSensorML, physical sensors are all modeled as physical processes that are "*process*" series and can be connected while the mathematical operations or functions in the software applications can be modeled as non-physical processes. In all, physical or non-physical input-processing-output (IPO) procedures can also be viewed as "*process*" series.

"*AggregateProcess*" is defined as a set of interconnected processes or aggregate processes with an explicit mapping of the input-output data flow among these processes. It is clear that an "*AggregateProcess*" can be viewed as a process network or process chain. "*AggregateProcess*" is equal to the processes with their own inputs, outputs, and parameters.

"*ProcessMethod*" is defined as the algorithms, behaviors, and interfaces of a "*Process*", especially for the social interaction functions, such as big data method, data fusion, and clustering algorithm. Note that "*ProcessMethod*" in SocialSensorML includes the methods and algorithms for social interactions, while not in SensorML.

The UML models, corresponding properties, and inner relationships of "*SocialSensor*", "*Process*", "*AggregateProcess*" and "*ProcessMethod*" are described in Fig. 6.9. The detailed explanations of these properties can be found in reference [22]. Specifically, "*AggregateProcess*" has the properties of components and connections, which indicates the component processes making up the "*AggregateProcess*" and the connection relationships among these component processes. Based on the UML models, the XML schemas can be automatically generated by applying the standard UML to XML schema encoding rules. Figure 6.10 illustrates an example of social sensor formal description by using SocialSensorML.

SocialSensorML addresses the discovery of social sensor, acquirement of social sensor's attributes and behaviors, description of social sensor's workflow, and handling of social sensor data. Besides, SocialSensorML enables the development of plug-and-play social sensor and processes, which may be seamlessly integrated to CPS for decision-making support. The SocialSensorML-based social sensor description also supports the development of an autonomous social sensor network where social sensors can interact with each other and publish alerts and tasks for other social sensors to subscribe and react.

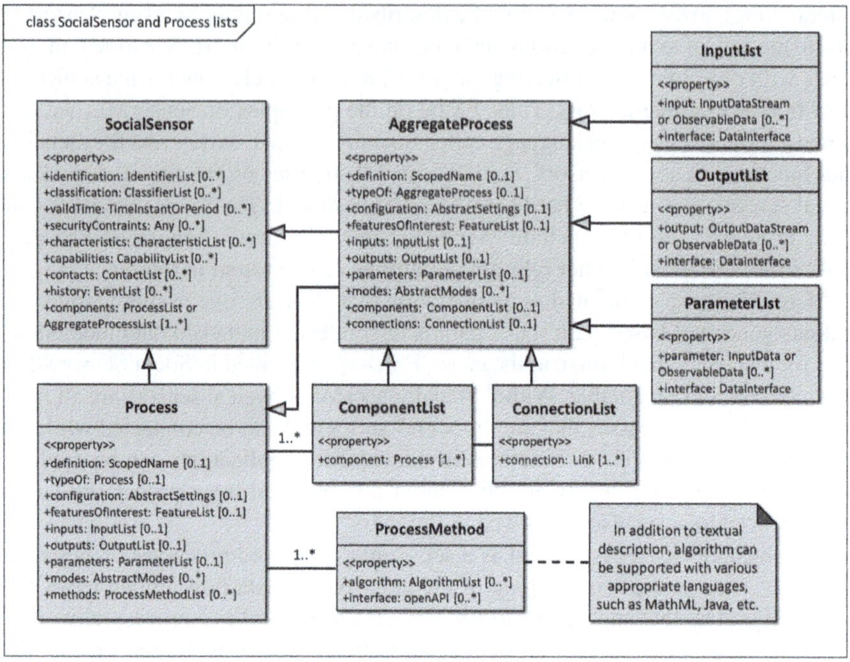

Fig. 6.9 UML model and relationships [15]

```xml
<SensorML xmlns="http://www.opengis.net/sensorML/1.0.1"
    xmlns:xsi="http://www.w3.org/2001/XMLSchema-instance"
    xmlns:swe="http://www.opengis.net/swe/1.0.1"
    xmlns:gml="http://www.opengis.net/gml"
    xmlns:xlink="http://www.w3.org/1999/xlink"
    xsi:schemaLocation="http://www.opengis.net/sensorML/1.0.1"
    http://schema.opengis.net/sensorML/1.0.1/sensorML.xsd version="1.0.1">
    <member xlink:arcrole="urn:ogc:def:process:OGC:detector">
      <ProcessModel>
        <gml:description>This is an MM-S2ensor</gml:description>
        <gml:name>MM-S2ensor</gml:name>
        <!--metadata-->
        <keywords>Machine-to-machine interaction,EB,Machine 1</keywords>
        <identification>http://115.154.191.25/S2ensors/MM-S2ensor/EB/1</identification>
        <classification>MM-S2ensor</classification>
        <characteristics>feature 1,feature 2</characteristics>
        <contact>EB manager</contact>
        <!--I/O-->
        <inputs>vibration,voltage,electric current,spindle speed,deviation,current task,...</inputs>
        <outputs>working-state,task-state,production-capacity,...</outputs>
        <parameters>A1,A2,α,γ,PList</parameters>
        <!--process method-->
        <method xlink:href="http://115.154.191.25/S2ensors/ProcessMethod/MM-S2ensor/Data-fusion"/>
      </ProcessModel>
    </member>
</SensorML>
```

Basic attributes

Input, Output, and Parameters

Social data fusion algorithms and methods

Fig. 6.10 An example of social sensor formal description using SocialSensorML

(2) ***Embedded-web-based configuration of social sensors by using REST architecture***

Social sensor integrates various physical sensors that are made with different communication protocols by different manufacturers. Thus, it is impossible for all the heterogeneous physical sensors to communicate with each other directly. Moreover, some physical sensors don't have Ethernet ports for accessing to the Internet. Therefore, a kind of middleware is used to shield the difference of physical sensors and bridge physical sensors and the Internet. The embedded device, called Raspberry Pi (RPi), is selected as the middleware to enable the REST-based infrastructure of a social sensor. An RPi is equipped with an ARM processor and a memory. It also has several USB slots, a RS232/485 port, a 10/100 Ethernet port, an HDMI and composite video output, and so on. The RPi owns four capabilities, including that it has various I/O ports so that HMI (e.g., screen, microphone and Web camera) and various physical sensors can be deployed on it, can connect various physical sensors or humans to the Web by assigning them IP addresses, has an embedded Web server so that software modules of social sensors can be developed, and possesses the capability of computation and data storage. RPi can be viewed as the cluster head of physical sensor group, which not only communicates with other cluster heads, but also communicates with the physical sensors from the same group.

REST architecture, defined by Roy Fielding [23], is a resource-oriented service access architectural based on the World Wide Web, whose performances includes scalability, simplicity, modifiability, visibility, portability, and reliability. According to the conception of REST, the social sensor is viewed as a set of Web resources by integrating different kinds of physical sensors and cluster head with URIs. The Web resources are carried out by combining URIs (specified by client) and HTTP verbs. In this chapter, REST architecture is adopted to implement the registration and mapping of cluster head and physical sensors. Firstly, physical sensors are registered and virtualized as physical sensor nodes (PSNs) (see Fig. 6.11a) to constitute the Web resource pool. Then the physical sensor drivers are programed with C language based on the communication protocol and data format. These drivers are coded as REST operation interfaces (POST, PUT, DELETE, and GET) that can be invoked by the software modules. Finally, various software modules (e.g., crawler module, chat bot module, and data processing module) are written by Tornado, a Python Web-based asynchronous networking library. Figure 6.11b shows the REST-based hierarchical model for the social sensor. Each social sensor is mapped as a three-layered model. The top layer, medium layer, and bottom layer correspond to the cluster head, elements of the social sensor and physical sensors, respectively. They are identified via URIs in the form of "*cluster head IP address+cluster head port+cluster head identifier+sensor node identifier*". The partial REST operation interfaces and the constitution of URIs are presented in Table 6.3.

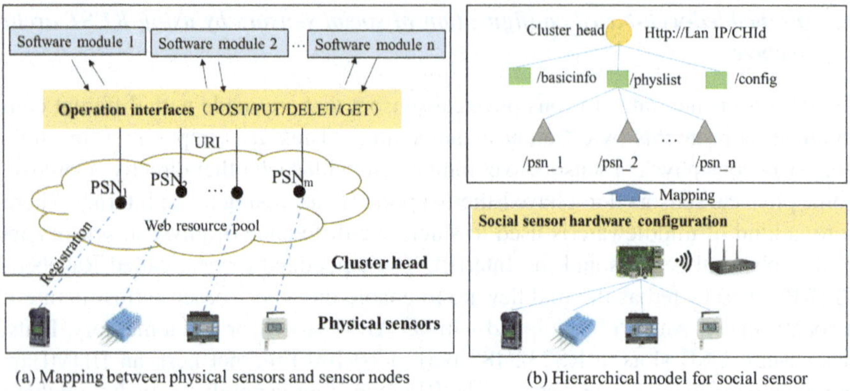

Fig. 6.11 Registration and mapping of cluster head and physical sensors

Table 6.3 Operation interfaces and the constitution of URIs [19]

URI	REST operation	Description
/{CHId}/sensorList	GET	Get a list of PSNs of the specified cluster head
{CHId}/basicInfo	GET	Get the basic information of social sensor
/{CHId}/sensorId	GET	Get the basic information of the specified PSN
/{CHId}/sensorId	PUT	Register a new PSN to the cluster head
/{CHId}/sensorId	DELETE	Delete the specified PSN
/{CHId}/sensorId	POST	Modify the information of the specified PSN
/{CHId}/sensorId/value	GET	Get the value of the specified PSN

6.3.3 Social Sensor Network for Manufacturing Workshop and Manufacturing Supply

There are different intra-enterprise-level and inter-enterprise-level production interactions under the context of SM. Thus, the data from heterogeneous social sensors should be virtualized and shared with different production participants. A social sensor network for both production workshop and manufacturing supply chain is a feasible solution. The functions of social sensor network provided by third-party service provider deal with four aspects, that is, providing a way to virtualize and register social sensors to the unified platform, integrating social media and communication interface mechanisms to enable H2H, H2M/M2H, and M2M social communication and information/command exchanging, enabling enterprises to form a private social sensor network, and self-organizing dynamically into social sensor sub-network for authorized sharing.

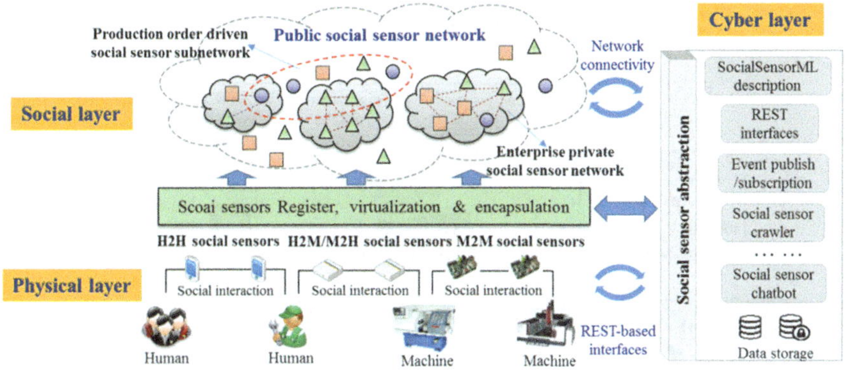

Fig. 6.12 Architecture of social sensor network [15]

The system architecture of social sensor network consists of three key elements which perform different roles, including social sensor hardware in the physical layer, social sensor abstraction in the cyber layer, and social sensor network in the social layer, as shown in Fig. 6.12.

In the physical layer, H2H social sensors, H2M/M2H social sensors, and M2M social sensors are deployed onto humans and machines. These social sensors are capable of collecting and handling real-time data, interacting with others, and reacting to transmitted production commands in an authorized plug-and-play way. All the social sensors are mapped into virtual nodes via registration and virtualization.

In the cyber layer, social sensor abstraction provides functions, application tools, data storage, and network access for humans and machines to communicate with each other. REST interfaces guarantee that physical sensors can be invoked by the software modules such as crawler and chat bot. The information generated from social interactions is stored in the local database and cloud database with different authorities for further handling. The event subscription and publish mechanism enables the data to be periodically or event-triggered collected during social interactions.

In the social layer, different social sensor owners register their social sensors to the Web so as to form the public social sensor network. Especially, the enterprise registers its own social sensors to the Web and forms an enterprise private social sensor network that can assist internal employees to control productions and make decisions in the production workshop level. In the manufacturing supply chain level, dynamically self-organized social sensor sub-networks are generated for different production orders so as to provide various services via social interactions and authorized sharing. When the production order is finished, the corresponding social sensor sub-network is dissolved.

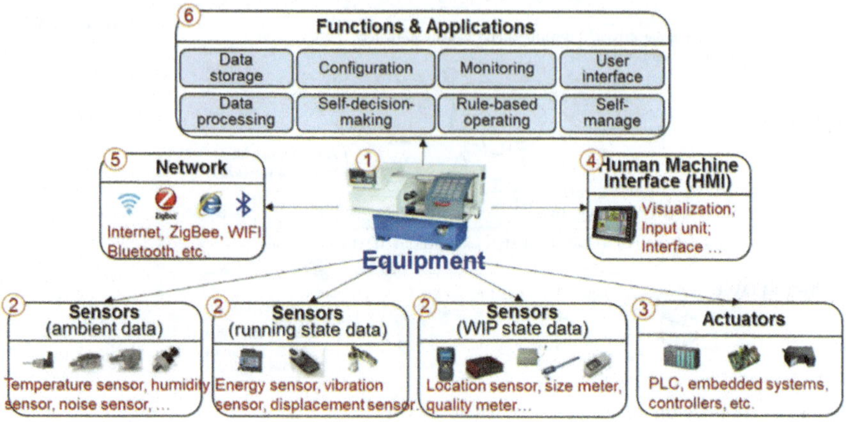

Fig. 6.13 Components of extended CPS nodes [18]

6.4 Extended CPS for Powering Social Manufacturing

6.4.1 *Integrating of RFID and Social Sensors with Existing CPS*

Existing CPS is a deeply integration of physical processes and software components. Combined with the RFID and social sensors mentioned above, the functions of an existing CPS can be enhanced and satisfy the needs of the intra- and inter-enterprise-level production cooperation and coordination under the context of SM. Accordingly, an extended CPS node (ECPSN) is defined as a kind of hardware-software-integrated mediator for machines to integrate different functions related to perception, communication, interactions and control feedback to realize the autonomous operations of single machine and the cooperation of multiple machines. Each ECPSN is configured with an URI, and different ECPSNs can interact for autonomous business coordination. The components of ECPSNs include several modules concerning equipment, sensors, actuators, HMI, network, and functions and applications, as shown in Fig. 6.13.

Here, "*equipment module*" is the base to build an ECPSN. Different machining equipment has different capabilities and skills.

"*Sensor module*" includes physical sensors with various targets, such as production environment data (e.g., temperature, humidity, noise), equipment's running states (e.g., vibration, displacement, energy consumption, spindle speed), and WIP states (e.g., RFID device, digital caliper, roughometer). Especially, RFID devices, including readers, antennas and tags, can be viewed as sensing system and have been integrated into the extended CPS so as to monitor and track machining processes.

"*Actuator module*" aims to execute the production commands. For example, PLC receives commands from the upper systems and drives the manipulators or machines.

The embedded system integrates the sensors, receives and preprocesses the sensor data, and further channels the data to the upper systems.

"*HMI*" acts as an intermediate between physical objects (i.e., equipment, sensors, and actuators) and humans (internal employees or external partners). Humans apply the HMI to communicate with physical objects and handle raw sensor data. Humans can input parameters or commands and exchange information with physical objects. In this case, H2M/M2H social sensors are integrated into ECPSNs and act as the HMI that deals with H2M/M2H interactions.

"*Network module*" adopts different communication protocols or interfaces (e.g., WIFI, ZigBee, Bluetooth, and Ethernet) to make the physical objects networked. Besides, the "*network module*" provides an application interface to realize plug-and-play equipment configuration for interoperability. Thus, different equipment can coordinate to make decisions. In this case, M2M social sensors are integrated into the extended CPS and act as the "*network module*" that deals with M2M interactions.

"*Function and Application module*" connects the physical space with the cyber space, and enables various functional applications of ECPSNs. Based on data capturing, handling and storing, ECPSNs can realize functions such as real-time monitoring, dynamic configuration, self-decision-making, rule-based operation, prognostics and health management (PHM), etc.

XML-RDF is specialized at describing the metadata of Web resources. As shown in Fig. 6.14, the formal description template of ECPSN is described as

$$ECPSN = \{URI, Operation, Config, Information, Function, CurSt\} \tag{6.14}$$

where URI is the URI of ECPSN in certain form (*http://public gateway IP address: port number/public gateway identifier/ECPSN identifier*). $Operation$ is the set of operation methods executed on ECPSN, such as HTTP GET, PUT, POST, and DELETE. $Config$ is the set of configuration information such as IP address and port number. $Information$ represents the basic parameters (e.g., location, bound equipment, list of embedded sensors and relevant actuators) and other information of ECPSN. $Function$ represents the computational applications of ECPSN, including middleware and gateway. $CurSt$ is the current status of ECPSN, i.e., available or occupied.

6.4.2 Extended CPS for Powering Social Manufacturing Nodes, Communities, and Network

(1) ***Framework of extended CPS platform for powering social manufacutring nodes, communities, and network***

After clarifying ECPSN, a systematic framework that interconnects ECPSNs from various enterprises is built for production collaboration. Figure 6.15 describes this

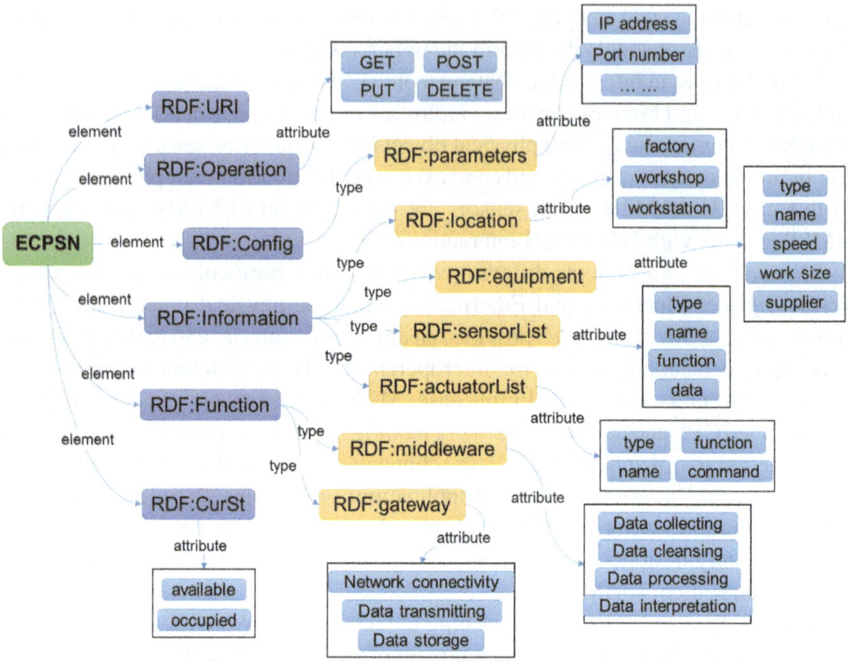

Fig. 6.14 XML-RDF template for extended CPS nodes [18]

framework. It can be simply divided into three layers, i.e., physical layer, cyber layer, and social layer, which are correspondent to the physical space, cyber space, and social space respectively.

Physical layer includes various enterprises that are regarded as social manufacturing nodes. Within a social manufacturing node, a public gateway is utilized to group all the ECPSNs to give data authority to upper-level systems. The public gateway is a Web server that abstracts the communication between the ECPSN and the upper-level systems. It offers its functionalities via a REST API, which makes the ECPSNs accessible to external suppliers. When a new ECPSN is plugged and activated, it will start an association process by sending its XML-RDF template to the public gateway. After the ECPSN passes the authentication, authorization and accounting (3A) checking, it will be automatically registered to the public gateway. Then, it can communicate with other group members. From the perspective of social manufacturing node, accordingly, ECPS-enabled production is endowed with the capabilities of self-configuration, self-coordination, and distributive control.

Cyber layer connects all the social manufacturing nodes to form a global social manufacturing network. In the network, some social manufacturing nodes self-organize themselves into a manufacturing community that aims to undertake a production order. During the order production, the social and industrial big data are collected and stored in the public or private database according to the data security

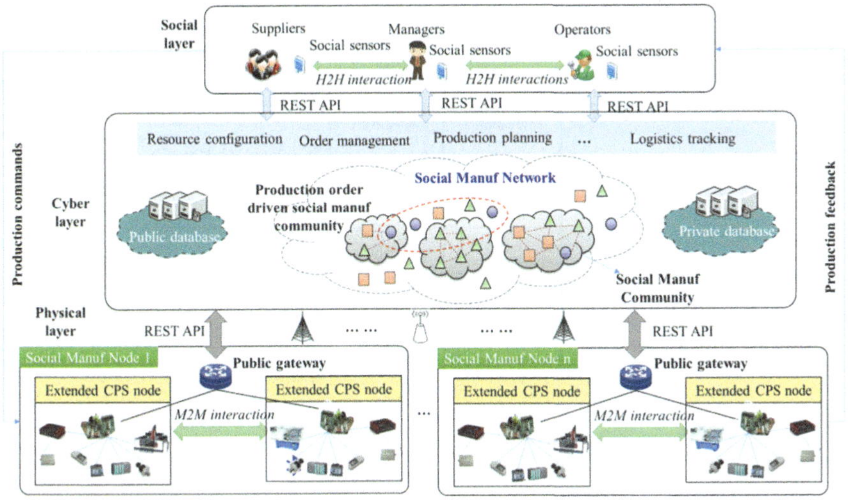

Fig. 6.15 Framework of extended CPS platform for powering social manufacturing nodes, communities, and network

grades. Besides, this layer provides Web applications such as resource configuration, order management, and production planning for production operations.

Social layer integrates various social media (e.g., instant messaging, live-streaming) via open interfaces, which facilitates the anytime and anywhere H2H interactions among enterprises and their suppliers. Furthermore, humans such as internal employees (managers and operators) and external suppliers can directly communicate with ECPSNs in the social manufacturing nodes via the H2M/M2H social sensors. Thus humans can send the production commands to the filed machines for production feedback and decision making.

(2) *Extended CPS-enabled production interactions*

Assume that the social manufacturing node A (SMN_A) is a core enterprise, SMN_B and SMN_C are two suppliers who undertake production orders from SMN_A. The managers of SMN_A, SMN_B, and SMN_C are equipped with H2H social sensors. SMN_B and SMN_C both have two machines, all of which are configured as ECPSNs. The abovementioned production interaction contexts include production commands transmission, production state uploading, production task reallocation, etc., which are related to different participant couples.

Scenario modeling method based on message sequence chart (MSC) is adopted to describe the production interactions. The message flow among them is illustrated according to the actual sequences of production interactions. The messages are triggered by different events, such as dialog initialization, message sending event, message receiving event, timer, and so on. The multi-role interactions during order production, as illustrated in Fig. 6.16, can be described as follows:

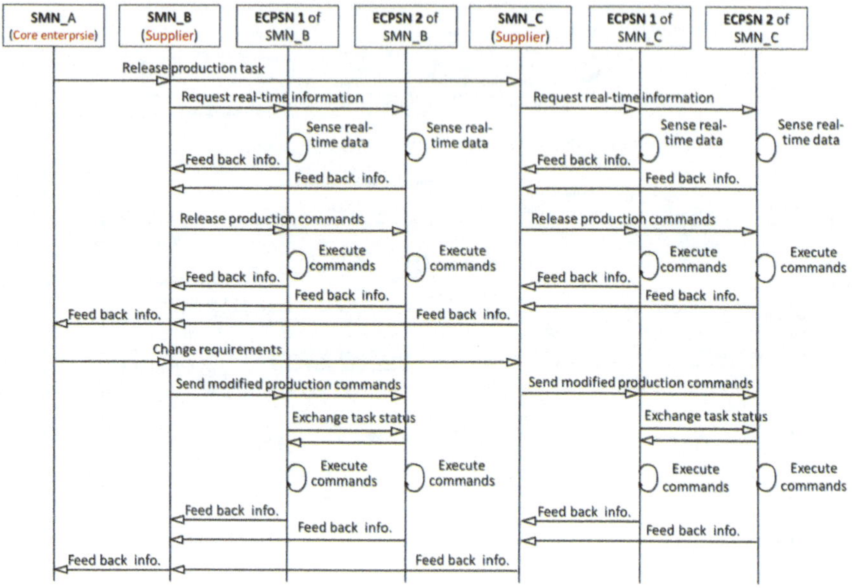

Fig. 6.16 Extended CPS-enabled production interactions [15]

- SMN_A releases production tasks to SMN_B and SMN_C through H2H social sensors,
- SMN_B and SMN_C receive real-time data from their machines periodically through the corresponding ECPSNs and can allocate appropriate amount of tasks to their machines,
- During the order production, ECPSNs collect the state data of SMN_B's machine 1 and 2 (and SMN_C's machine 1 and 2) in real-time and can coordinate and react to the unexpected events. For example, if machine 1 breaks down, machine 2 will take over its tasks after interactions,
- If changes need to be made, SMN_A will describe its dynamic demands through H2H social sensors. After receiving these sensory input data and relevant objective data, SMN_B and SMN_C will handle these data into engineering requirements by applying H2H social sensors,
- After that, SMN_B and SMN_C interact with their machines to adjust the production commands through H2M/M2H social sensors integrated into ECPSNs, and
- Finally, finished products are delivered from SMN_B and SMN_C to SMN_A.

It can be seen that the extended CPSs integrates different kinds of social manufacturing nodes, and self-organize into manufacturing communities for order production. Through H2H, H2M/M2H, and M2H interactions, the order production processes are transparent to enterprises, and make the inter-enterprise-level production coordination in close-loop control manner.

6.5 Concluding Remarks

Integrating RFID and social sensors with an existing CPS has paved the way for the production interactions and mass collaboration under the context of social manufacturing. For RFID application scenarios in both the inter-enterprise level and intra-enterprise level, an RFID-based graphical deduction model is developed for tracking and monitoring machining-process material flows and manufacturing supply chain. For social sensors aiming at bridging human-machine interactions, the concept clarification, operational logics, and functional implementation are addressed. The social sensor network for production workshop and manufacturing supply chain is constructed to enable production interactions and cooperation among humans and machines. Despite of the potential significance, RFID and social sensors are isolated and it is much necessary to incorporate them into a suitable framework. The current CPS framework is selected to integrate RFID and social sensors, which is called as extended CPS. The extended CPS framework breaks through the barriers of production interactions, production tracking and monitoring, and production control, which benefits for building an easy-to-deploy social manufacturing nodes, communities and network.

References

1. Jiang PY, Cao W (2013) An RFID-driven graphical formalized deduction for describing the time-sensitive state and position changes of work-in-progress material flows in a job-shop floor. J Manuf Sci Eng-Trans ASME 135(3):1–13
2. Qu T, Yang HD, Huang GQ et al (2012) A case of implementing RFID-based real-time shop-floor material management for household electrical appliance manufacturers. J Intell Manuf 23(6):2343–2356
3. Jiang PY, Fu YB (2012) RFID based e-quality tracking in service-oriented manufacturing execution system. Chin J Mech Eng 25(5):974–981
4. Chongwatpol J, Sharda R (2013) RFID-enabled track and traceability in job-shop scheduling environment. Eur J Oper Res 227(3):453–463
5. Guo ZX, Ngai EWT, Yang C et al (2015) An RFID-based intelligent decision support system architecture for production monitoring and scheduling in a distributed manufacturing environment. Int J Prod Econ 159(2015):16–28
6. Cai H, Andoh AR, Su X et al (2014) A boundary condition based algorithm for locating construction site objects using RFID and GPS. Adv Eng Inform 28(4):455–468
7. Sari K (2010) Exploring the impacts of radio frequency identification (RFID) technology on supply chain performance. Eur J Oper Res 207(1):174–183
8. Vlachos IP (2014) A hierarchical model of the impact of RFID practices on retail supply chain performance. Expert Syst Appl 41(1):5–15
9. Sridhar L, Prabha VL (2013) RFID based access control protection scheme for SRAM FPGA IP cores. Microprocess Microsyst 37(6–7):629–640
10. Ding K, Jiang PY, Sun PL et al (2016) RFID-enabled physical object tracking in process flow based on an enhanced graphical deduction modeling method. IEEE Trans Syst Man Cybern Syst PP(99):1–13
11. Wang C, Jiang PY, Ding K (2015) A hybrid-data-on-tag-enabled decentralized control system for flexible smart workpiece manufacturing shop floors. Proc Inst Mech Eng Part C: J Mech Eng Sci 231(4):764–782

12. Wang C, Jiang PY (2017) Deep neural networks based order completion time prediction by using real-time job shop RFID data. J Intell Manuf, Published online, 1–16
13. Lee CKM, Ho W, Ho GTS et al (2011) Design and development of logistics workflow systems for demand management with RFID. Expert Syst Appl 38(5):5428–5437
14. Ding K, Jiang PY, Su SL (2018) RFID-enabled social manufacturing system for inter-enterprise monitoring and dispatching of integrated production and transportation tasks. Robot Comput-Integr Manuf 49(2018):120–133
15. Ding K, Jiang PY (2017) Social sensors (S^2ensors): a kind of hardware-software-integrated mediators for social manufacturing systems under mass individualization. Chin J Mech Eng 30(5):1150–1161
16. Kagermann H, Wahlster W, Helbig J (2013) Securing the future of German manufacturing industry: recommendations for implementing the strategic initiative Industrie 4.0; final report of the industrie 4.0 working group. Forschungsunion 1–84
17. Monostori L, Kádár B, Bauernhansl T et al (2016) Cyber-physical systems in manufacturing. CIRP Ann Manuf Technol 65(2):621–641
18. Ding K, Jiang PY (2017) Incorporating social sensors, cyber-physical system nodes, and smart products for personalized production in a social manufacturing environment. Proc Inst Mech Eng Part B: J Eng Manuf, Publish Online
19. Liu C, Jiang PY (2016) A cyber-physical system architecture in shop floor for intelligent manufacturing. Procedia CIRP 56(2016):372–377
20. Lee J, Bagheri B, Kao HA (2015) A cyber-physical systems architecture for industry 4.0-based manufacturing systems. Manuf Lett 3:18–23
21. Seitz KF, Nyhuis P (2015) Cyber-physical production systems combined with logistic models—a learning factory concept for an improved production planning and control. Procedia CIRP 32(2015):92–97
22. N.N. (2018) Sensor Model Language (SensorML). OGC® SensorML, available at: http://www.opengeospatial.org/standards/sensorml
23. Fielding RT, Taylor RN (2002) Principled design of the modern Web architecture. ACM Trans Internet Technol (TOIT) 2(2):115–150

Chapter 7
Social Factory and Interconnections

Chao Liu and Pingyu Jiang

7.1 Introduction

The great and collaborative progress on computer science (CS), information and communication technologies (ICTs), and manufacturing science and technology (MST) not only enhances the role of socio-technical systems in depth but also refines the contents of customer needs in height. It also drives manufacturing factories to transfer their original enterprise-centric working mode to a new customer-centric one. Here, the original working mode indeed runs as a closed system while the new one just aims to build an open system that focuses on the production interactions and collaboration with customers and suppliers. Under the context of social manufacturing (SM), on the one hand, distributive socialized manufacturing resources (SMRs) possessing similar interests and common goals tend to autonomously self-organize into manufacturing communities (MCs) to fulfill production orders and enhance collaboration efficiency in the form of value co-creation and knowledge share [1]. Thus, manufacturing factories can be considered as production nodes of MCs that aim to complete the production tasks via inter-enterprise-level collaboration. On the other hand, the technology advances in the aspects of the Internet of Things (IoT), cyber-physical system (CPS), cloud computing, and social networking have paved the way to a foreground of the ubiquitous interconnections among intra- and inter-enterprise resources, data/information, processes, machines and humans. These interconnections are also presented respectively in the form of IoT, Internet of Data (IoD), Internet of Services (IoS) and Internet of People (IoP) [2], which can transform smart objects into social entities that are capable of bridging human-to-object interactions [3].

Under the context of SM, product manufacturing activities are becoming more and more complex, diverse, and personalized. Thus, a manufacturing factory has to be open, sharing, collaborative, flexible, and intelligent to meet the new requirements. Different from traditional manufacturing factories which are operated in rigid and closed manner, the new factory mode in the context of SM makes it possible that product orders together with customer requirements are directly sent to the factory,

P. Jiang, *Social Manufacturing: Fundamentals and Applications*, Springer Series in Advanced Manufacturing, https://doi.org/10.1007/978-3-319-72986-2_7

and the orders and requirements can be responded in no time. At the same time, the factory collaborates with other production nodes or its suppliers to get raw materials, semi-finished parts and finished parts to finish its own order tasks. Thus, production information such as order progress, part quality and material flows is sharable and transparent to all the participators. Furthermore, because the product orders are undertaken and completed by a group of production nodes, both unoccupied production capability in such a factory and order tasks can also be shared among the collaborative production nodes. Within the factory, all the physical devices are interconnected and virtualized as corresponding digital objects. The expected and unexpected production events are captured and handled in real time based on which production decisions can be made. We define this future factory mode as social factory [4]. Actually, social factory is a kind of production node involved in accomplishing part machining and product assembling tasks in the production stage of a product life cycle. Its fundamental goal is to orchestrate humans, smart machines, physical processes, and software components in an optimal manner to fulfil order-driven production tasks via the intra-enterprise-level and inter-enterprise-level interactions and collaboration.

The "*social*" features of social factory are reflected from two perspectives. First, from the inter-enterprise-level perspective, each machining equipment can be accessible to its customers, other social factories and suppliers. It means that external participators know which machines are responsible for the production orders and what are the better production schedules. Thus social factory can interact and collaborate with its customers, other social factories and suppliers to complete the production tasks via the social network powered with social sensors [5]. Second, from the intra-enterprise-level perspective, all the factory objects such as machine tools, automatic guided vehicles, conveyer belts, industrial robots, and smart workpieces are interconnected and equipped with computing and communication capabilities to perform human-like behaviors. Thus these smart objects are capable of sensing the production environment, interacting and cooperating with each other, and making intelligent decisions.

According to its working principles, social factory at least has these important characteristics, such as open and shared production services, distributive customers and suppliers, rapid resource configuration, flexible production, intelligent decision making, reconfigurable operation processes, etc. As one of the most important technological bases of *Industry 4.0*, CPS has presented its great potential in the aspect of constructing future manufacturing factories from hardware configurations and data acquisition, handling and visualization, to eventually knowledge acquisition and learning [6, 7]. However, current IoT/CPS-enabled manufacturing systems are mainly centralized and hierarchical control architecture [8], which cannot fit well with the upcoming challenges of social factory, such as decentralization, customers' participations, Internet-based social interactive behaviors, mass collaboration, fast changing production requirements, various production disturbances, flexible resource configuration, transparent production, etc.

The extended CPS (ECPS) discussed in Chap. 6 can be applied as the solution to address these issues. ECPS is characterized by the tight integration of physical entities (e.g., machines, tools, sensors, and actuators) from physical space with computational entities from cyber space where physical entities and computational entities coordinate with each other through cyber-physical interconnections [9]. It has two main functions. The first one is the cyber-physical interconnection which is responsible for real-time data collection from the physical space and command information feedback from the cyber space. And the second one is large scale data management, analysis, and computation in the cyber space. In this chapter, some enabled technologies such as social network analysis, multi-agent approach and digital twin technology will be applied to designing and implementing the social factory, together with the above ECPS technology. Therefore, a decentralized production control framework for ECPS-enabled social factory is proposed. The purpose of this framework is to develop a collaborative and reconfigurable production system that not only supports small batches production and product diversity, but also guarantees high quality and low cost in terms of the customer orders. In this production system, physical objects including sensors, actuators, control components and intelligent embedded devices are installed on machining equipments (e.g., machine tools, industrial robots, and conveyor belt) which are connected with each other through fieldbus technology and/or industrial Ethernet. These physical objects constitute the physical implementation of ECPS node (ECPSN). And then these physical objects are digitalized and represented as virtual objects. Some mechanisms or algorithms such as self-organization mechanism, state transition and update mechanism, learning mechanism, and exception prediction and handling algorithms are integrated into the virtual objects so as to constitute the virtual implementation of ECPSN. Each ECPSN can be regarded as a smart agent that is able to autonomously interact and cooperate with humans or other agents to achieve adaptation, autonomy and decentralization. It means that the social implementation of ECPSN is reached in a higher level.

7.2 Some Definitions

7.2.1 Smart Workpiece

Smart workpiece (SW) is a kind of physical entity that can be uniquely identified [10, 11]. It has certain level of intelligence to recognize how to produce itself, record its historical and current status, check its own next working process, and communicate with other machining equipments (in the form of ECPSNs) for dynamic and decentralized production control so as to achieve its production goals.

As shown in Fig. 7.1a, the components of SW include a "*physical workpiece*", an "*active RFID tag*", an "*RFID antenna/reader*", an "*embedded device*", and a "*carrier*".

"*Physical workpiece*" is a kind of part to be machined and linked to "*active RFID tag*", "*embedded devices*", and "*carrier*".

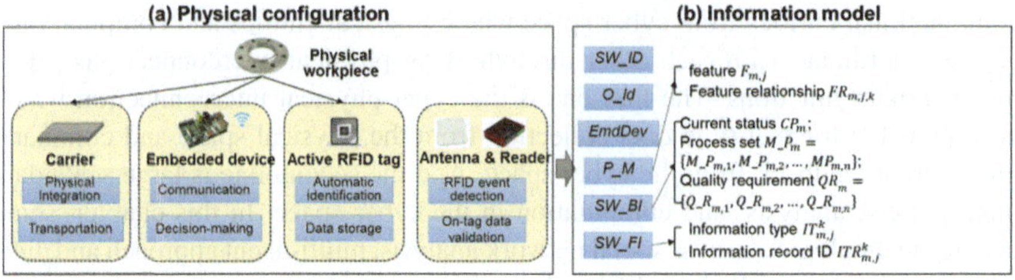

Fig. 7.1 Components of smart workpiece [10]

"*Active RFID tag*" is the unique identifier for real-time and automatic object identifying. It is attached to either the "*physical workpiece*" or its "*carrier*" and invokes the communication modules of "*embedded device*" to communicate with ECPSNs. Besides, the "*active RFID tag*" stores some very important parameters from the product model (e.g., product BOM) and production model (e.g., machining parameters, production processes) to provide basic machining information for dynamic production control.

"*RFID antenna/reader*" is responsible for RFID event detection when the "*active RFID tag*" enters the signal coverage area of RFID antennas.

"*Embedded devices*" are responsible for external communication, data computing and decision-making (e.g., duplicate data filtration, basic machining information transmission).

"*Carrier*" is a physical entity that loads and transports "*physical workpiece*", and binds the "*active RFID tag*" to "*physical workpiece*" if necessary. Actually, it also acts as a key object to enable SW in a WIP materials and machining process flow.

As shown in Fig. 7.1b, the aforementioned SW is described as follows:

$$SW = \{SW\_ID, O\_ID, EmdDev, P\_M, SW\_BI, SW\_FI\} \tag{7.1}$$

where SW_ID and O_ID represent the unique identifier of RFID tag and production order. $EmdDev$ stands for the embedded device. P_M expresses the product model of SW. SW_BI and SW_FI represent the basic and detailed information of SW, respectively. SW_BI facilitates SW to communicate with the control system directly, while SW_FI stores the index of process-related data (e.g., machine ID, operator ID, and cutting tool ID) generated from the process:

$$SW\_BI_m = \{STA_m, MP_m, CP_m, QR_m\} \tag{7.2}$$

where STA_m is the current status of SW (i.e., work-in-progress—WIP, finished workpiece—FW, waste workpiece—WW), $STA_m \in \{WIP, FW, WW\}$; MP_m is the machining process set of SW m, $MP_m = \{M\_P_{m,1}, M\_P_{m,2}, \ldots, M\_P_{m,n}\}$; CP_m is the current process of SW m, $CP_m \in \{M\_P_{m,1}, M\_P_{m,2}, \ldots, M\_P_{m,n}\}$; QR_m is the quality requirements of each machining process, $QR_m = \{Q\_R_{m,1}, Q\_R_{m,2}, \ldots, Q\_R_{m,n}\}$.

SW_FI is used to acquire detailed machining process information which is stored in the backend database. It can be described as follows:

$$SW\_FI_m = \left\{IT^1_{m,1}, ITR^1_{m,1}, \ldots, IT^p_{m,1}, ITR^p_{m,1}, \cdots\cdots, IT^1_{m,n}, ITR^1_{m,n}, \ldots, IT^q_{m,n}, ITR^q_{m,n}\right\} \quad (7.3)$$

where $< IT^k_{m,j}, ITR^k_{m,j} >$ represents a pair of information ($IT^k_{m,j}$—information type, $ITR^k_{m,j}$—information record ID in the database) that indicates k-th information pair of j-th process of m-th SW. For example, $\langle ECPSN\#1, record\#12\rangle$ stands for ECPS node #1 which is used in a machining process, and the index of ECPS node #1 is defined as $record\#12$. Detailed information of ECPS node #1 can be referred according to the mapping relationship between $record\#12$ and the data sheet.

Based on the aforementioned discussion, the on-tag data of SW m can be formalized as follows:

$$OTD_m = \{SW\_ID_m, SW\_BI_m, SW\_FI_m\} \quad (7.4)$$

where OTD_m is the on-tag data of SW m.

Based on the formalized description of SW, all the machining process data are correlated. Thus, SWs can cooperate with ECPSNs to decide process routings according to the real-time automated execution context. In this way, the decentralized production control in a social factory is facilitated and there is no more requirements for enterprise-level control decisions.

7.2.2 Digital Twin-Based Extended Cyber-Physical System Node

As shown in Fig. 7.2, an ECPSN deals with its physical and virtual implementations and further social implementation in a higher level.

The physical implementation of an ECPSN is responsible for sensor perception and actuator execution, and includes physical components such as one core equipment, sensors, actuators, networking devices, human-machine interface (HMI), and embedded devices. The core equipment is the foundation of this ECPSN. Sensors are the gateway of equipment to sense the surrounding physical environment (e.g., temperature, humidity, noise) and running states (e.g., vibration, spindle speed). Actuators are responsible for the execution of production commands generated from the virtual implementation of the ECPSN. Networking devices including RS232, RJ45 and routers connect equipment with other physical objects through various network protocols such as WIFI, ZigBee, Bluetooth, and industrial Ethernet. HMI deals with interactions between ECPSNs and humans. On one hand, HMI allows humans to examine, operate and control the ECPSN. On the other hand, HMI offers the current status of ECPSN and the feedback information that aid operators to make

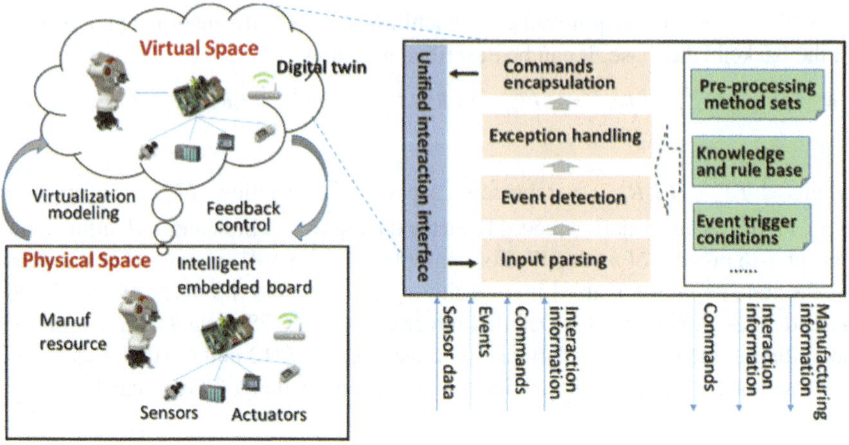

Fig. 7.2 Components of digital twin-based extended cyber-physical system node

decisions. Embedded devices are utilized to integrate all the physical devices and act as the operational environment for the virtual implementation of ECPSN.

The virtual implementation of an ECPSN is responsible for high-level decision making and can be seen as a digital representation of the physical objects, which is widely known as digital twins [12]. In fact, a digital twin is the exact virtual counterpart of a physical object and can express the characteristics, functionalities and performances of the physical object in a digital way. Here, these characteristics, functionalities and performances deal with time-variation attributes of the physical object and are related to its shape, position, topologic, status, kinematics, dynamics, etc. In general, a fundamental digital twin model can be created referring to a physical equipment and ECPSN physical components around the equipment. Whenever production events occur (e.g., disturbances, exceptions), a physical sensor updates the current status to its digital twin hosted in the embedded device. And then, the digital twin handles the sensory information through input parsing, event detection, exception handling, and commands encapsulation. Furthermore, it sends commands to the physical objects for feedback control. Some algorithms and rules such as pre-processing methods, rule base, and event trigger conditions are embedded in the digital twin to improve the intelligence and autonomy of the ECPSN. Note that the input may come from the physical sensors, commands issued by humans, or interaction information generated by other ECPSNs. The output includes commands, interaction information, and production information that can be sent respectively to the physical objects, other ECPSNs, and backend database.

To sum up, the main characteristics of ECPSNs include "*reactivity*", "*autonomy*", "*interoperability*", "*diversity*", "*adaption*" and "*flexibility*". Here, "*reactivity*" means that an ECPSN can sense its own working conditions, and react to disturbances and exceptions. "*Autonomy*" implies that an ECPSN can autonomously interact with other entities (humans or other ECPSNs) and make decisions by itself without direct control and intervention from these entities because it knows its capabilities and status. "*Interoperability*" indicates that the interactions and cooperation between ECPSNs can be achieved with proper agent communication language, ontologies or interaction protocols, etc. "*Diversity*" declares that ECPSNs can be categorized as machine tool ECPSN (MT_CPSN), industrial robot ECPSN (IR_CPSN), conveyer belt ECPSN (CB_ECPSN), etc., and each ECPSN has its own knowledge, skills and objectives based on different machining equipments. "*Adaption*" just means that each ECPSN can learn from history experience and behaviors, and become more intelligent due to its learning mechanism. "*Flexibility*" shows such a fact that a lot of ECPSNs can be interconnected to build a cooperative network for a certain production task and each ECPSN may move in and out freely without damaging the entire network, i.e., the plug-and-produce configuration.

7.2.3 Classification of Social Factory

As mentioned in Chap. 2, social factory can be further defined as a kind of ECPSN-interconnected, data-driven and intelligent factory model which collaborates with customers and suppliers, and provides production-order-driven machining and assembling services. In the inter-enterprise level, a social factory builds the ubiquitous connections with its customers and suppliers through the production orders that are fulfilled by the collaboration and share mechanism among social factories, customers, and suppliers. It is also a production node which aims to finish either part machining or product assembling tasks under the context of SM. In the intra-enterprise level, a social factory aims at orchestrating humans, machines and smart objects in an appropriate and optimal manner to realize autonomous decisions-making and decentralized production control.

Social factory can be classified into three categories according to its accepting production-order types, i.e., outsourcing-order-driven social factory, crowdsourcing-order-driven social factory, and order-prediction-and-allocation-driven social factory. Either outsourcing-order-driven or crowdsourcing-order-driven social factory enables itself through its role as a manufacturing service provider related to outsourcing or crowdsourcing production orders. While order-prediction-and-allocation-driven social factory enables itself either as a manufacturing service provider mentioned above or as a core producer who holds the intellectual property of the product to be produced. Actually, there is no obvious difference for these three kinds of social factories to organize their production activities except social business interactions, and cooperation and share mechanism related to Internet-based connecting and communicating behaviors in business. Such "*social-issue-related*" different points are just concerned with using social sensors inside ECPSNs in both intra-enterprise level and inter-enterprise level.

7.3 Framework and Operational Logic of Social Factory

7.3.1 Framework of Social Factory

An easy-to-deploy and simple-to-use social factory framework which enables manufacturing resource self-organization, distributive production control, and intelligent decision making is introduced in this section. As shown in Fig. 7.3, it consists of four layers, that is, physical connection layer, computing layer, application layer, and interaction layer.

Physical connection layer aims at connecting all the involved machining equipments, physical sensors, controllers, and actuators. According to the customer requirements and specific production logic, different types of sensors, actuators, controllers, HMIs, and embedded devices are deployed on the machine tools, AGVs, workpieces, etc., and further construct different kinds of ECPSNs to realize plug-and-produce configurations. Various ECPSNs are interconnected through factory bus and the Internet so as to enable data exchanges between physical space and cyber space. ECPSNs have the capabilities of sensing the real-time production

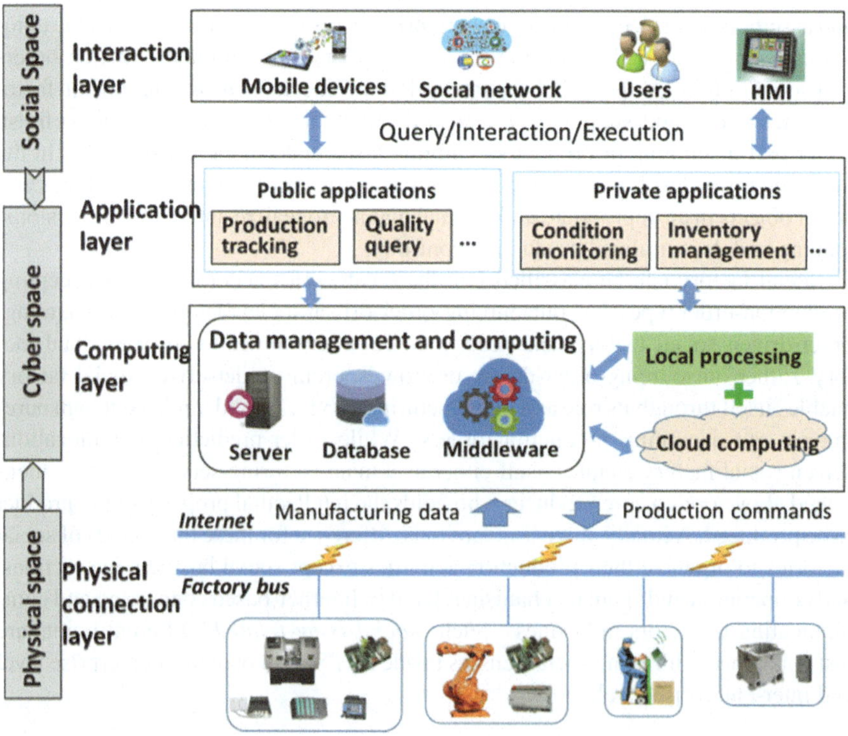

Fig. 7.3 Framework of social factory

environment, interacting with humans and other ECPSNs proactively, and making decisions intelligently and autonomously.

Computing layer consists of three components: middleware, production database and Web server. Middleware is responsible for virtualizing different types of ECPSNs as the corresponding digital twins, and providing the unified data access interfaces. Production database is used to store real-time production data, social contexts, and production rules/knowledge. Web server guarantees the stable running of middleware and production database. With the aid of these components, this layer aggregates real time sensor data from physical connection layer, production commands and social contexts from application layer. The aggregated production information is further handled through local handling capability in the ECPSN or cloud computing in the cloud server.

Application layer contains different applications used for production coordination, production monitoring, and quality control. The public applications can be accessed by external partners such as customers and suppliers. The private applications are mainly utilized to control and execute production operations by internal operators or managers.

Interaction layer allows customers and suppliers to interact with factories anywhere and anytime via social sensors (e.g., mobile devices, social media tools and HMIs). Social data or social contexts are collected and locally handled by the social sensors and then transferred to the backend database for further analysis.

7.3.2 Operational Logics of Social Factory

Generally, a product order is decomposed into several sub-orders according to the product bill of materials (BOM). Some of these sub-orders together with customer requirements are presented as production orders, and shared by or assigned to a group of suitable social factories based on outsourcing and crowdsourcing service mechanisms. Here, a social factory as production order accepter will complete the sub-order by means of collaborating with its customers, other social factories and suppliers with the aid of social sensors. While other correlated social factories and suppliers mentioned above just provide required materials, parts, and components to the above social factory according to the product BOM. Furthermore, this social factory can also share its own unoccupied production capability with external production nodes or social factories which accept their own production orders so as to increase the resource utilization rate.

Within the social factory, as shown in Fig. 7.4, all the machining equipments and the corresponding add-on sensors, controllers, and actuators are interconnected through factory bus and virtualized as ECPSNs with digital twin technology. Diverse ECPSNs constitute the ECPSN network. Based on a production order the social factory accepts, a production order-driven ECPSN sub-network can be shaped from the above ECPSN network. The operation of the sub-network is under the support of hardware, software, process, and service resources. In order to complete this pro-

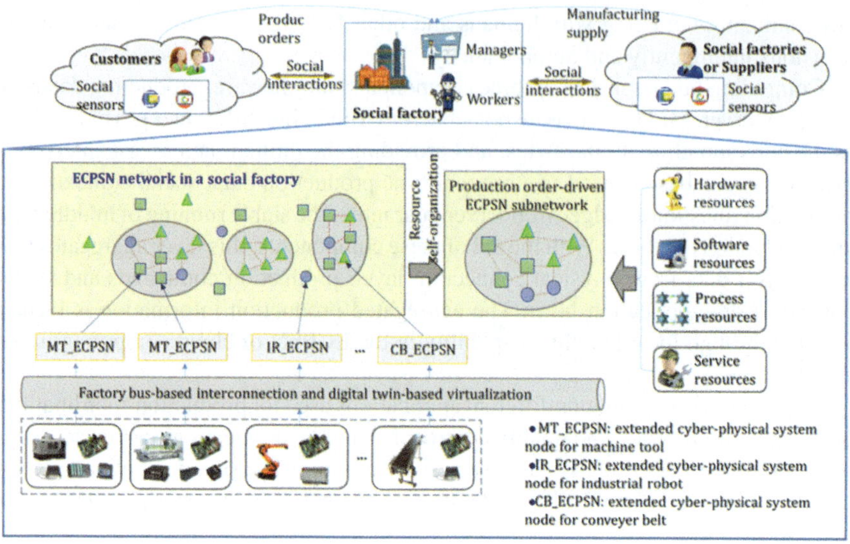

Fig. 7.4 Operational logics of social factory

duction order well, customers need to interact with the involved entities (workers, managers and ECPSNs) in the social factory via social sensors. On one hand, the production order requirements are delivered to the social factory and then transformed into process- or operation-level production commands which are sent to ECPSNs for production share, coordination and control. On the other hand, the real-time production information (e.g., the production schedule, machining quality) is sent to customers for requirements feedback. Besides, the requirements for materials, parts, and components can be directly sent to suppliers. In some cases, customers can select specific suppliers and learn something form the whole production processes. In addition to the inter-enterprise-level interactions, share and collaboration among customers, suppliers and social factory, there is also an autonomous interacting, sharing and collaborating implementation among ECPSNs, operators, managers, and smart workpieces in the intra-enterprise level. The interactions and connection are supported by multi-mode social sensors, such as H2H social sensors, H2M/M2H social sensors, and M2M social sensors.

7.4 Decentralized Production Control of Social Factory

7.4.1 Decentralized Control Node Model

For a decentralized control node model, each ECPSN can be considered as a control node. The physical connection inside a control node is shown in the right side of

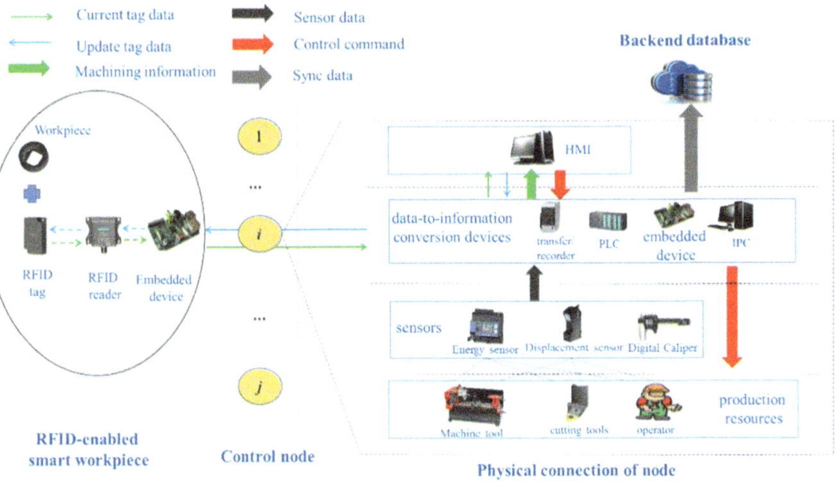

Fig. 7.5 Decentralized control node model [11]

Fig. 7.5. Here, production resources such as machine tools, cutting tools, measuring tools, operators, etc. are monitored by sensors via wired or wireless network. Those sensors connect the node controller with the data-to-information conversion devices such as transfer/recorder, embedded device, industrial personal computer (IPC), programmable logic controller (PLC), etc. In order to receive the control commands, the node actuator is directly connected to the node controller, and controls the operations of production resources. There are three kinds of communications related to ECPSNs, including communication between ECPSN and smart workpiece, communication between ECPSN and machine operator, and communication among ECPSNs. The Chapter 6 has demonstrated that the communication between ECPSN and humans is realized via the H2M/M2H social sensors that take the form of HMIs or chart bots, and the communication among ECPSNs via the predefined communication protocols or rules. Note that RFID-enabled smart workpiece can be considered as a special ECPSN that is capable of sensing its own status, and directly communicating with other ECPSNs for production share and cooperation. So the communication between ECPSN and smart workpiece is just a special case related to the communication among ECPSNs.

Taking into consideration of the communication between ECPSN and smart workpiece further, the control node of ECPSN requires interfaces to communicate with smart workpieces. For a RFID-enabled smart workpiece, the "*active RFID tag*" which contains a very limited temporary storage is bundled with the "*physical workpiece*" and thereby acts as a tiny information model of the workpiece. The reader can read the on-tag data through a non-contact way, and then sends them to the embedded device for analysis. After that, the embedded device sends the preprocessed data to the control node for further handling. For the control node, there are some different signals in the physical control entities of the control nodes. One is the sensor data

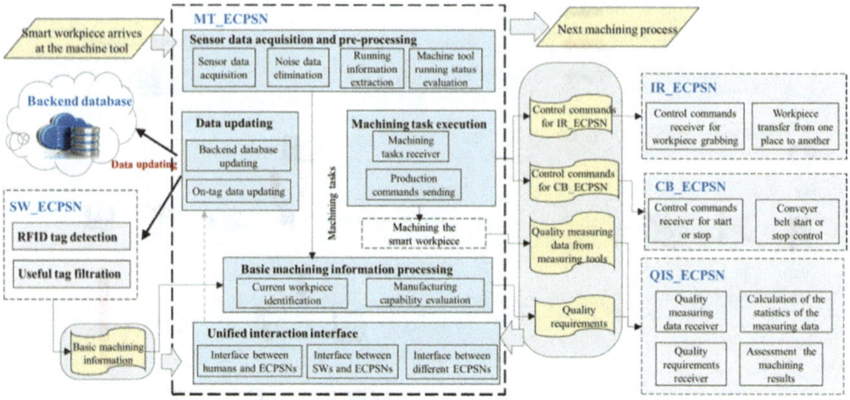

Fig. 7.6 Functional modules of extended cyber-physical system nodes and their inner relationships

from production resources, which is ultimately converted to machining information by the data-to-information conversion devices. The other is control commands which are directly sent to node actuator from node controller. The third one is the sync data which represent the running status of machining equipment. And the data should be stored on the backend database. The last one is the communication information between workpieces and control nodes.

7.4.2 Functional Modules of Extended CPS Node

As shown in Fig. 7.6, different virtual-implementation-related functional modules are related respectively to correspondent ECPSNs and can be presented according to the decentralized control node model. Here, at least five kinds of ECPSNs are involved, including ECPSN for smart workpiece (SW_ECPSN), ECPSN for machine tool (MT_ECPSN), ECPSN for industrial robot (IR_ECPSN), ECPSN for conveyer belt (CB_ECPSN), and ECPSN for quality inspection station (QIS_ECPSN). The correspondent functional modules include:

- RFID tag detection (TD) module,
- useful tag filtration (UTF) module,
- basic machining information analysis (BMIA) module,
- sensor data pre-processing (SDP) module,
- machining task execution (MTE) module,
- unified interaction interface (UII) module, and
- data updating (DU) module.

RFID tag detection (TD) module acquires the RFID tag data through the RFID reader, which is deployed on the machine tool, when a smart workpiece arrives. Then, it judges whether the tag is valid or not by checking the data format according to the predefined RFID data structure, and then sends the valid on-tag data to the UTF module. Smart workpiece m entering the radio frequency signal coverage area of machine tool i could be treated as an event, which can be described as:

$$e\_Tag_{i,m} = \{e\_ID_{i,m}, e\_T_{i,m}, e\_Info_{i,m}, t_{i,m}, T_{i,m}\} \tag{7.5}$$

where $e\_Tag_{i,m}$ and $e\_ID_{i,m}$ represent the event and event type respectively, $e\_T_{i,m} \in \{e_{in}, e_{pass}, e_{duplicate}\}$ represents the enter event, pass-through event and duplicate read event respectively, $e\_Info_{i,m}$ is the event information, $e\_ID_{i,m}$ is the ID code of smart workpiece, $t_{i,m}$ is the event occurrence time, and $T_{i,m}$ is the deviation of adjacent captured time of the same tag. All the captured RFID tags by TD module can be described as:

$$E\_Tag_i = \{M\_CP_i, e\_Tag_{i,1}, e\_Tag_{i,2}, \cdots\} \tag{7.6}$$

where M_CP_i stands for the current process of machine tool i.

The output of TD module can be formalized as:

$$TD_i = \sigma_{e\_Info_{i,m}=OTD_m}(E\_Tag_i) \tag{7.7}$$

Useful tag filtration (UTF) module is responsible for filtering duplicate on-tag data. When the tag enters the radio frequency signal coverage area of the reader, the tag signal would be captured by the reader continually until it leaves the area. Thus the data are reported repeatedly by the RFID reader at a regular periodicity. The periodicity is usually set as one second, and it can be reconfigured through the RFID management software. The module filters the data based on a given timeout. If the deviation of adjacent captured time of the same tag is smaller than the given timeout, the data would not be sent to the MT_ECPSN. And if the deviation of adjacent captured time of the same tag is longer than the given timeout, it is interpreted as a new entrance of the tag, and the captured data would be sent to the MT_ECPSN. The filtered useful on-tag data that would be sent to BMIA module can be formalized as:

$$UTF_i = \sigma_{T_{i,m} \geq T_{gv}}(TD_i) \tag{7.8}$$

where UTF_i indicates the output of UTF module, and T_{gv} is the lower threshold of $T_{i,m}$.

Basic machining information analysis (BMIA) module is used for identifying whether the smart workpiece passing through a machine tool needs to be processed there. If the completion status of smart workpiece is WW or FW, the tag would be directly ignored and the smart workpiece would be moved to the next machine. If the completion status of smart workpiece is WIP, BMIA module would judge whether its current process should be executed at this machine tool. Besides, BMIA module would evaluate whether the current machine tool satisfies the required machinability according to the machine tool running status collected from the sensor data pre-process module. If the current machine tool is able to handle the current workpiece,

BMIA module will send the machining tasks to the machining task execution module. Furthermore, this module will send the quality requirements to the QIS_ECPSN via the unified interaction interface module. The decision procedure can be described as:

$$
\begin{aligned}
M\_T_i &= \prod_{M\_P_{m,i}} (UTF_i) \\
&= \prod_{M\_P_{m,i}} (OTD_m)\,,\text{if } (OTD_m \supset STA_m, CP_m) \\
&\quad \wedge (STA_m = WIP) \wedge (CP_m = M\_CP_i)
\end{aligned} \tag{7.9}
$$

and

$$
\begin{aligned}
Q\_R_i &= \prod_{Q\_R_{m,i}} (UTF_i) \\
&= \prod_{Q\_R_{m,i}} (OTD_m)\,,\text{if } (OTD_m \supset STA_m, CP_m) \\
&\quad \wedge (STA_m = WIP) \wedge (CP_m = M\_CP_i)
\end{aligned} \tag{7.10}
$$

where M_T_i is the machining tasks of current machine tool i, and Q_R_i represents the corresponding quality requirements.

Sensor data pre-processing (SDP) module eliminates the incorrect, redundant and noisy data from sensor data, and then extracts the running information of machine tool based on the sifted sensor data. After that, the current status (available, out of service, or overloaded) of machine tool would be evaluated depending on the running information. SDP module would notify MTE module that the machine tool can keep on working until the process is finished. That is,

$$
Run\_Info_i = f_{SDP}(Sensor_i) \tag{7.11}
$$

where Run_Info_i is the running information of machine tool i, $Sensor_i$ represents the monitoring data from different sensors, $Sensor_i = \{sensor_{i,1}, sensor_{i,2}, \ldots\}$.

Machining task execution (MTE) module collaborates with the external ECPSNs such as IR_ECPSN and CB_ECPSN by sending production control commands to these ECPSNs according to the received machining tasks. For example, after receiving the machining task, MTE module notifies CB_ECPSN to stop the conveyer belt for workpiece transportation. At the same time, MTE module notifies IR_ECPSN to transfer the current workpiece to the in-buffer of machine tool, waiting for the workpiece to be machined.

Unified interaction interface (UII) module provides interfaces for the communication with external entities (humans or other ECPSNs). Interface between humans and ECPSNs enables the direct access between humans (customers, suppliers, managers, and operators) and machine tools. Interface between smart workpieces and

ECPSNs enables the basic machining information attached on the RFID tag to be transmitted to the machine tool for further analysis. Interface between different ECPSNs enables ECPSNs (e.g., IR_ECPSN, and CB_ECPSN) to communicate with each other for production share and cooperation. In addition, IR_ECPSN and CB_ECPSN aims to collaborate with MT_ECPSN to transfer the workpiece from one place to another (e.g., from the conveyor belt to the in-buffer of machine tool, from out-buffer of machine tool to conveyor belt) during the production process. After the current machining process is finished, the smart workpiece will be transferred to QIS_ECPSN for quality inspection. QIS_ECPSN would calculate the statistics of quality measuring data, and compare the statistics with quality requirements. If the statistics are within the range of quality requirements, it means the current process is qualified. Otherwise, the current process is unqualified. The evaluation process can be formalized as:

$$Q\_Measuring_i = f_{QIS}(Measuring\_Data_i) \tag{7.12}$$

$$PR_i = f_{PRJ}(Q\_Measuring_i, Q\_R_i), PR_i \in (WIP, WW, FW) \tag{7.13}$$

where $Q\_Measuring_i$ is the statistics of quality measuring data; $Measuring\_Data_i$ is the quality measuring data, $Measuring\_Data_i = \{measuring\_data_{i,1}, measuring\_data_{i,2}, \ldots\}$; PR_i is the process result.

Data updating (DU) module has two functions, that is, backend database updating function and on-tag data updating function. Backend database updating is used for saving the machining information derived from a machining process (such as machining results, quality measuring data, quality statistics, sensor data, cutting tools, operator, etc.) to the backend database. On-tag data updating is responsible for updating the on-tag data after finishing the current process. It changes the completion status of smart workpiece to a new level depending on the process result PR_i. Meanwhile, it leads the current process to the next step. The above new on-tag data of completion status in the current process are further written back to the tag with RFID reader. The on-tag data updating rules include that:

- the completion status in the current process is marked as WW if the machining process is unqualified, and
- the completion status in the current process is marked as WIP if the machining process is qualified and the current process is not the last one, otherwise the completion status is marked as FW.

These updating rules are formalized as:

$$STA_m = PR_i \tag{7.14}$$

$$CP_m = \begin{cases} M\_P_{m,i+1} \; if PR_i = WIP \\ M\_P_{m,n} \quad if(PR_i = WW) \vee (PR_i = FW) \end{cases} \tag{7.15}$$

It should be noted that most of the functional modules mentioned above work universally in different ECPSNs. For example, UII module and DU module exist in other kinds of ECPSNs such as IR_ECPSN, CB_ECPSN, and QIS_ECPSN.

7.4.3 Production Interactions and Cooperation

After clarifying the internal functional modules of ECPSNs, the production interactions and cooperation in both the inter-enterprise level and the intra-enterprise level are illustrated in Fig. 7.7. In the inter-enterprise level, social factory communicates with customers and other social factories to generate product orders that are further decomposed into several sub-orders including production orders. These sub-orders are undertook and shared by a group of social factories. For a certain social factory, it finishes the sub-order tasks by collaborating with customers, other social factories and suppliers. During the production process, real-time data on production process and workpiece flows are offered to customers, other social factories and suppliers via unified data interfaces to respond to any potential sub-order requirement changing (e.g., quantity, quality, and lead time). If a sub-order requirement changing does happens, all the required materials, components and parts would be timely supplied to make sure that the changing can be satisfied, even if it is proposed when the sub-order is almost done. In some cases, participators such as customers, other social factories and suppliers may even have direct access and specific control authority to the production-order-related machining equipments under the authorization mechanism. In the intra-enterprise level, social factory begins to execute the production order via the decentralized interactions and cooperation among ECPSNs after the RFID-enabled smart workpieces (SW_ECPSN) are configured and delivered to the shop floor in which all the machine tools are connected by the conveyer belts, and a smart workpiece is transported from the first machine tool to the last one along the conveyer belts. The production process has been illustrated in the right side of Fig. 7.7.

The production interactions and collaboration can also be explained in detail. Firstly, the smart workpiece arrives at the first machine tool via the conveyer belt and queries the available MT_ECPSN whether it is capable and available to execute the current machining process. After receiving the machining task information from the SW_CPSN, the MT_CPSN evaluates whether its machining capability and production capacity can meet with the machining requirements through considering the current task queue, and informs the SW_CPSN about the current status (out of service, overloaded, and available). If the current machine tool can't handle the machining task, the MT_ECPSN would ask the CB_ECPSN to transport the smart workpiece to the next MT_ECPSN. The evaluation and mapping process repeats until the available MT_ECPSN is eventually selected.

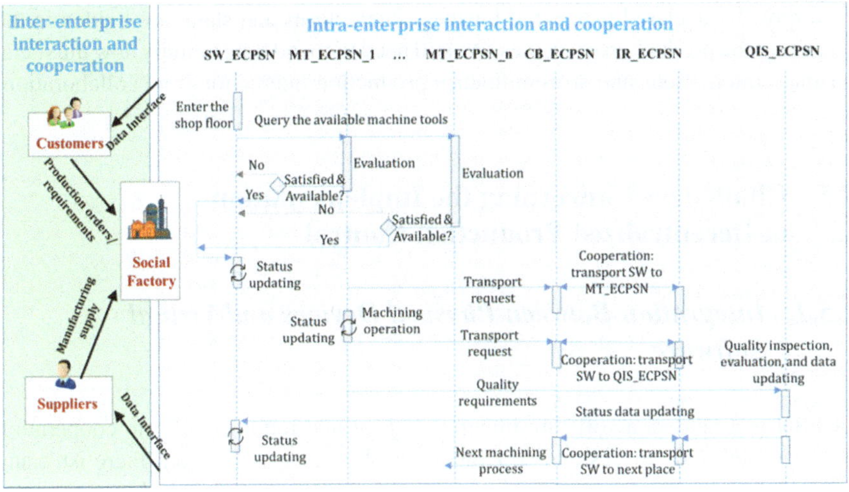

Fig. 7.7 Production interaction and cooperation at the inter-enterprise level and intra-enterprise level

Secondly, the selected MT_ECPSN interacts with CB_ECPSN so as to request CB_ECPSN to transport the workpiece to the designated machine tool. At the same time, the IR_ECPSN is informed to grab the workpiece from the conveyer belt to the in-buffer of machine tool for further execution.

Thirdly, the machining operation is conducted by the collaboration between human operator and the selected machine tool. The real-time production data are collected and processed locally, and the production status is monitored and updated to the MT_ECPSN. Assume that an unexpected production disturbance (such as machine tool breakdown) happens during the production process, once the disturbance is detected, the virtual modules of the MT_ECPSN will halt the machine tool immediately to prevent further damage. The current status will be updated and the MT_ECPSN will try to identify and eliminate the fault. The production progress will continue afterwards, and the current status will be updated. If the MT_ECPSN isn't able to handle the disturbance itself, it will interact with other MT_CPSNs to search for alternative machine tool.

Fourthly, when the current machining process is finished, the MT_ECPSN asks the IR_ECPSN to grab the workpiece from the out-buffer of machine tool to the conveyer belt. And the CB_ECPSN transports the workpiece to QIS_ECPSN for quality inspection and evaluation. At the same time, the MT_ECPSN sends quality requirements extracted from the smart workpiece to the QIS_ECPSN. After the quality inspection and evaluation, process results are updated to the SW_ECPSN.

Finally, QIS_ECPSN asks IR_ECPSN and CB_ECPSN to cooperate with each other to transport the workpiece to the next designated location.

It must be declared that social business interactions and share are hidden in the context of the production processes. We still need to spend time to study how the social business interactions and share influence production interactions and collaboration.

7.5 Challenges Concerning the Implementation of Decentralized Production Control

7.5.1 Integration Between Physical Devices and Virtual Software

ECPSN is a hardware-software integrated agent and requires delicate cooperation between physical devices and software control modules. However, there isn't any methodology to support a convenient, fast, transparent, and plug-and-produce integration of physical and automatic devices. In addition, the heterogeneity of these physical devices increases the difficulty for integration. As a result, there isn't any effective method to seamlessly integrate the physical devices into the software agents to form an ECPSN. From the physical device perspective, different hardware devices have different characteristics, specifications and communication protocols. It means that connecting them in an efficient form is still a challenge [7]. Some low-cost and energy-efficient embedded devices (e.g., *Raspberry* Pi), equipped with processor, RAM, network interfaces and peripherals, seem to give us an answer to this challenge. They can be utilized to integrate the heterogeneous devices so as to shield the difference between hardware platforms. From the cyber network perspective, digital twin technology can be employed to virtualize the physical devices and simulate the machining operations [13]. Technologies such as Web service, 3D modeling, virtual reality, and augmented reality can also be integrated into embedded devices to construct a digital representation of physical devices. Here, digital twin technology makes it possible to connect physical devices with virtual software. Through creating a digital twin model related to machining operations, for example, real-time monitoring data in the physical world can be updated to the virtual model simultaneously and the control commands from the virtual model would be transmitted to the physical devices.

7.5.2 Resource Self-organization Mechanism

According to specific production tasks, related ECPSNs are interconnected to shape an ECPSN-interconnected network in which ECPSNs negotiate with each other and try to achieve a global goal. However, performances such as production efficiency, production load, resource allocation are generally not good enough because each ECPSN can only make local decisions and cannot consider something in a global viewpoint [14]. Furthermore, these ECPSNs have to share a common understanding of the syntax and semantics of the exchanged information to realize the seamless communication among different ECPSNs [15]. Thus a resource self-organization mechanism needs to be developed to make ECPSNs automatically match production tasks in a global viewpoint. Key enabled technologies related to resource self-organization mechanism include manufacturing capability modeling, manufacturing task modeling, requirement-capability matching, machine tool optimal configuration, and learning mechanism. The modeling of capability and requirements means that both capability and requirements are formalized as readable formats for machine tools, for example, in the form of ontology-based XML format. The requirement-capability matching means that SW_ECPSNs and MT_ECPSNs negotiate with each other to identify available machine tools. The optimal configuration of machine tools ensures that the optimal machine tools can be picked out from the large-scale solution space with the constraint of cost, quality, lead time, etc. The learning mechanism contributes to the intelligence of an ECPSN so as to improve the efficiency of resource organization.

7.5.3 Event-Driven Control Mechanism

Production disturbances may lead to the deviations of production plan even the breakdown of the whole production system, so high-level monitoring and control solutions are required. Such solutions include two phases. The first phase is expected to detect and recognize different events occur in the physical world (e.g., production process, machine tools and production logistics), and then these events are autonomously reflected in the cyber world (e.g., digital twins, various production applications). In this phase, the production event model for ECPSNs, which deals with the temporal and spatial properties of events, event-trigger conditions and event classifications, must be explored [9]. The second phase is expected to handle, with certain limits, unexpected or unforeseen production disturbances following the detection of production events. In this phase, the logical operators such as AND, OR, and NOT can be used to combine different types of event conditions to capture composite events, and then the useful production information can be extracted and some predefined operations can be carried out. In order to achieve intelligent capabilities, artificial intelligence (AI) methods especially machine learning algorithms can be applied. For example, for a group of similar ECPSNs, their behaviors and reactions to the production events can be recorded and trained through the embedded algorithms such as deep neural learning algorithms in a collaborative way, thus ECPSNs can acquire disturbance handling capabilities. Under this topic, some useful enabled technologies like production event modeling, event or disturbance detection, identification and diagnosis, disturbance prediction and forecasting, disturbance handling mechanism, etc., can be used. In short, the event-driven control mechanism enables the production control from traditional "*fail and recover*" practices into "*predict and prevent*" practices and makes it possible to consequently improve the efficiency of the ECPSN production network.

7.6 A Demonstrative Example of Social Factory

7.6.1 Hardware and Software Configuration

To exemplify the advantages of distributive production control architecture mentioned in this chapter, a small-scale flexible smart production line has been developed as an ECPS-enabled social factory prototype in our *Manufacturing System Lab*, By means of using a set of middleware, the main goal to develop this prototype is to demonstrate the principles and models including the organizations, configurations and runtime logic of:

- physical space that deals with machine tools, various physical and social sensors, RFID, their ECPSNs, etc.,
- cyber space concerning digital twins that contains a variety of functional modules, 3D geometrical models of physical objects, etc., and
- social space which is concerned with social business interactions, share and collaboration related to Internet-based connecting and communicating behaviors in the intra- and inter-enterprise levels.

In the physical space, the built production line consists of three machine tools, three robotic arms, and four conveyer belts, as shown in the top left side of Fig. 7.8. Three machine tools include a CNC lathe, a CNC milling machine, and a small machining center. Each machine tool is deployed with an RFID reader, three RFID antennas, and some add-on sensors (e.g., vibration, power, temperature). Three robotic arms are installed near the machine tools to load and upload the smart workpieces. The four conveyer belts connect the machine tools and transport the workpieces from one machine tool to another. Each workpiece is attached with an "*active RFID tag*" which stores machining process information (e.g., machining demands and workpiece status). The machine tools, robotic arms, and conveyer belts are integrated and connected to the factory bus through *Rasberry Pi* (RPi), which is a single-board computer equipped with an ARM processor with 700 MHz, a memory with 256 MB, and some hardware interfaces (e.g., four USB slots, a RS232/485 port, a 10/100 Ethernet port). Through the interconnections of heterogeneous devices, the physical implementation of MT_ECPSN, IR_CPSN, CB_CPSN, and SW_CPSN can be finished.

In the cyber space, the digital twin models of the corresponding physical devices are developed. In order to digitalize the behavior of the physical devices, firstly, we utilize Web Graphics Library (WebGL) to construct the 3D model, plan the motion relationship and write the control script. Secondly, communication interfaces between physical devices and digital models are designed and developed via the middleware embedded in the RPi. Thirdly, some basic inference mechanisms, such as production rules and intelligent algorithms, are embedded into the RPi to endow ECPSNs with intelligence and negotiation ability. Fourthly, these CPSNs are endowed with Web access capability using REST architecture so that anyone can access the CPSNs in any places, under the control of trust and security mechanism.

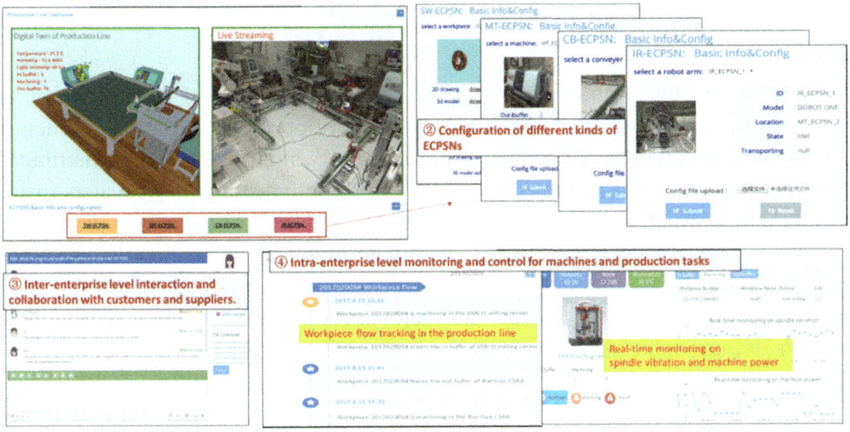

Fig. 7.8 An example of SF configuration and operation

In order to achieve flexibility, finally, *Resource Description Framework* (RDF) which is a semantic web data model and can be applied for conceptual description with resource-attribute-value labels is used for virtualizing machining equipments and their add-on sensors as resource nodes. Here, RDF descriptions are a series of configuration files in the form of XML to make sure that they are discoverable and machine-readable. Users only need to modify and upload the configuration files to the Web server hosted in the RPi to fulfill rapid and flexible configuration. Different kinds of ECPSNs have different configuration items, as shown in the top right side of Fig. 7.8. For example, the ID and machining process information are written to SW_ECPSN while the conveyor speed is configured to CB_ECPSN. Based on the steps mentioned above, digital twins of ECPSNs are developed. The real-time status of the physical device can be synchronized with the digital twins, and the production commands are sent to the physical devices for feedback control. It should be noted that certain mechanisms are embedded into the Production Line Management Center (PLMC) to achieve overall scheduling and coordination.

In the social space, participants can orchestrate the functionalities and algorithms embedded in various ECPSNs in the form of public or private *WebApps* under the support of REST interfaces provided by PLMC. The public *WebApps* (e.g., interaction center, production tracking, and quality query) allow all the registered users including managers and operators inside social factory, customers and suppliers outside social factory to visit and use them via Web browser. The private *WebApps* indicate that only the authenticated and authorized workers from social factory can visit and use them. These *WebApps* enable users to interact and coordinate with ECPSNs so as to create and execute production tasks in the context of social manufacturing.

7.6.2 *Decentralized Production Control*

As soon as finishing a hardware and software configuration of social factory, as shown in the bottom side of Fig. 7.8, we can develop a decentralized production control mechanism in the demo platform. In the inter-enterprise level, the distributed customers, other social factories and suppliers communicate with managers inside social factory which holds the production order, and control the whole production process via the *WebApps* developed for remote monitoring. Customers query and track the status of the production order, other social factories and suppliers deliver parts and components to the above social factory according to outsourcing and crowd-sourcing requirements issued by the above social factory.

In the intra-enterprise level, each workpiece is attached with "*active RFID tag*" in which very important machining process information is stored and thus turned into SW_CPSN. The detailed information of the workpiece such as the 3D model and 2D drawing is stored into backend database and can be found through its unique identifier. After entering the production line, we can use the SW_ECPSN to send its machining demands to three MT_ECPSNs. PLMC will coordinate these three MT_ECPSNs and plan a machining path according to matching results between machining demands and machining capability. The CNC lathe, the CNC milling machine, and the small machining center undertake the first, the second, and the third machining process respectively. Interactions and negotiation among ECPSNs (SW_ECPSN, IR_ECPSN, MT_ECPSN, and CB_ECPSN) during each machining process are similar. Here, we take the first machining process as example.

First, the IR_ECPSN binding to the CNC lathe is informed to grab the workpiece. And the workpiece is put into the in-buffer of the CNC lathe and its status is updated as "*in-buffer*". Then the operator begins to produce the workpiece and the workpiece status is updated as "*machining*". After finishing the current process, the operator unloads the workpiece, puts it into the out-buffer of the CNC lathe, and changes the workpiece status as "*out-buffer*". Simultaneously, the part machining quality is inspected and then updated to the SW_ECPSN. After that, the IR_ECPSN is informed to grab the workpiece from the out-buffer and put it on the conveyer belt. Finally, the CB_ECPSN is informed to transfer the workpiece to the next machine tool. During the entire executing process, physical operational activities such as workpiece flow and robotic arm grabbing happened in the physical world are synchronously reflected in the virtual digital twin models. The running states of machine tools are also monitored and updated to the backend database.

7.7 Conclusion

To sum up, our current work provides a blueprint for the distributive intelligent control of social factory. However, the implementation architecture is still facing great challenges, especially in the aspect of social space that enable Internet-based connecting and communicating behaviors in business although some progresses on difficult problems related to both physical and cyber spaces have been made.

It can be said that, in general, there is still a long way to go before we achieve the autonomy, intelligence, and distributive share and coordination by using digital twin-based ECPSNs under the context of social factory. The future work will be focused on ECPSN improvement and social factory implementation, including social business interactions and their impacts on production processes, production order sharing mechanism, the universal access and connectivity of physical objects, the digital representation of physical objects, more efficient communication mechanism among physical space, cyber space and social space, eventually the decentralization features embodiment of physical objects, i.e., autonomy, self-organization, self-learning, and self-negotiation, etc.

References

1. Jiang PY, Ding K, Leng JW (2016) Towards a cyber-physical-social-connected and service-oriented manufacturing paradigm: Social Manufacturing. Manuf Lett 7(2016):15–21
2. Yao XF, Jin H, Zhang J (2015) Towards a wisdom manufacturing vision. Int J Comput Integr Manuf 28(12):1291–1312
3. Atzori L, Iera A, Morabito G (2014) From 'smart objects' to 'social objects': the next evolutionary step of the internet of things. IEEE Commun Mag 52(1):97–105
4. Jiang PY, Liu C (2017) Discussion on the implication, architecture and operation logic of social factory. Aeronaut Manuf Technol 526(7):26–30 (in Chinese)
5. Ding K, Jiang PY (2017) Social sensors (S^2ensors): a kind of hardware-software-integrated mediators for social manufacturing systems under mass individualization. Chin J Mech Eng 30(5):1150–1161
6. Lee J, Bagheri B, Kao HA (2015) A cyber-physical systems architecture for industry 4.0-based manufacturing systems. Manuf Lett 3(2015):18–23
7. Liu C, Jiang PY, Zhang CY (2017) A resource-oriented middleware in a prototype cyber-physical manufacturing system. Proc Inst Mech Eng, Part B: J Eng Manuf. Publish Online
8. Lepuschitz W, Zoitl A, Vallee M et al (2011) Toward self-reconfiguration of manufacturing systems using automation agents. IEEE Trans Syst Man Cybern Part C: Appl Rev 41(1):52–69
9. Monostori L, Kádár B, Bauernhansl T et al (2016) Cyber-physical systems in manufacturing. CIRP Ann Manuf Technol 65(2):621–641
10. Ding K, Jiang PY (2017) Incorporating social sensors, cyber-physical system nodes, and smart products for personalized production in a social manufacturing environment. Proc Inst Mech Eng, Part B: J Eng Manuf. Publish Online
11. Wang C, Jiang PY, Ding K (2015) A hybrid-data-on-tag-enabled decentralized control system for flexible smart workpiece manufacturing shop floors. Proc Inst Mech Eng, Part C: J Mech Eng Sci 231(4):764–782
12. Grieves M (2014) Digital twin: manufacturing excellence through virtual factory replication. White Paper
13. Schleich B, Anwer N, Mathieu L et al (2017) Shaping the digital twin for design and production engineering. CIRP Ann 66(1):141–144
14. Zhang YF, Qian C, Lv JX et al (2016) Agent and cyber-physical system based self-organizing and self-adaptive intelligent shopfloor. IEEE Trans Ind Inform
15. Zuehlke D (2010) SmartFactory—towards a factory-of-things. Ann Rev Control 34(1):129–138

Chapter 8
Product Service Systems for Social Manufacturing

Wei Guo and Pingyu Jiang

8.1 Product Service System in Social Manufacturing

This section explores the basic concepts of product service system (PSS) and its advantages on integrating product with services. Under the context of PSS, many enterprises changed their business models from providing tangible products to integrated product services. However, this transition has been confronted with challenges from enterprises themselves, the society, and the customers. Hence, a novel approach, called as PSS for SocialM, is proposed in our work to solve the problems.

8.1.1 Useable PSSs in Product Life Cycle

The prototype of PSS was originally proposed in the report for the Dutch government in 1999 [1]. In fact, PSS can be understood as a business system which consists of products, services, supporting networks and infrastructure, is able to satisfy customer requirements and brings lower impact on the environment than the traditional business models [2]. One of the most obvious characteristics of PSS is the transformation from product selling to product-service providing. In this way, customers can obtain the services they need without paying for tangible products, and enterprises or service providers can obtain sustainable profits by serving the customers continuously.

On a life cycle basis, the implementations of PSS can be classified into three categories, that is, product-oriented services, use-oriented services and result-oriented services [3, 4]. Product-oriented services simply provide additional services to their original product, such as after-sales service and maintenance repair operating (MRO) service. For use-oriented services, the product is still owned by the service provider but shared to a number of other users, such as product renting, sharing, and pooling. Result-oriented services are pure customer requirements driven PSS that the service provider develops a subversive way to provide services to the customer, such as gas

P. Jiang, *Social Manufacturing: Fundamentals and Applications*, Springer Series in Advanced Manufacturing, https://doi.org/10.1007/978-3-319-72986-2_8

service [5], professional printing service [6] and civil aerospace service [7]. In this chapter, PSS refers to result-oriented PSS by default, and services or product-services refer to the result-oriented product-services by default.

Nowadays, enterprises are aware of such a fact that PSS is a sustainable and powerful tool for getting profits and promoting customer satisfactions. However, most of enterprises still remain in the stages of either product-oriented services or use-oriented services. Transformation to product-service providers (PSPs), which provides integrated product-services to customers, is facing with various challenges and obstacles from enterprises themselves. For example, there exist high labor costs, extreme service complexity, different culture background, and so on [8, 9]. On the basis of the internal barriers of enterprises, we know that the transformational challenges would come from the following three aspects.

The first aspect is about challenge on the enterprises' organization. To provide product-services, PSPs must shift their focus on enhancing their service departments and simplify other departments like manufacturing and sale departments. This may lead to enterprise instability which is difficult to eliminate.

The second aspect is about challenge on labor demands. Services must be consumed as they are provided because they cannot be saved, stored, returned, or carried forward for later use or sale [10]. Due to the variation in labor requirements, too much labor reserve may increase the economic burden of enterprises, and insufficient labor reserve may not be able to satisfy the requirements when more labors are required.

The third aspect is about challenge on complex services. Due to the incensement of product-service complexity and expertise, PSPs have to manage and control all the detailed processes of product-services and cannot concentrate on the core services that represent the core competency of the PSPs.

All these challenges are hindering the development and transformation of the PSPs to a higher level.

8.1.2 PSS for Social Manufacturing as a New Way to Run Products and Attached Services

According to the challenges mentioned above, a flexible and efficient mode should be introduced to enhance the current PSS. Facing the similar challenges in manufacturing, we proposed a new manufacturing paradigm called social manufacturing [11]. This new manufacturing paradigm starts from Internet-based connecting and communicating behaviors in business, focuses on self-organizing socialized manufacturing resources into manufacturing communities, and runs under the support of extended CPS and social factory model so as to implement mass collaboration and share during product manufacturing activities covering the whole stages of a product life cycle [12]. Enterprises in the same communities should have the same or similar manufacturing resources and capabilities. Based on the concept of SocialM, products are produced by gathering the socialized manufacturing resources into manufacturing communities

In service area, there are also many micro-and-small-scale service enterprises (MSSEs) which specialize in providing services. However, their sizes and strategies hinder their possibilities to further develop into very strong PSS providers [13]. In this chapter, two types of MSSEs sources are dealt with. The first one is the traditional MSSEs which can be distinguished by the number of employees, quantity of fixed assets, etc. And the second one is formed by MSSEs decomposed from the large-scale manufacturing enterprise.

In this chapter, the above MSSEs which contains socialized service resources (SSRs) can be cognized as independent PSPs which are gathered into service communities (SCs) to provide product-services. The core PSP only concentrates on the key phase of product-services and manages the others PSPs. In order to solve the problem in PSS under the context of SocialM, we proposed a novel business model called PSS for SocialM that gathers the SMSEs into SCs so as to provide product-services collaboratively. In the context of PSS for SocialM, the MSSEs share their service resources and capabilities on a specified platform, and then the service resources are aggregated into different SCs according to their similarity. Once a customer proposes a service order to the platform, suitable SCs will be selected and arranged according to their service capabilities and established a service community network.

8.2 Concepts and Implementing Architecture of PSS for Social Manufacturing

In the previous section, integrating of PPS and SocialM has been proposed to form a novel business model to tackle the challenges of PSS. Therefore, concepts and characteristics of this new business model should be defined and declared in detail so as to establish the implementing architecture and the operational logic of PSS for SocialM.

8.2.1 Definitions

The following key definitions explain basic concepts and research boundary of PSS for SocialM.

Definition 1: SSRs are defined as a set of the equipments and human resources involved in the processes of service providing, such as products themselves, diagnosis and maintenance tools, maintenance staffs, service packages that are presented as a series of services, etc. Resource sharing is one of the most core characteristics of SSRs to ensure the efficient and effective use of available service resources under the ever-increasing competitiveness [14]. PSS for SocialM aggregates and clusters the decentralized service resources and capabilities into SCs. In this way, problems caused by information islands [15] are solved, such as low resource utilization and low service efficiency.

Definition 2: Service capability is defined as a kind of the ability of using SSRs for operating and completing a specific service to satisfy the service requirements of the customer. Generally speaking, service capability includes stable equipment service capability and ever-increasing labor service capability.

Definition 3: MSSE is defined as a kind of social micro-and-small-scale service enterprise which provides product-services to customers. In a generic meanings, an MSSE also implies that it organizes and uses its own SSRs for product-services under considering its implementing structure and operational logic. To the extreme form, individuals like makers and individual serviceman who provide product-services can also be seen as a kind of virtual "*MSSE*".

Definition 4: PSP is defined as a kind of role an MSSE acts as and also as the representative or agent of an MSSE. It emphasizes the type of SSRs and the versatility of service capabilities.

Definition 5: Service Community (SC) can be defined as an aggregation of inter-related MSSEs and their SSRs and inter-connection relationships among MSSEs, their PSPs and their SSRs. Within an SC, its MSSEs have common or similar benefits and collaboratively complete product-services to satisfy customer requirements. As the core content of PSS for SocialM, SC has two obvious natures. The first is that relationships among SSRs include not only collaboration but also competition within an SC. Since every PSP wants to realize high profits and low costs, correspondent MSSEs need to change their strategies on SSRs to increase competitiveness. The second is that the organization procedure of SC is of self-organization and virus-like propagation. It means that whether an MSSE joins or leaves an SC is up to PSP's own strategy.

Definition 6: SC network is defined as the inter-connection among SCs that have various service types and service capabilities. In fact, PSS for SocialM uses an SC network responsible for different product-service phases and processes. Relationships among SCs are fuzzy and vague. It means that any MSSE together with its PSS and partial or all SSRs can be a member of different SCs. SC network is a typical relationship network which can be used to express the relationships among SCs and MSSEs, such as sequential relationship, supporting relationship, etc.

Definition 7: Service order can be defined as official service requests from customers guaranteed by contracts. It includes service type, service quantity, service capability, order allocation, response time, etc., and depends on customers' requirements analysis. Due to the intangible nature of services, the operation procedure of service order and service consumption is synchronous.

8.2.2 Characteristics of PSS for Social Manufacturing

Since PSS for SocialM integrates the core ideas of both PSS and SocialM, it has its own unique characteristics, which determine how to enable the implementing architecture and operational logic of PSS for SocialM.

The first characteristic is about service relationships. In the context of PSS for SocialM, SSRs from distributed PSPs self-organize into various SCs for collaborative services. Since an SC network may contain vast SCs and SSRs inside them, relationships among SCs and inside an SC are complicated. Generally speaking, there are only two kinds of relationships inside an SC, that is, cooperative relationship and competitive relationship [16]. PSPs either collaborate with each other to complete the product-services or compete with one another to obtain more profits. Therefore, relationships among SCs not only include the above two typical relationships, but also attachment relationship, sequenced relationship, etc.

The second characteristic is service-order-oriented. Due to the intangible nature of services, services cannot be stored like products and cannot have inventory [17]. They must follow a service order and be consumed as they are provided. The operation of a service order relies on triggering and executing a series of service events. Here, a service event is used for changing service state from A to B. Besides the above service order, service contract, service flow, service result and service evaluation are also different from products [18].

The third characteristic is about lean services. In traditional PSS, only the core enterprise provides product-services for customers, this makes it difficult to satisfy the detailed requirements in the service processes and leads to possible economic waste. In the context of PSS for SocialM, a mass of professional service MSSEs act as PSPs for specialized product-service providing and one PSP only needs to concentrate on one kind of service. Since product-services can be divided into different service phases, a lean-philosophy-driven optimal matching among PSPs and service phases can be realized through so-called "*lean services*" [19].

8.2.3 Implementing Architecture of PSS for Social Manufacturing

In order to clarify the concepts and structure of PSS for SocialM, a four-layer implementing architecture of PSS for SocialM is illustrated, as shown in Fig. 8.1. From the bottom to top, there are resource layer, community layer, organization layer and operation layer. A functional platform of PSS for SocialM is created to manage and control the realization of the implementing architecture.

In the resource layer, massive SSRs that belong to different PSPs are socialized and virtualized. A PSP may have one or more SSRs, and each SSR can complete a specific service without assistance from others. However, the SSRs are chaotic and disordered in the resource layer, and unable to provide integral product-services

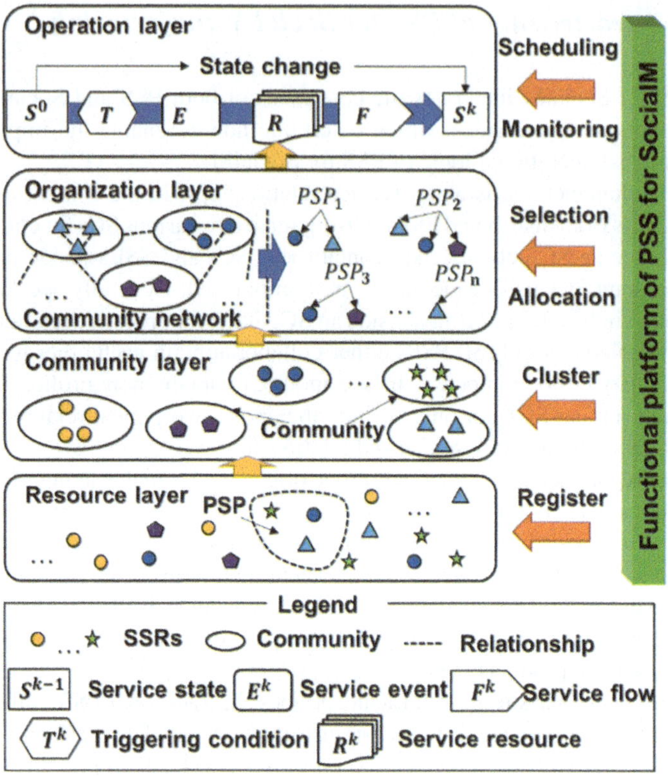

Fig. 8.1 The implementing architecture of PSS for SocialM

for customers. In order to increase the service capability, the SSRs are clustered into different SCs according to their service capabilities in the community layer. However, the results of clustering demonstrate that the prototype of SC is in unstable situations. The SSRs need to change their strategies to form an ordered and stable community structure through self-organization mechanism. Generally speaking, an SC may have the service capabilities for a specific product-service and several SCs may work together to provide integrated PSS. Hence, dynamic SCs may be combined further into an SC network in the organization layer according to the service order and detailed customer requirements. Here, the SSRs are identified by the capabilities of service resources, but the service strategies of SSRs are determined by PSPs. Therefore, the SSRs should be correlated with the PSPs. Different service strategies may affect the service costs and quality of service (QoS), and then affect the service order allocation to the SSRs. The service processes can just be described as that customers acquire services from service providers with service resources through service flows or service channel, and need to consequently change their states through event-based operations in the operation layer, as shown in Fig. 8.2 [20]. As the output of PSS for SocialM in Fig. 8.1, service flows change the states of the customer in the

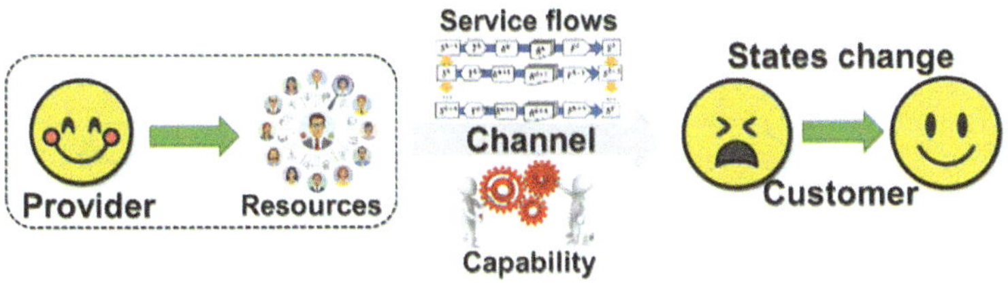

Fig. 8.2 The purpose and definition of a service

operation layer. When the customer is in a state of S^{k-1}, the triggering condition T^k is activated. According to the content of service event E^k, the functional platform organizes the selected communities with service resources R^k to provide service flow F^k to change the customer state to S^k. This four-layer architecture illustrates the service processes from SSRs organization to product-service output.

As the basic for PSS for SocialM, the functional platform of PSS for SocialM can be seen as a web-based platform integrated with various intelligence algorithms to realize the functions of each layer, such as deep learning for resources clustering, genetic algorithm for resources selection, game theory for service order allocation, Petri net for service flows design, etc. The platform is also integrated with instant messaging tools for customer participation and communication among all the PSPs. And it is also integrated with resource management tools to enable SSRs and structured tools for service process visualization.

The implementing architecture constructs an ecological environment for PSPs to mainly increase their service resources and service capabilities to provide the customer with satisfaction of product-services. The other things will also be handled appropriately by the functional platform.

8.2.4 Operational Logic of PSS for Social Manufacturing

Based on the implementing architecture of PSS for SocialM, its operational logic can be organized, as shown in the Fig. 8.3. The operational logic of PSS for SocialM can be explained by means of declaring four issues, that is, modeling of service capability, modeling of service flow, service monitoring and scheduling, and service quality evaluation. The four issues cover the whole operational processes of PSS for SocialM, including service contract establishing (service capacity and capability), product-service providing (service flow), service process controlling (monitoring and scheduling) and service resulting (service evaluation). The detailed steps of these four issues are described as follows.

As to the first issue concerning PSS for SocialM, distributed PSPs release their SSRs for clustering into different SCs according to the similarity of service capabilities. Therefore, the service capability modeling method needs to be proposed so as to describe and analyze the service capabilities of each SSR. Based on the similarity

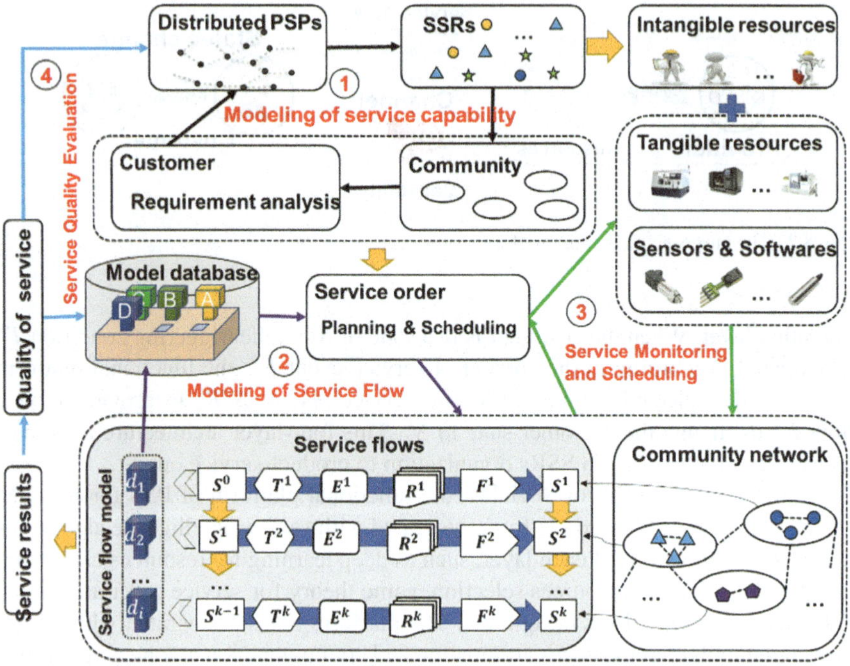

Fig. 8.3 Operational logic of PSS for SocialM

analysis results, SSRs with similar capabilities will be clustered into the same SCs. In order to provide lean services, SCs match with the analysis results of customer requirements and suitable SCs are selected. These matching results are fed back to correspondent PSPs who are representatives of SSRs, based on which PSPs would optimize their service resources and service strategies to fit with the changing market.

According to the matching results, for the second issue concerning PSS for SocialM, service orders and contracts are created between customers and PSPs. SSRs inside an SC or among SCs either collaborate or compete with other each to complete services specified by the above service orders. And service orders allocation must be tackled among SSRs. Generally speaking, a complex service package may be divided into several independent services and each service may be satisfied by an SC. The service flow modeling includes service triggering condition T^k, service event E^k, service resources R^k and the service flows F^k, all of which are the core contents of every service package or its services. The designed service flow may be stored in a model database for reuse when a new service is similar with the stored model. The model database will be updated if a new service flow is designed and is better than the old one in the database.

In order to provide lean services, for the third issue concerning PSS for SocialM, the service processes of PSS for SocialM must have a feedback mechanism to realize closed-loop control. In the context of PSS for SocialM, the services are product-services or product-based services. Therefore, the embedded sensors and corresponding software for products need to be developed to monitor the states of SC network and service flows. Service monitoring mainly includes two aspects. The first is to monitor service flows and schedule service resources according to the service orders. The second is to monitor service states when customers are in specific states, and events would be triggered with the condition to start service flows and change the customers' state to the next step.

As to the fourth issue concerning PSS for SocialM, it should be pointed out that the outputs of PSS for SocialM are product-services and their evaluation criterions are different from products. For products, we usually focus on production costs and product function implementations and neglect benefits from product operations. Since the outputs of product-services are service results of satisfying customer requirements, evaluation criterions need to include not only service costs and service functions, but also service time, service efficiency, service value creation, etc. which can be summarized as QoS. Evaluation results would be fed back to PSPs for assisting them to optimize resources allocation and service strategies. Moreover, the results are used to update the model database if the QoS of the new service flow is better than the stored model.

According to the above four issue, the operational logic of PSS for SocialM becomes clear and its corresponding key enabled technologies will be proposed in the next section.

8.3 Key Enabled Technologies

8.3.1 Modeling of Service Capability and Costs Attached to Products

Service capacity and capability description and modeling of an SSR are the primary work of PSS for SocialM. An SSR includes intangible labor resources and tangible product resources, and can be described from four aspects, that is, service functions, service performances, service structure and service activities. The definitions of the four aspects are declared as follows. Examples of logistics services are provided for better understanding.

Service functions are determined by the basic service properties of an SMR, including service types, service attributes and service contents to be able to provide to customers, physical products and their functions to be used for services, etc. For example, service function of the logistics services includes transporting service types, transport carrier, transportation route, etc.

Service performances describe service quality, service efficiency, service reliability and response time based on service functions, these indexes can be sum up as QoS. For instance, transporting time and on-time rate are the key indicators of logistics services. Service performances directly determine whether an SSR can complete the service order or not.

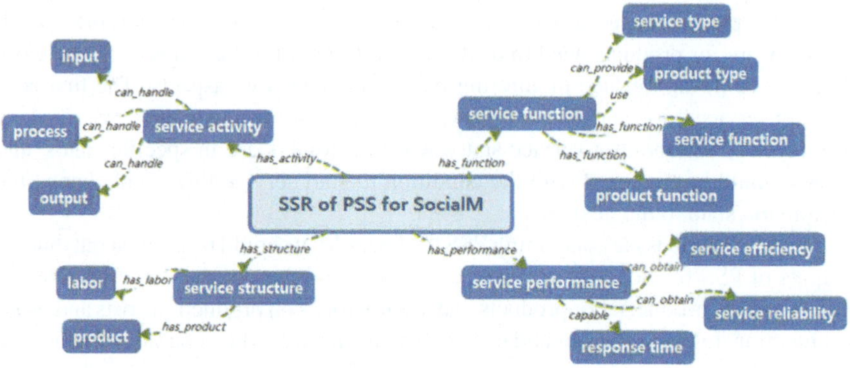

Fig. 8.4 The ontology structure for SSR (top-level)

Service structure denotes the organizational form of an SSR, including the type and quantity of labor, product devoting for services, etc. Service structure represents service strategies of a PSP who devotes more labors and physical products to an SSR or not. It is because that benefiting service performances would increase service costs and vice versa. For example, the number of truck drivers and trucks put into a transportation route are related to the service structure which is decided by the PSP.

Service activities represent a series of nodes in a service flow. Each service activity is correspondent with a node in the service flow and contains three elements, that is, "*service input*", "*service process*" and "*service output*". The detailed service flow will be designed and optimized according to the specific customer requirements. For logistics services, service activities deal with constructing a transportation route from city A to city B, loading and unloading service processes, etc.

Service functions depend on the capacity and capabilities of an SSR. With the help of service structure and service activities, a PSP is able to provide different service capabilities with specific service performances to customers. There exist many models and methods to describe an SSR from the above four aspects. In order to decrease ambiguity and make a more effective explanation of an SSR, ontology method is often used to understand the terms within the service domain and define formal specification [21]. Actually, ontology is commonly defined as an explicit formal specification of describing terms, relations among the terms, rules to generate terms and their relations in a specific domain [22]. The ontology structure for SSR is shown in Fig. 8.4 (top-level abstract classes of an SSR) and the detailed description can be expressed in the form of structural knowledge using Ontology Web Language (OWL) [23].

Whether a PSP can get service orders or not is partly decided based on its service capacity, capabilities and the service costs related to its SSRs. Profits and costs are important concerns for both the PSPs and customers, they all want to reduce costs and improve profits simultaneously. There are various methods for service cost evaluation, such as performance-based contracting [24], pay-for-performance [25],

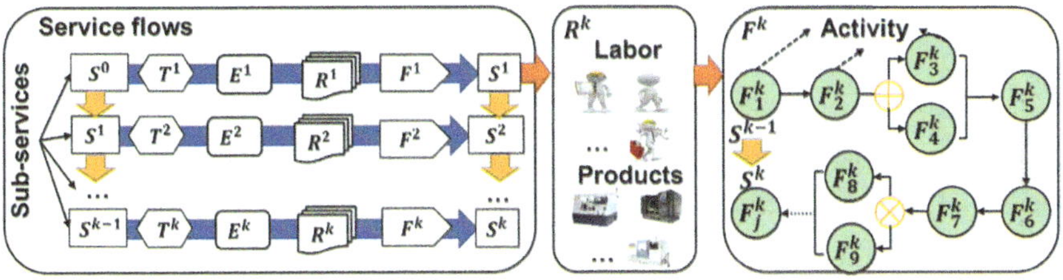

Fig. 8.5 Activity-based costing method for PSS for SocialM

Table 8.1 The parameters of service costs

Parameters	Remarks
$C_{k-1,k}^{n}$	The costs of PSP *n* for changing customer from *k* − *1* to *k*
$C_{k-1,k}$	The costs of the sub-service changing customer from *k* − *1* to *k*
C	The total costs of the service
λ_i^n	Service order allocated to SSR *i* of PSP *n*
i_n	The number of SSRs in the PSP *n*
k_j	The number of activities in the sub-service
α	The probability of selecting this SSR in select relation
$C_{i,j}^{n}$	The costs of SSR *i* of PSP *n* for activity *j*
N	The number of PSPs for the sub-service
K	The number of sub-services in the PSS for SocialM

pay per service unit [26], activity-based costing [27], etc. Considering the PSS for SocialM, a service order which often consists of a service package is completed by various SCs and SSRs, the evaluation of service costs will depend on costing different sub-service flows related to a service task inside the above package. As shown in Fig. 8.5, we propose a refined activity-based costing (ABC) method for the costs estimating of PSS for SocialM.

A service task in PSS for SocialM is usually divided into several sub-services and may be completed by an SC with various SSRs. A service flow includes different kinds of relations, including sequential relation (represented by '→'), concurrent relation (represented by '⊗') and selective relation (represented by '⊕'). The service costs of PSS for SocialM can be calculated as follows. The parameters in the calculation formulas are listed in Table 8.1.

$$C_{k-1,k}^{n} = \sum_{i=1}^{i_n} \lambda_i^n \cdot \sum_{j=1}^{k_j} \alpha \cdot C_{i,j}^{n} \tag{8.1}$$

$$C_{k-1,k} = \sum_{n=1}^{N} C_{k-1,k}^{n} \tag{8.2}$$

$$C = \sum_{k=1}^{K} C_{k-1,k} \tag{8.3}$$

The three formulas illustrate the costs of PSPs, sub-services, and services of PSS for SocialM respectively. The service costs are one of the significant evaluation indicators of QoS.

8.3.2 *Modeling of Order-Driven Service Flow*

Product-services are intangible and cannot be stored like physical products. Hence, the make-to-stock strategy for physical products cannot be applied to services. Only the make-to-order or order-driven strategy can be applied. To satisfy customer requirements, service flows design must be treated carefully. From macroscopic design to microscopic design, Shimomura developed a service modeling method consisting of four models, that is, "*flow model*", "*scope model*", "*scenario model*", and "*view model*" [28]. It emphasizes that service flow design should include service participators, state changes, outline and detailed service flows. In the context of PSS for SocialM, a complex service can be divided into several simpler sub-services which have the elements of service state ($\boldsymbol{S}^k$), triggering condition ($\boldsymbol{T}^k$), event ($\boldsymbol{E}^k$), service resource ($\boldsymbol{R}^k$) and service flow ($\boldsymbol{F}^k$), as shown in the left side of Fig. 8.5.

Petri nets are widely studied and successfully applied in workflow designing, process modeling and flow modeling for discrete-event dynamic systems [29]. In this section, we use the Petri net to build a service flow model because it has a well-defined mathematical foundation and a clear graphical feature [30]. A typical Petri net can be defined as a directed graph with three structural components, "*Places*", "*Transitions*", and "*Arcs*". "*Places*" represents states or conditions of the system, "*Transitions*" describes events that may modify system states, and the relationships between places and transitions are connected by "*Arcs*". "*Places*" can contain "*Tokens*" with which the number and position may be described during the Petri net execution. A Petri net is a 3-tuple $N = \langle P, T, F \rangle$, where:

$P = \{p_i : i = 1, \ldots, |P|\}$ is a finite set of "*Places*",
$T = \{t_i : j = 1, \ldots, |T|\}$ is a finite set of "*Transitions*", $P \cap T = \emptyset$,
$F \subset (P \times T) \cup (T \times P)$ is the set of directed "*Arcs*" representing flow relations, connecting "*Places*" and "*Transitions*" together.

When applying Petri net to service or sub-service flow design, the three tuples would have specific physical meanings, as shown in Table 8.2. Right side of Fig. 8.5 shows the sequence relation, concurrent relation, selective relation of service and sub-service flow. In fact, a Petri net can be used to explain the logic for ordering and selecting different relations, as shown in Fig. 8.6. Obviously, the Petri net is able to satisfy the requirements of service flow design and can be applied to design the required services and their sub-services. Services of PSS for SocialM emphasize

Table 8.2 Meanings of the three tuples in different contexts

	Typical	Service	Sub-service
Place (*P*)	States	States	States
Transition (*T*)	Activity	States change	States change
Arc (*F*)	Between two places	Between two sub-services	Between two service points
Token	Resources	A community	Resources set
Identification (*ID*)		Community ID	SSRs ID set
Type (*TY*)		Labor and product type for service	Labor and product type for sub-service
Quantity (*Q*)		Quantity of labor and product	Quantity of labor and product
Workload (*W*)		Orders allocate to the community	Orders allocate to each SSR

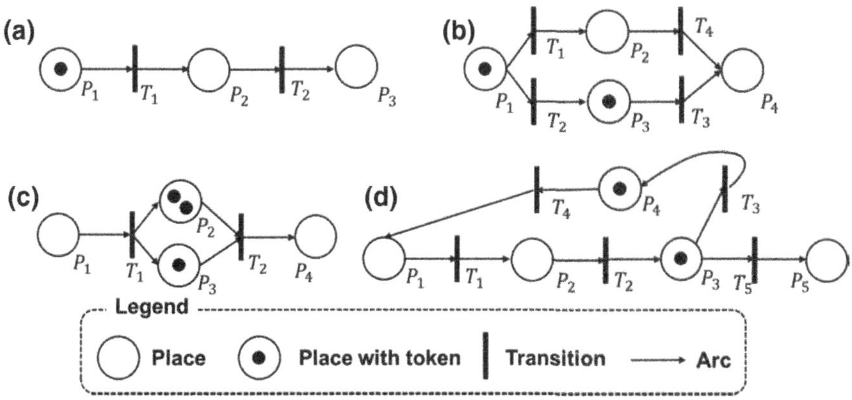

Fig. 8.6 The relationships in a Petri net. **a** Sequence relation, **b** selective relation, **c** concurrent relation, **d** selective relation

utilizing SSRs and collaboration to provide services. "*Tokens*" in typical Petri net is used to represent resources for "*Transitions*". However, "*Tokens*" only emphasizes on existence of the resources and neglects characteristics and quantity of the resources.

Based on the advantages of resource-aware Petri net [31], we proposed an SSR-aware Petri net with 7 tuples, $N = P, T, F, ID, TY, Q, W$. The extended four parameters are defined to assist and support the "*Tokens*" contents, as shown in Table 8.2. A simplified example of logistics services is illustrated in Fig. 8.7. The SSR-aware Petri net and its corresponding explanations are shown in Table 8.3.

To sum up, in this section, an RSS-aware Petri net for service flow modeling is presented. It highlights the SSRs description and service order allocation in the context of PSS for SocialM, and can express service flows clearly and accurately.

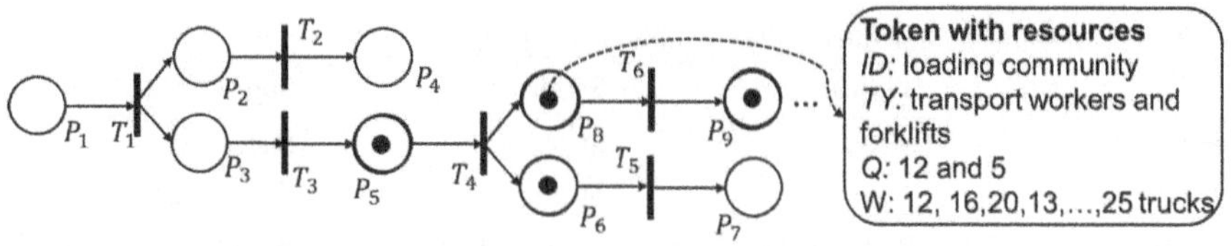

Fig. 8.7 A simplified example of logistics service with Petri net

Table 8.3 The explanation of places and transitions in logistics Petri net

Places	Remarks	Transitions	Remarks
P_1	Logistic order	T_1	Acceptance approval
P_2	Not undertake	T_2	Chargeback
P_3	Undertake	T_3	Goods consolidation
P_4	End	T_4	Goods inspection
P_5	Goods for shipment	T_5	Returned goods
P_6	Disqualified goods	T_6	Loading
P_7	End		
P_8	Qualified goods		
P_9	Loaded trucks		

8.3.3 Planning, Scheduling, and Monitoring of PSS for Social Manufacturing

Planning, Scheduling, and Monitoring are the core contents of PSS for SocialM. The operational logic of these contents is shown in Fig. 8.8. For an order-driven service mode, a service order with customer requirements and service contents is the input of the system. According to the service order that often consists of a service package and SSRs that belong to correspondent PSPs, service planning decides which SSRs should be devoted to the service package, and then selects the appropriate SCs to form the corresponding SC network. Each SC in charge of a sub-service flow of a service and the SC network will complete the entire service flow related to the service package. To monitor the service and its sub-service flow, some sensors are deployed and loaded in SSRs, such as global positioning system (GPS) modules, radio frequency identification (RFID) tags, and web-cameras. They are embedded into physical products to track routes and monitor labors' activities during service processes, and monitoring data will be fed back to a scheduling center. If one or more SSRs cannot complete their tasks with the required time or costs, the original service planning will be changed to schedule or reschedule the SSRs to be consistent with the new service planning.

For the participators of product-services, profits and costs are the most focused points. Such participants always want to reduce costs and improve profits at the same time. Therefore, service planning and scheduling should consider the economic factor

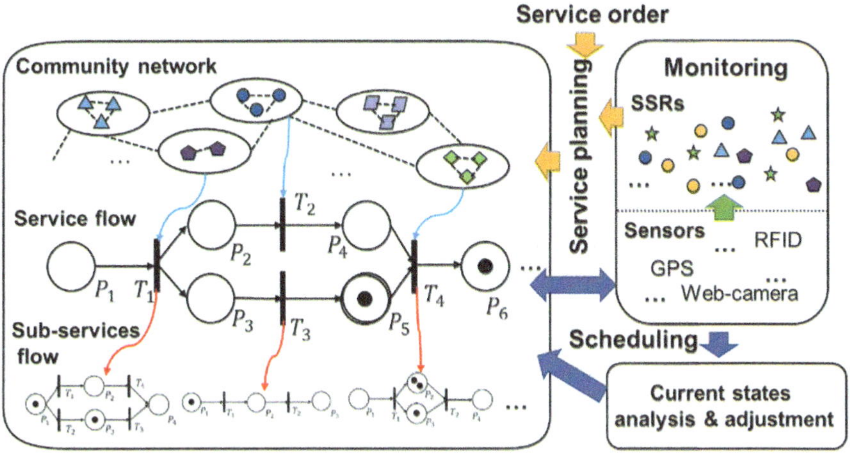

Fig. 8.8 Operational logic of service planning, scheduling, and monitoring

of services. After an SC network establishment and service flow design, the service planning and scheduling can be mapped into a service order allocation problem with criterions of costs or profits. In essence, service order allocation is an optimization problem of allocating the optimal order quantities to PSPs. Focusing on order allocation methods, scholars proposed many kinds of optimization algorithms. For example, Demirtas and Üstün proposed an integrated approach to analyze network process and multi-objective mixed integer linear programming under the consideration of the cost factor [32]. Kannan et al. introduced a set of fuzzy multi-criteria decision-making method and multi-objective programming approach for green supply chain order allocation [33]. Çebi and Otay mainly considered quantity discounts and lead time as main factors that influence order allocation problem, and solved the problem with a fuzzy multi-objective model [34]. Jain et al. introduced the chaotic bee colony algorithm for order allocation with different discounting policies [35].

In PSS for SocialM, there are two types of participators, i.e., customers and PSPs. They can change their strategies to get better payoffs independently. For customers, they can change the quantity of orders allocating to a PSP. And for PSPs, they can change the quantity of labors and physical products devoted to the services in response. If a customer changes the strategy first, correlated PSPs will also change their strategies according to the customer's move and then the customer will response to the PSPs' strategies. The iterations will continue until reaching an equilibrant state.

In order to solve the problem related to the service order allocation, we use a Stackelberg non-cooperative game model in which the leader moves first and then the follower moves sequentially [36]. In this game model, a customer is mapped as the leader and PSPs are mapped as the followers in Fig. 8.9. Based on the concept of Stackelberg game model and PSS for SocialM philosophy, the customer and the correspondent PSPs have the payoffs on costs represented by *LC* and *FC* respectively.

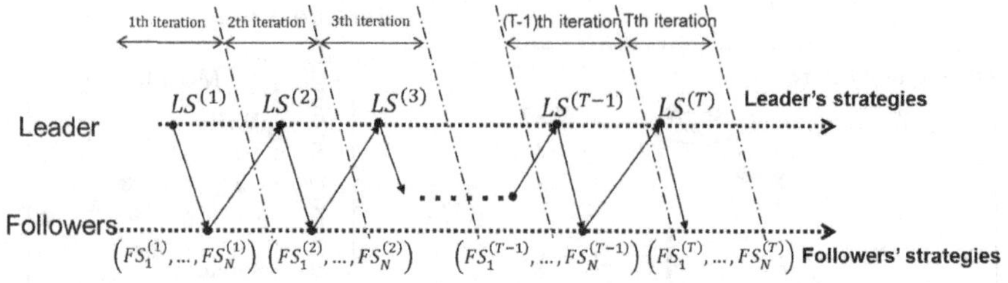

Fig. 8.9 Iterations between leader and followers of Stackelberg game

The payoffs strategies are represented by *LS* for the customer and $FS_1, FS_2, \ldots, FS_n$ for the PSPs, and the gaming goal is to minimize the costs of PSPs and customer.

In order to find the equilibrium of the game (the best *LS* and *FS* for *LC* and *FC*), a modified hierarchical Bird Swarm Algorithm (HBSA) was proposed based on the Bird Swarm Algorithm (BSA) [37]. It mimics the foraging behavior, vigilance behavior and flight behavior of birds. As to foraging behavior, each bird searches for food according to its previous experience and the swarms' experience. This operator aims at searching for feasible solutions and finding dominant solutions. As to vigilance behavior, birds try to move to the center of the swarm for foraging and would inevitably compete with each other according to foraging behavior. To avoid this phenomenon, some birds would not directly move towards the center of the swarm and keep vigilance to avoid trapping in local optimum. As to flight behavior, birds may fly to another site on a frequency *FQ*. When arrived at a new site, some birds acting as producers would search for food patches, while others acting as scroungers would follow the producers.

The BSA can be applied to solve single level problems, but the problem in this section has two levels, i.e., leader level and follower level. According to the core idea of BSA, an HBSA algorithm to solve multi-objective Bi-level programming is proposed. Here, the HBSA consists of two BSAs, one is for solving the leader-level problem and the other for follower-level problem. The flowchart of the HBSA is shown in Fig. 8.10 and the corresponding parameters are demonstrated in detail in Meng's research [37].

According to iteration results, customer and PSPs dynamically adjust the service strategies to minimize the service costs. In the real case, however, the evaluation indicators between customers and PSPs include not only the costs but also the indicators of QoS. Thus the Stackelberg game model will turn into a multi-objective Bi-level optimization problem, which requires further study in depth.

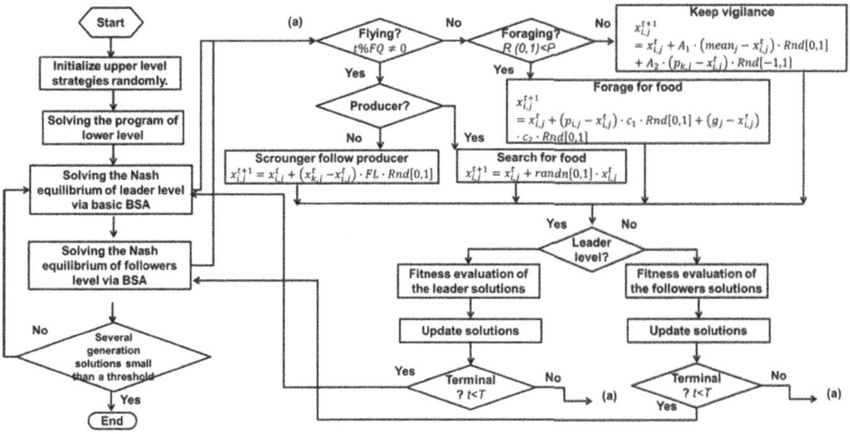

Fig. 8.10 The flowchart of HBSA

8.3.4 *Service Quality Evaluation*

After modeling the service capacity, service capabilities, service flow, and operation principles, an integrated PSS for SocialM is proposed. However, it is still unknown that whether the PSS for SocialM has the ability to provide high-efficient and sustainable services. It is quite necessary to evaluate its QoS from different perspectives with various criterions. Actually, QoS is the description or measurement of the overall performance of a service that originally applied to telephony computer network services. It considers service response time, packet loss, transmission delay and so on. Recently, QoS is used to evaluate common services and product-services based on its principle [38].

For a kind of product-service, the QoS evaluation can be classified into two categories. The first is to make an evaluation before service operations to predict unexpected failure and adjust unexpected service operations accordingly in advance. The second is to do an evaluation after service operations to provide a reference or guidance for future work by analyzing the service results. The common QoS evaluation processes are illustrated in Fig. 8.11. Qu et al. conducted an evaluation analysis through three aspects, that is, customer value, sustainability and trade-offs between them [39]. Yoon et al. pointed out that evaluation must be considered from two aspects including evaluation from the viewpoint of PSS providers to find the potential risk, and evaluation from the viewpoint of customers based on their satisfaction degree [40]. Key performance indicators (KPI) can be used for measuring and evaluating a service with criterions on service production, customer requirements and so on [41]. The relative weights of the criteria are determined by using fuzzy analytic hierarchy process [42].

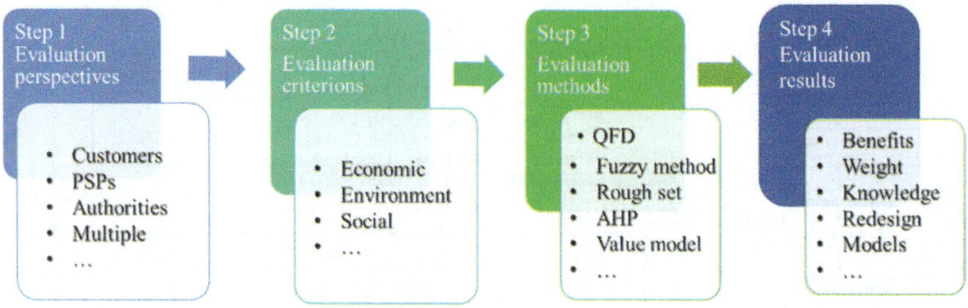

Fig. 8.11 The common processes of QoS evaluation

On the basis of synthesizing the above viewpoints, a Quality Function Deployment (QFD) based evaluation method is applied to evaluate the effects of a service concerning customers served [43, 44]. The four evaluation steps are detailed as follows.

The first step is about evaluation perspectives. QoS can be evaluated from various perspectives based on the focus of stakeholders. Customers always pay attention to value creation, service result, and effect. While PSPs care more about service sustainability, economic income, and competitive power. The authorities keep a watchful eye on social and environmental benefits which are provided by the product-services. Generally speaking, scholars choose one or two perspectives as entry points to evaluate whether a product-service can satisfy the criterions or not, and draw lessons from the evaluation results.

The second step is concerned with evaluation criterions. Different evaluation perspectives require corresponding evaluation criterions. Product-service evaluation criterions can be classified into three categories, that is, economic criterion, environment criterion, and social criterion. For economic criterion, added value, consumption and price are the most significant indicators. For environment criterion, energy consumption, hazardous materials, and emissions of pollutants, etc. are the main indicators. And for social criterion, different service scenarios have different indicators, such as health and safety, customers' culture, job creation, etc. An evaluation criterion may have multiple evaluation indicators and an evaluation indicator can be applied to multiple evaluation criterions.

The third step is about evaluation methods. Before calculating the values of evaluation indicators, the weight of each indicator should be tackled first. Since different weights of indicators have varied effects on QoS, greater weight indicator has greater impact on evaluation criterion, and vice versa. Fuzzy computation and analytic hierarchy process (AHP) methods are always applied to evaluate the weights [42]. In fact, AHP is a structured technique to organize and analyze complex decisions. To improve the precision of AHP, fuzzy computation is integrated with AHP, such as triangular fuzzy numbers and so on. Even though the evaluation criterions and indicators have been determined, different calculation formulas still need to be considered for service evaluation. There are no consensus formulas for evaluation indicators, which should be adjusted according to the characteristics of the service.

The fourth step is concerned with evaluation results. QoS evaluation can be classified into two categories, i.e., evaluation before service operations and evaluation after service operations. Hence, the evaluation results also have corresponding attributes for different kinds of product-services. Due to lack of related knowledge about product-services, PSPs may suffer from unexpected failure during the services. Therefore, necessary service evaluation before service operations should be used to assist the PSPs to optimize and adjust their service strategies. When a service is completed, the service results should be saved into knowledge base as references for future service design and operations. Some product-service models can be abstracted from the knowledge base. In this way, new product-services can be redesigned rapidly by changing the parameters of the models.

The evaluation of QoS plays a role in instructing product-service design and operations. With the above four steps, a product-service may have various evaluation results from different perspectives. It should be pointed out that different evaluation formulas should be applied to different types of product-services according to their natures.

8.4 Examples

In this section, two typical commercial applications are utilized to verify the effectiveness and efficiency of PSS for SocialM. One is from a logistics company where it outsources its logistics orders to individual truck drivers so as to realize wide coverage and high-efficiency logistics services. Another is from an air-conditioner manufacturing company where it outsources its transportation service, installation service and MRO service to related product-service SCs, and the company only focuses on the assembling operations which are the core work of producing air-conditions.

8.4.1 Electronic Product Trucking Services Through Individual Truck Drivers

Nowadays, logistics services have become an important content to support product manufacturing activities. However, there are still many problems to be solved. The first one is low degree of logistics resources intensiveness. When goods are sent from city A to city B in the form of "*Express*", they need to be transferred among many locations. It means that the logistics company has to set up lots of stations and hire many drivers to handle the goods. The second one is asymmetric information between senders and truck drivers. The processes of logistics are complicated and unstable due to the various natures and transportation routes. Because of the asymmetric information of every logistics process, it is difficult for logistics companies to select the optimal transportation route, transportation mode and transportation strategy so

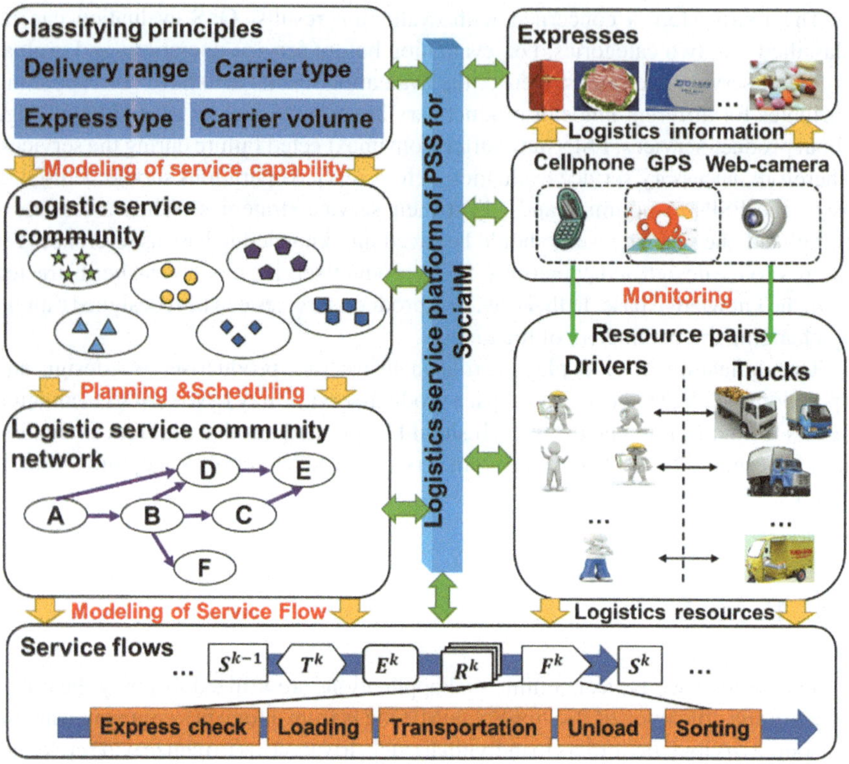

Fig. 8.12 The operational logic of the logistics service platform

as to increase the logistics costs and waste their transportation capacity. The third one is lack of professional logistics. The types of goods are different, including fragile goods, frozen goods, large-scale goods, etc. Different goods should be transported in specialized transportation methods which require specialized transport vehicles.

To solve aforementioned problems, the logistics company, called as RRS, established a logistics service platform based on the theory of PSS for SocialM. The operational logic of the logistics service platform is shown in Fig. 8.12.

The platform is based on the instant interactive information network to attract individual truck drivers, and then truck drivers are classified into SCs for various logistics services. The classifying principles include delivery range, carrier type, "*Express*" type and carrier capacity, as shown in Fig. 8.13. Based on the classification principles, logistics resources can be classified into different logistics SCs. The platform selects and organizes suitable SCs to form the logistics SC network through matching the resources' capabilities with customer requirements. In the procedure of network forming, the platform applies artificial intelligence algorithms to do service planning and scheduling so as to maximize transportation capacity and minimize the logistics costs and logistics time. During the transportation process, cellphone, GPS,

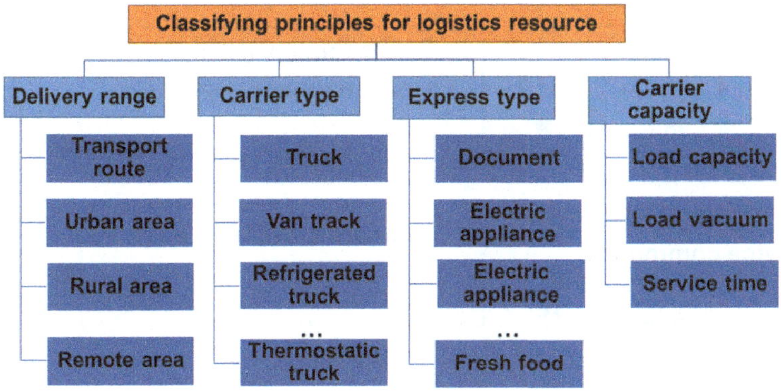

Fig. 8.13 Classifying principles for logistics resource

and web-camera are equipped to ensure the real-time monitoring of the conditions of trucks and goods, and send instant messages to the truck drivers about the weather and road conditions. On the other hand, customers can use the APP or website to obtain the goods information about their locations and expected arrival time.

Besides logistics services, RRS has proposed the installation services especial for household appliances. The company has been training the truck drivers on how to install the common household appliances, this indirectly increases the incomes of truck drivers and satisfy the needs of customers at the same time.

In general, this logistics company takes advantage of PSS for SocialM to form a logistics service platform which integrates individual truck drivers into SCs based on customer requirements. Some advantages can be drawn. The first advantage is to utilize logistics resources with high efficiency. With optimal logistics routes, truck loading and order allocation, the platform builds a bridge between goods and trucks to maximize the logistic capacity of the company. The second advantage is to share information. The platform provides instant communication tools, so truck drivers can share logistics information with each other and eliminate asymmetric information. The third advantage is to provide professional logistics services and add-value services. For some special goods, such as fresh food, medicines and electronic products, professional truck drivers and carriers are required. The platform classifies logistics resources into different SCs to provide specific professional logistics services for certain goods. Installation services and some others add-value services are proposed for better service experiences.

8.4.2 Air-Condition Service System for Social Manufacturing

Air conditioner is a common electric appliance in our daily life and we usually buy it from the market. Currently, people from a lot of public places such as office spaces and dormitories want to buy "*cooling*" and "*heating*" services instead of buying air

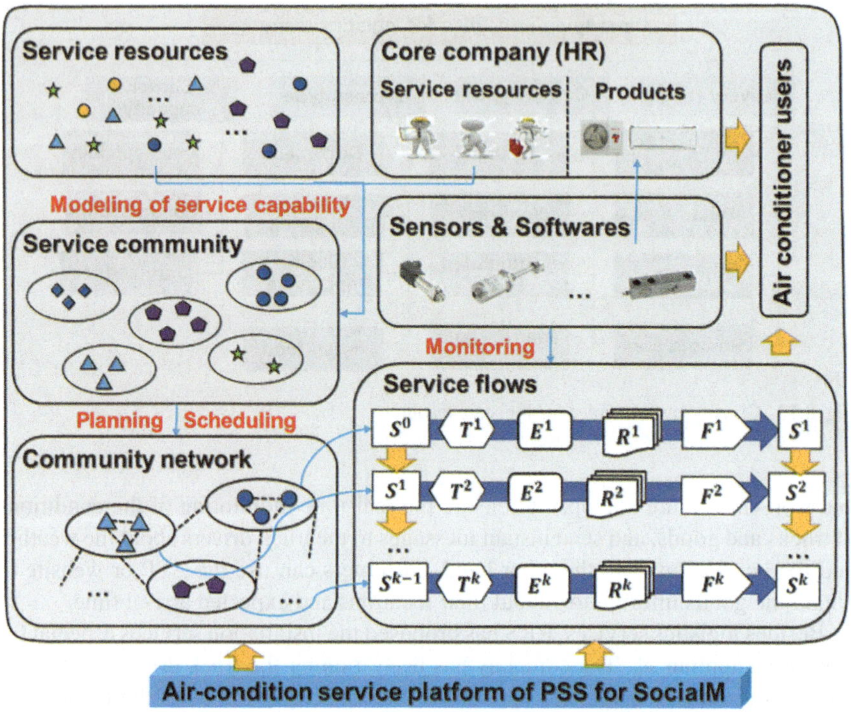

Fig. 8.14 The operational logic of the air-conditioner service platform

conditioners so as to save money. About from 2016, an air conditioner manufacturer, called as HR, started a financial leasing program in which HR rents its products to customers to provide "*cooling*" and "*heating*" services and add-value services around an air conditioner, such as installation, cleanout, MRO, etc. The customers just pay for the "*cooling*" and "*heating*" services without buying any air conditioner. To realize this result-oriented PSS, HR decomposed its resources into several service sectors. Sometimes, HR's service capacity cannot satisfy the customer requirements, the external PSPs should join in the service processes with their service resources. In this context, HR, as a core enterprise, builds an air-conditioner-service platform based on PSS for SocialM philosophy, manage and control it. The operational logic of the air-conditioner-service platform is shown in Fig. 8.14.

Since this product-service mode is core enterprise driven PSS for SocialM, the service resources of external PSPs serve as the assistant for the core company. Within the financial leasing program, HR manufactured customized air conditioners and rent them to the customers without payment concerning products. The customers just pay for the "*cooling*" and "*heating*" services of the air conditioners and the service fees can be calculated according to the followed formulas:

$$T = K_t \times C \times (K_h \times T_h + K_m \times T_m + K_l \times T_l) \quad (8.4)$$

$$P = T \times E \times \alpha \tag{8.5}$$

where T is the equivalent running time of an air conditioner, and K_t is the running time of an air conditioner. C denotes the operation mode of the air conditioner, refrigeration or heating. T_h, T_m and T_l are the running time of an air conditioner in high-grade, middle-grade and low-grade respectively. T_h, T_m and T_l represent the grade coefficients of high-grade, middle-grade and low-grade respectively. E is the unit price of the air condition service and α represents the loss coefficient of an air conditioner.

The core enterprise HR provides the most important "*cooling*" and "*heating*" services and monitors correspondent service processes, the external PSPs assistant the core company to complete add-value services around an air conditioner. The SSRs that belong to external PSPs and service resources from HR are classified into SCs according to their service types and service capacity. Then a SC network is established. It should be pointed out that the product-services around an air conditioner are heterogeneous. It means that different SCs provide different product-services. According to the continuity of the product-services, SCs can be classified into three types, that is, one-time service SC, intermittent service SC and continuity service SC.

After service planning, each SC in the SC network completes one or several services divided from the entire air-conditioner-service package. The sensors and corresponding software are embedded in air conditioners to collect operating data, monitor service processes and operate states. The operating data will be transferred to the platform and feedback the analysis results to HR. Based on the results, HR can optimize the product-service strategies to increase product-service quality and maintain continuous relationships with customers. The customers can monitor the service processes and product states with APPs installed in their cellphones, with which the customers can also participate in the service processes and send their new requirements to HR.

In 2017, for example, HR has installed almost 70,000 air conditioners to dormitories of 12 universities and provides 200,000 times "*cooling*" and "*heating*" services including repair services. In this way, HR has occupied the air-conditioner market of the university by providing "*cooling*" and "*heating*" services and has accumulated enough product-service experiences to compete with others. The air-conditioner-services create a solution that benefits everyone. On the one hand, selling air-conditioner-services bring continuous profits for HR, On the other hand, customers can pay less money to acquire the services they need.

8.5 Concluding Remarks

In this chapter, a novel service mode called PSS for SocialM is proposed to realize mass service collaboration. This mode addresses MSSEs within an SC by collaboration and competition to satisfy customers' requirements and realize product-service value adding. By means of analyzing the implementing architecture and key enabled technologies, we believe that PSS for SocialM can assist MSSEs to realize and adapt product-services. The two practical cases also confirm the feasibility and effectiveness of PSS for SocialM, even though there still are some unsatisfactory aspects. As a newly proposed method, our researches on PSS for SocialM are mainly from theoretical perspective. Hence, more practical applications should be carried out to fully develop and verify this novel business mode.

References

1. Goedkoop M, Halen V (1999) Product service systems, ecological and economic basics
2. Mont OK (2002) Clarifying the concept of product–service system. J Clean Prod 10(3):237–245
3. Tukker A, Tischner U (2006) Product-services as a research field: past, present and future—Reflections from a decade of research. J Clean Prod 14(17):1552–1556
4. Tukker A (2004) Eight types of product–service system: eight ways to sustainability? Experiences from SusProNet. Bus Strategy Environ 13(4):246–260
5. Gao J, Yao Y, Zhu V, Sun L, Lin L (2011) Service-oriented manufacturing: a new product pattern and manufacturing paradigm. J Intell Manuf 22(3):435–446
6. Visintin F (2012) Providing integrated solutions in the professional printing industry: the case of Océ. Comput Ind 63(4):379–388
7. Kerley W, Wynn DC, Eckert C, Clarkson PJ (2011) Redesigning the design process through interactive simulation: a case study of life-cycle engineering in jet engine conceptual design. Int J Serv Oper Manage 10(1):30–51
8. Vezzoli C, Ceschin F, Diehl JC, Kohtala C (2015) New design challenges to widely implement 'Sustainable Product-Service Systems'. J Clean Prod 97:1–12
9. Martinez V, Bastl M, Kingston J, Evans S (2010) Challenges in transforming manufacturing organisations into product-service providers. J Manuf Technol Manage 21(4):449–469
10. Kundu S, McKay A, de Pennington A, Moss N, Chapman N (2007) Implications for engineering information systems design in the product-service paradigm. In: Advances in life cycle engineering for sustainable manufacturing businesses. Springer, London, pp 165–170
11. Jiang PY, Ding K (2012) Social manufacturing: a new way to support outsourcing production. In: Proceedings of the 2nd International Conference on Innovative Design and Manufacturing, Taipei, Taiwan, 20–23 December 2012
12. Jiang PY, Leng JW, Ding K, Gu P, Koren Y (2016) Social manufacturing as a sustainable paradigm for mass individualization. Proc Inst Mech Eng, Part B: J Eng Manuf 230(10):1961–1968
13. Lelah A, Mathieux F, Brissaud D, Vincent L (2012) Collaborative network with SMEs providing a backbone for urban PSS: a model and initial sustainability analysis. Prod Plann Control 23(4SI):299–314
14. Samaddar S, Rabinowitz G, Zhang GP (2005) An experimental analysis of solution performance in a resource sharing and scheduling problem. Eur J Oper Res 165(1):139–156
15. Tao F, Cheng Y, Zhang L, Nee AYC (2017) Advanced manufacturing systems: socialization characteristics and trends. J Intell Manuf 28(5):1079–1094
16. Ding K, Jiang PY, Leng JW, Cao W (2016) Modeling and analyzing of an enterprise relationship network in the context of social manufacturing. Proc Inst Mech Eng, Part B: J Eng Manuf 230(4):752–769
17. Meier H, Roy R, Seliger G (2010) Industrial product-service systems—IPS2. CIRP Ann Manuf Technol 59(2):607–627

18. Tukker A (2013) Product services for a resource-efficient and circular economy—a review. J Clean Prod 97:76–91
19. Resta B, Powell D, Gaiardelli P, Dotti S (2015) Towards a framework for lean operations in product-oriented product service systems. CIRP J Manufact Sci Technol 9:12–22
20. Arai TŒSY (2004) Proposal of service CAD system—A tool for service engineering. CIRP Ann Manuf Technol 53(1):397–400
21. Vasantha GVA, Roy R, Lelah A, Brissaud D (2012) A review of product-service systems design methodologies. J Eng Des 23(9):635–659
22. Gruber TR (1993) A translation approach to portable ontology specifications. Knowl Acquisition 5(2):120–199
23. Kulvatunyou B, Ivezic N, Lee Y, Shin J (2014) An analysis of OWL-based semantic mediation approaches to enhance manufacturing service capability models. Int J Comput Integr Manuf 27(9):803–823
24. Selviaridis K, Wynstra F (2015) Performance-based contracting: a literature review and future research directions. Int J Prod Res 53(12):3505–3540
25. Langbein L (2010) Economics, public service motivation, and pay for performance: complements or substitutes? Int Public Manage J 13(1):9–23
26. Williams A (2007) Product service systems in the automobile industry: contribution to system innovation? J Clean Prod 15(11):1093–1103
27. Kimita K, Hara T, Shimomura Y, Arai T (2008) Cost evaluation method for service design based on activity based costing. In: Manufacturing systems and technologies for the new frontier. Springer, London, pp 477–480
28. Sakao T, Shimomura Y, Sundin E, Comstock M (2009) Modeling design objects in CAD system for service/product engineering. Comput Aided Des 41(3):197–213
29. Dong M, Chen FF (2001) Process modeling and analysis of manufacturing supply chain networks using object-oriented Petri nets. Robot Comput-Integr Manuf 17(1–2):121–129
30. Salimifard K, Wright M (2001) Petri net-based modelling of workflow systems: An overview. Eur J Oper Res 134(3):664–676
31. Pla A, Gay P, Meléndez J, López B (2014) Petri net-based process monitoring: a workflow management system for process modelling and monitoring. J Intell Manuf 25(3):539–554
32. Demirtas EA, Üstün Ö (2008) An integrated multiobjective decision making process for supplier selection and order allocation. Omega 36(1):76–90
33. Kannan D, Khodaverdi R, Olfat L, Jafarian A, Diabat A (2013) Integrated fuzzy multi criteria decision making method and multi-objective programming approach for supplier selection and order allocation in a green supply chain. J Clean Prod 47:355–367
34. Çebi F, Otay I (2016) A two-stage fuzzy approach for supplier evaluation and order allocation problem with quantity discounts and lead time. Inf Sci 339:143–157
35. Jain V, Kundu A, Chan FTS, Patel M (2015) A Chaotic Bee Colony approach for supplier selection-order allocation with different discounting policies in a coopetitive multi-echelon supply chain. J Intell Manuf 26(6):1131–1144
36. Mu H, Jiang P, Leng J (2017) Costing-based coordination between mt-iPSS customer and providers for job shop production using game theory. Int J Prod Res 55(2):430–446
37. Meng X, Gao XZ, Lu L, Liu Y, Zhang H (2016) A new bio-inspired optimisation algorithm: Bird Swarm Algorithm. J Exp Theor Artif Intell 28(4):673–687
38. Sun H, Wang Z, Zhang Y, Chang Z, Mo R, Liu Y (2012) Evaluation method of product-service performance. Int J Comput Integr Manuf 25(2):150–157
39. Qu M, Yu S, Chen D, Chu J, Tian B (2016) State-of-the-art of design, evaluation, and operation methodologies in product service systems. Comput Ind 77:1–14
40. Yoon B, Kim S, Rhee J (2012) An evaluation method for designing a new product-service system. Expert Syst Appl 39(3):3100–3108
41. Waltemode S, Aurich JC (2013) Productivity and quality assessment of services within technical product-service systems In: The philosopher's stone for sustainability. Springer, Berlin, pp 125–130

42. Allen Hu H, Chen SH, Hsu CW, Wang C, Wu CL (2012) Development of sustainability evaluation model for implementing product service systems. Int J Environ Sci Technol 9(2):343–354
43. Sakao T, Lindahl M (2012) A value based evaluation method for Product/Service System using design information. CIRP Ann Manuf Technol 61(1):51–54
44. Shimomura Y, Hara T, Arai T (2008) A service evaluation method using mathematical methodologies. CIRP Ann Manuf Technol 57(1):437–440

Chapter 9
Blockchain Models for Cyber-Credits of Social Manufacturing

Jiewu Leng, Jiajun Liu and Pingyu Jiang

9.1 Blockchain in Social Manufacturing

With the coming of sharing economy [1], increasing manufacturing enterprises tend to share their socialized manufacturing resources (SMRs) and crowd-source/outsource their non-core service to reduce cost and gain faster market response. The SMRs with similar capabilities and interests aggregate into self-organized manufacturing community (MC) [2] to improve their bargaining power and competence. In this situation, a new manufacturing paradigm, called social manufacturing (SocialM) [3–7], is proposed. Under the circumstances of SocialM, a third-party platform is utilized to integrate distributed SMRs into decentralized manufacturing network called social manufacturing network (SMN). SMN provides an effective approach for the organization of MC and interaction among SMRs, which maximizes SMRs utilization efficiency and facilitates SMRs cooperation with each other for providing more precise and professional manufacturing services for personalized demands from prosumers [2, 8]. In SMN, prosumers can also be regarded as a kind of SMRs who provide product innovative ideas and personalized demands, and participate in the entire manufacturing process [9]. Around these demands, different SMRs owning corresponding manufacturing capacity focus on different phases of product lifecycle and autonomously organize into MCs. Within the MC, SMRs establish seamless cross-enterprise collaborations with each other to accomplish demand order.

Due to the peer-to-peer characteristic of SMN, the cross-enterprise collaborations have enormous potential risks because there is no supervising in production activities from the trusted third party [10]. For instance, enterprises can abort on-going partnerships casually and customers can default, even unpaid order payments intentionally. In short, the peer-to-peer cyber-credits are not guaranteed without a trusted third party or mechanism. Cyber-credits are the foundation for the cross-enterprise collaborations under SocialM context. It is closely related to vital interests of prosumers and MC/SMRs. Crediting MC/SMRs' competence and honesty is

P. Jiang, *Social Manufacturing: Fundamentals and Applications*, Springer Series in Advanced Manufacturing, https://doi.org/10.1007/978-3-319-72986-2_9

a vital factor for prosumers to select trustworthy MC/SMRs to cooperate with, and MC/SMRs must credit that prosumers can provide their with high quality manufacturing services. Apparently, the establishment of manufacturing cyber-credits is an important aspect for cross-enterprise collaborations under SocialM context for keeping the operation of SMN smoothly and reliably. Generally speaking, guaranteeing credit among MC/SMRs depends on the trusted third party, but there is no such role in SMN. Consequently, there is a gap among MC/SMRs in establishing reliable partnership because they know nothing about each other. Hence, guaranteeing the cyber-credits among MC/SMRs becomes an important problem. Furthermore, in SMN, mass cross-enterprise collaborations involve numerous interest disputes among MC/SMRs. This requires a highly trustworthy technology to guarantee the manufacturing cyber-credits and the balance of profits among MC/SMRs in SMN.

To solve the above problems, the distributed blockchain technology [11–13] is introduced into SocialM. The basic theory of the distributed blockchain technology is originally proposed by Satoshi [14], which is utilized to guarantee Bitcoin payments in an electronic cash system based on peer-to-peer network instead of a trusted third party. Utilizing Bitcoin system, Andrychowicz et al. [15] constructed some decentralized protocols to implement fairness among multi-party transaction. Richard and Gareth [16] presented a reputation system based on a decentralized smart contract system. Aaron and Primavera [10] elaborated the benefits and challenges brought by the distributed blockchain and discussed that blockchain can enable decentralized organization with codified governance rules by smart contracts. Hence, the distributed blockchain technology is a promising tool to guarantee manufacturing cyber-credits in SMN without the trusted third party.

9.1.1 Concepts of Blockchain

The organization and operation of traditional enterprises are centered on large equipment producer. But, nowadays, some emerging open source products (e.g., *RepRap* open source 3D printer) are developed by manufacturing network composed of peer-to-peer small and micro enterprises (SMEs). In that case, intellectual property, data privacy and credit mechanism are increasingly important so that some central manufacturing platforms can't adapt to the decentralized peer-to-peer manufacturing network any more. In addition, current service interactions between SMEs are offline, peer-to-peer and direct under market adjustment without auxiliary methods and tools. Considering these trends, this chapter proposes a blockchain technology-based manufacturing community self-organizing technology for SMEs.

Figure 9.1 shows the evolution of organization and operation of enterprises. Current emerging networked manufacturing paradigm (e.g., Cloud Manufacturing) depends on centralized platform, server system and whole management model. And SMRs owing great manufacturing capacity are organized and operated by connecting to cloud server. When the scale of SMRs increases gradually, the disadvantages of the networked manufacturing will appear:

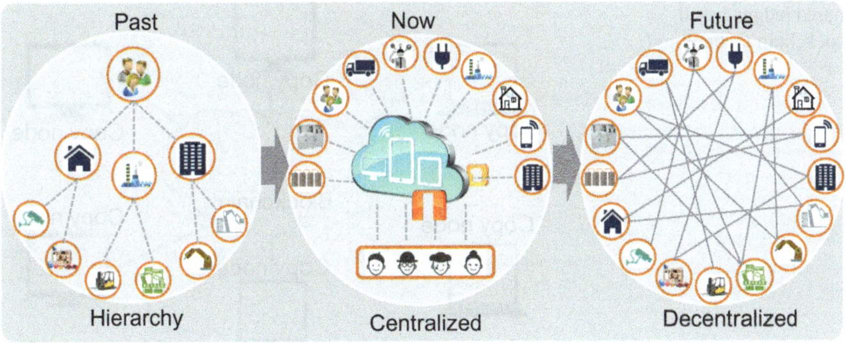

Fig. 9.1 Evolution of organization and operation of enterprises

(1) Coordination of upstream and downstream SMRs is weak and traceability of manufacturing in-formation is poor. And the factors affecting product quality are complex and diverse, including design, manufacturing, materials, assembly and link between productions.
(2) The gradually increasing SMRs requires larger scale cloud severs from more large-scale server cluster and network equipment, which will be confronted with huge risk that breakdown of one node will greatly collapse the function of whole manufacturing network.
(3) Heterogeneous of equipment and diverse personalized server demands obstruct peer-to-peer inter-action between SMRs, and equipment from different SMRs can't meet the requirements of interoperability and compatibility.

The above disadvantages stem from that organization and interconnection of the current networked manufacturing stay on the surface of application layer, and the interconnection and manufacturing logic between underlying equipment cannot be implemented effectively by social media merely.

This chapter applies the blockchain technology to construct self-organizing structure of SMEs under social manufacturing context [17, 18], and the structure covers not only application layer but also interconnection and management of underlying equipment. This enables the interconnection between SMEs to become decentralized and improves the robustness of large-scale manufacturing network.

Blockchain technology is a new decentralized architecture and computing paradigm firstly used in Bitcoin. To guarantee credit under peer-to-peer network circumstance, blockchain technology adopts a decentralized shared ledger to record interaction information from network nodes.

Figure 9.2 depicts the significant characteristics of blockchain technology: (1) blockchain technology is a decentralized scheme to accomplish storage, validation, delivery of data depending on network nodes not the trusted third party; (2) blockchain is an open source technology and all data stored on blocks are accessible to everyone at anytime and anywhere; and (3) blockchain runs on consensus

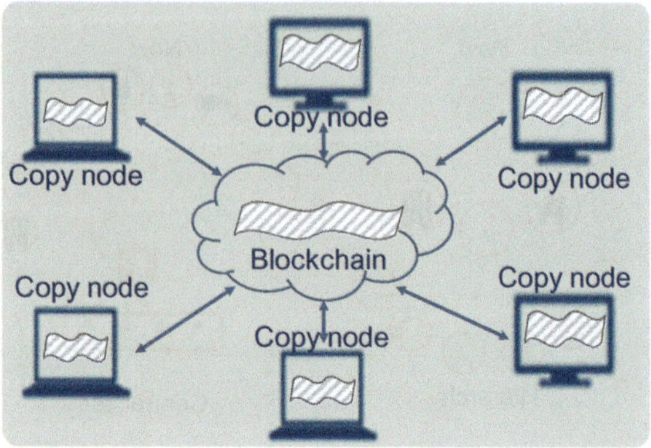

Fig. 9.2 Decentralized shared ledger-based blockchain technology

specifications and protocols (e.g., hash algorithm) which enable network nodes verify and exchange data securely by encryption technique without management and intervention from the trusted third party.

Blockchain technology applies decentralized shared ledger encryption technology, distributed consensus algorithm and economic incentive mechanism to accomplish peer-to-peer trade, coordination and cooperation in decentralized system, which provides an effective strategy to avoid high cost, low efficiency, and unsafe data storage in centralized platform.

9.1.2 Characteristics of Blockchain Models for Cyber-Credits in Social Manufacturing

In SMN, SMRs scattering over different regions are aggregated by DSMP to cooperate to satisfy personalized demands. To assist SMRs in selecting reliable partner without the trusted third party, modeling of SMR credit is necessary. As shown in Fig. 9.3, firstly, blockchain data of SMRs is mapped to a comprehensive indicator system containing service capacity, assets, innovation, teamwork, delivery, etc. Then, a set of assessment algorithms are developed to calculate credit of SMR and it can be calculated as

$$\mathrm{C} = \sum_{i=1}^{n} \lambda_i I_i \tag{9.1}$$

where C denotes credit of enterprise, I_i denotes quantification of inference indicator i and λ_i denotes the weight of the indicator i, n denotes the mount of indicators. And, a self-change mechanism is established to regulate credit according to daily activity information from SMR in SMN, for example if an SMR can't deliver as contracted, its credit will be decreased. Moreover, some classification algorithms (e.g., artificial

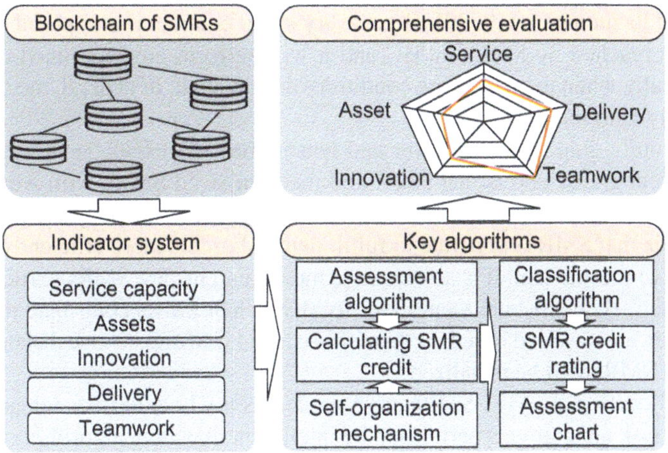

Fig. 9.3 Modeling of manufacturing cyber-credits

neural network (ANN), support vector machine (SVM), decision trees) are applied in SMR credit rating from different perspectives in concerns of different outsourcing demands. Finally, assessment charts are given to provide a much more indicative view of SMR credit for SMRs.

Manufacturing cyber-credits is a codified compulsory, deterrent, and incentive mechanism governed autonomously by a smart contract system based on the distributed blockchain technology. In SMN, cyber-credits act as the trusted third party to prevent defaults and frauds during manufacturing service interactions and bridge the credit gap among SMRs. In SMN, prosumers, as initiators of demand orders, must publish reasonable crowdsourcing/outsourcing demands in SMN and bear the manufacturing service charges as contracted. MC/SMRs are the performers who satisfy the crowdsourcing/outsourcing demands, and their processing capability, product quality, product delivery time must be reliable. A decentralized social manufacturing platform (DSMP) is used as the implementer of cyber-credits to ensure SMN operate smoothly and responsibly, including supervising production activities, rating the credit of MC/SMRs based on their historical transaction information, balancing interests among MC/SMRs and so on.

9.2 Implementing Architecture of Blockchain Models for Cyber-Credits

In SMN, many SMRs form decentralized and peer-to-peer MC for certain purpose by self-organizing. MC is an ecosystem living on SMN and exists autonomously. Community autonomy is a strategy that organization and management of MC depend on a series of smart contracts. In MC, community decisions are mutually made and performed its members [19, 20]. Here, every smart contract is composed of a series of clauses reached by SMRs in MC and it will perform predetermined operations automatically when the triggering conditions are satisfied. In general, the strategy of community autonomy includes:

Community goal. It is the driver and reason that MC exists. In SMN, there are two different goals: one is that small and medium-sized SMRs with weak service capability aggregate together to promote their competence and bargaining power, the other one is that SMRs cooperate to fulfill demand orders more efficiently.

Incentive mechanism. It is an incentive measure to inspire SMRs participating in MC and actively contribute themselves to MC. When SMRs contribute themselves to MC, they are awarded based on their efforts. And if SMRs behave dishonestly, the malicious SMRs will be penalized.

Task allocation. In MC, SMRs are regarded as a whole to take demand orders, after that, task allocation is performed by intelligent matching algorithms to achieve the optimal configuration of SMRs.

Voting mechanism. It is a foundational and key decision-making means in community autonomy. Voting is a process that SMRs establish consensus about community decisions [21]. The process is performed by voting smart contract based on PoS, which provides an efficient and reliable implement of voting. The weight of vote from SMR depends on its credit in SMN.

9.2.1 Implementing Architecture of Blockchain in Social Manufacturing

The framework of manufacturing cyber-credits is illustrated in Fig. 9.4. It can be clearly seen that there are four layers in this framework, i.e., social manufacturing resources layer, credit guarantee layer, decentralized social manufacturing platform (DSMP) layer, and application layer.

In social manufacturing resources layer, SMRs are configured by cyber-physical-social systems (CPSS). An SMR is a set of manufacturing resources shared in SMN by an enterprise, including manufacturing shop, enterprise information system and social media. Firstly, manufacturing shop of enterprise is equipped with a variety of sensors and radio frequency identification devices (RFID) to carry out self-perception to real-time production activities. And then, manufacturing shop, enterprise information system and social media are integrated by WAN/LAN to accomplish connection of physical space, cyber space and social space. Finally, the manufacturing resources are encapsulated as an SMR with information interaction interfaces. By the interfaces, SMR can implement information interaction with other SMRs in SMN and its confidential information can be encrypted for that specified SMRs have permission to access to. This can extremely promote service interaction efficiently among SMRs.

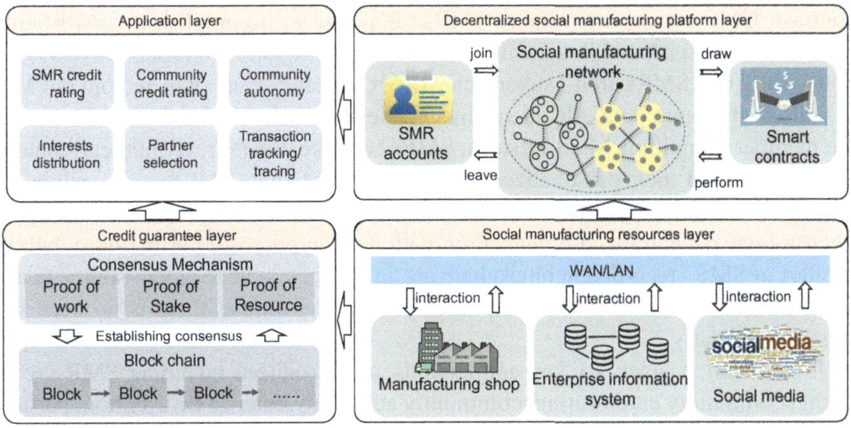

Fig. 9.4 Framework of manufacturing cyber-credits

In credit guarantee layer, the decentralized blockchain technology, a distributed shared encrypted-database serving as an irreversible and incorruptible public repository of information, is introduced to SMN as a guarantor of manufacturing cyber-credits. It facilitates SocialM in fully accomplishing decentralization and peer-to-peer. The blockchain can be compared as a chain in chronological order of block where SMR information especially manufacturing cyber-credits of SMRs/CMs in SMN is stored on. And a consensus mechanism composed of proof of work (PoW), proof of resource (PoR) and proof of stake (PoS) is utilized to establish consensus among SMRs/CMs and verify block information. If the block information is correct and not tampered, the block will be added to blockchain. PoW is a voucher how much SMRs contribute themselves to MCs and it is related to interest distribution among SMRs in MCs. PoR is a comprehensive cyber-credits proof of SMRs property including assets, manufacturing capacity, etc. And, PoS is applied to achieve consensus in MC decision-making according to votes from members whose weight of voting is determined by their cyber-credits.

In decentralized social manufacturing platform (DSMP) layer, an admittance condition is developed to create a safe, convenient, and worry-free environment for cross-enterprise collaborations in SMN. When a new SMR applies to join SMN, if it satisfies the admittance standard, it will be formally represented as a resource node of SMN and assigned an SMR account with an initial cyber-credit evaluated by PoR from other SMRs. SMRs admitted autonomously build connection relations and form different MCs by self-organization. The organization modes of SMRs exist as three patterns, i.e., one is called decentralized manufacturing community (DMC) with similar service capability, another is collaborative manufacturing community (CMC) around personalized demands, and the last one is dissociative individual temporarily in-dependent of the other SMRs, for example an SMR newly join SMN and haven't established business relations with any other SMRs. In DMC, SMRs with similar manufacturing capacity compete against each other for self-profit. By contrast, CMC forms when many SMRs cooperate to accomplish a crowdsourcing/outsourcing task and it is a temporary MC breaking up with order delivery. It is noteworthy that SMRs aren't subordinate to MC and they can opt in and opt out MC at will. At the same time, MC aren't independent of each other and an SMR can be the member in multiple MCs. Besides, SMRs can take orders freely but legitimately and put forward their personalized demands as prosumers. All these activities of SMRs are regulated by smart contracts which are programmable. It allow any MC to customize decentralized organization with its own purpose. Furthermore, public profiles of SMRs recorded in blockchain are easily accessible to any re-source node, which settles the problem of information asymmetry so that there is no need for trusted third party.

In application layer, six key applications are introduced including SMR credit rating, community credit rating, community autonomy, interest distribution, partner selection and transaction tracing/tracking.

9.2.2 Organizational and Operational Logic of Blockchain in Social Manufacturing

Figure 9.5 describes self-organizing logic of blockchain-based manufacturing service. SMRs, owning a certain manufacturing capacity, are encapsulated as service nodes to register in blockchain-based manufacturing service network. After that, they supply manufacturing service for prosumers and prosumers will pay virtual coin for their service. The virtual coin circulates in trading market by a transaction platform. Blockchain technology solves interest conflict and dispute among SMRs by codified contracts and guarantees SMRs having equal right and interest in manufacturing network. It enables that SMRs self-organize into dynamic manufacturing community, which promotes integration efficiency of SMRs by transforming serial production mode into parallel one. The prerequisite for blockchain-based manufacturing service is to establish a consensus computational algorithm for manufacturing service capacity and trading system.

Self-organizing logic of blockchain-based manufacturing service is reflected in two aspects: one is data registry, blockchain is credible and traceable so that it is used as a reliable distributed database for storing lifecycle information of manufacturing service; the other is self-organizing of manufacturing service, blockchain provide flexible script code system to support SMRs establishing diverse decentralized applications upon application layer for service demands, including information publish, supply and demand matching, credit evaluation, etc.

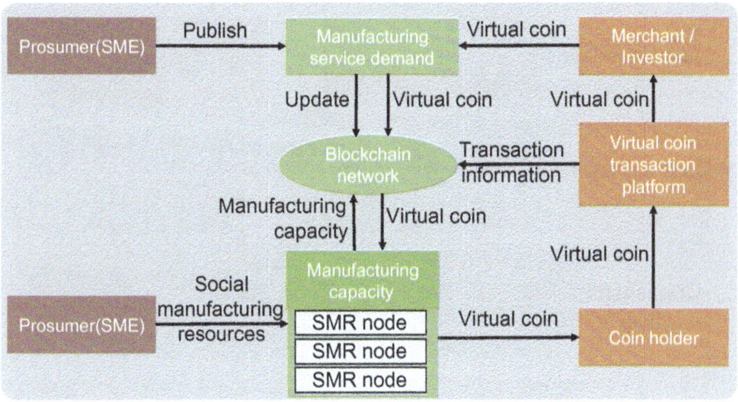

Fig. 9.5 Self-organizing logic of blockchain-based manufacturing service

9.3 Configuration of Blockchain in Social Manufacturing Contexts

9.3.1 Three Implementation Ideologies of Blockchain

According to application scenarios and service demand, blockchain-based manufacturing service can be divided into application pattern, namely, public service blockchain, consortium service blockchain, and private service blockchain.

(1) Public Service Blockchain

Public Service Blockchain is completely decentralized and maintained by encrypted currency system. It is implemented by economic incentive based on contribution of PoW and all nodes can participate in consensus of manufacturing service.

(2) Consortium Service Blockchain

Consortium Service Blockchain is partially decentralized or multi-centered so that it is applicable to organization or alliance of industrial park and enterprise group. Consensus of manufacturing service subjects to predefined node set. In this situation, block is verified by nodes from the set.

(3) Private Service Blockchain

Private Service Blockchain is completely centralized. It can be used to manage internal data of SMRs where SMRs is charge of the authorization of permission to write and read data. In contrast to Public Service Blockchain, Private Service Blockchain has three advantages: the first is high flexibility of discipline, SMRs can be modified blockchain rules easily according to their own requirements when they implement

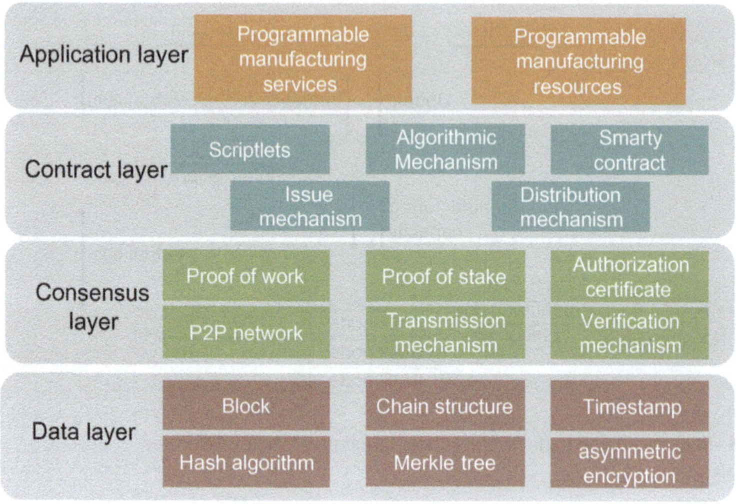

Fig. 9.6 Framework of blockchain-based manufacturing service

Private Service Blockchain; the second is high efficiency, verification of manufacturing service is simple that only small amount of nodes take part in that process; the last is authority protection, permission of write and read block data are controlled by SMRs which is beneficial to privacy protection. Based on the above points, the Private Service Blockchain is more suitable for implementation within SMRs. The Consortium Service Blockchain and Private Service Blockchain are not fully complied with the blockchain model because of the degree of decentralization, so there is no necessity of economic incentives.

9.3.2 *Formulation of Decentralized Trustiness in Social Manufacturing Contexts*

Figure 9.6 shows framework of blockchain-based manufacturing service. It is composed of data layer, consensus layer, contract layer and decentralized application layer.

Data layer encapsulates with data block, data encryption and timestamp technology. It functions as a virtual machine of blockchain to implement expression and encryption of block data, which provides support for upper layers to validate and transmit data. In blockchain network, every node has right to package interaction data into block by Merkle tree-based hash algorithm, and the block will be attached to the longest main chain with a timestamp. All blockchain data is recorded on the chain structure to guarantee its traceability. Figure 9.7 depicts the structure of block including block header and block body. The block header contains current version,

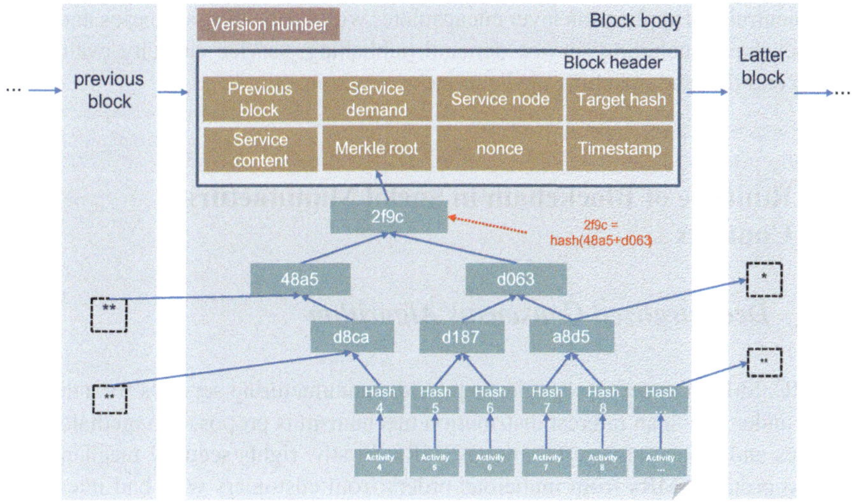

Fig. 9.7 Data structure of block

previous block address, service demand, service node, service content, hash value of current block, nonce, Merkle tree root and timestamp. Timestamp is a verification of block to ensure block data tamper-resistant. Merkle tree stores whole transaction data by hash algorithm and is convenient to check and query the data quickly.

Consensus layer consists of peer-to-peer network, data dissemination mechanism, data validation and consensus algorithm. SMR nodes is organized into peer-to-peer network by flatten topology and every node undertakes network routing, verifies/propagates data and discovers new node. Once a node mine block with manufacturing service data, the block will be broadcast to other SMR nodes to verify by the node. After other nodes receive the block, they will verify whether the block is valid or not. If the block is valid, the receiving node will disseminate the block in the network, otherwise the block will be discarded. Consensus algorithm build foundation for flexible programming, data operation, manufacturing server self-organizing and interaction. Consensus of blockchain-based manufacturing service accomplishes service demand, published on de-centralized shared ledger, by congregating service capacity of SMR nodes. It can be regarded as manufacturing task crowdsourcing among SMRs.

Contract layer includes manufacturing service demand publishing and matching, transaction mechanism, and diverse smart contract scripts and algorithms. In decentralized network, because SMR nodes attempt to maximize their own interests when they participate in manufacturing service, a reasonable crowdsourcing mechanism must be designed to regulate SMRs behavior and ensure blockchain network achieve overall goals. This also guarantees stability of blockchain consensus from SMRs nodes. Later section will focus on the above content.

Decentralized application layer encapsulates with application scenarios and cases of blockchain, including service demand publishing, service capacity evaluating, supply and demand matching, and service transaction, etc.

9.4 Runtime of Blockchain in Social Manufacturing Contexts

9.4.1 *Decentralized Consensus Algorithm*

In CMC, SMRs cooperate together to provide manufacturing services for a demand order, under which an interest distribution mechanism is proposed to mediate profit disputes and balance interests among SMRs. Firstly, rights security mechanism is used to protect SMRs from malicious orders from customers with bad intentions. After that, a performance index system is developed to evaluate the contributions of SMRs to CMC in manufacturing services. At last, a strategy of interest distribution is proposed and it can be formatted as

$$E_k = M \times \sum_{i=1}^{n} \left(\tau_k^i \times A_k^i\right) \tag{9.2}$$

where E_k denotes earnings of SMR_k, M denotes gross earnings of CMC, n denotes the amount of influence factors,A_k^i denotes the ith factor of profits of SMR_k,τ_k^i denotes the ith weight of factor A_k^i. It is noted that profit distribution is encapsulated as smart contracts. At the beginning of demand order, all SMRs participating in the order including prosumers need to cooperate to draw a set of smart contracts that elaborate principle of interest distribution. After product delivery, the smart contracts will be triggered and interest distribution will be implemented automatically based on service tasks of SMRs, which greatly reduce profit disputes among SMRs. Simple protocols of the smart contracts can be described as follows.

All order participants including SMRs and prosumers make up CMC and sent some guarantee deposit to a specified contract account as penalty for default. Moreover, the prosumers have to send extra charges of manufacturing service to the contract account to prevent from prosumers repudiation. The participants draw interest distribution smart contract (IDSC) together by a voting mechanism based on PoS. Once an SMR accomplishes its manufacturing task, it publishes a proposal about manufacturing information, and members of the CMC will verify the proposal. If over 51% members agree the proposal, a PoW will be allocated to the SMR account and the proposal will be stored on blockchain. When the demand order is delivered, the IDSC will verify PoWs of SMRs and distribute charges to corresponding SMR accounts. If a default occurs, the IDSC will perform indemnity contracts to confiscate guarantee deposit from confiscate account as compensation for the honesty SMRs.

Under self-organizing of blockchain-based manufacturing community, it is essential for SMRs to establish a unified agreement of economic incentives and participate in forming stability consensus of blockchain. An operation method is presented for blockchain-based manufacturing community from consensus and contract layers based on the underlying data layers of blockchain.

The consensus process of blockchain is a crowdsourcing process of manufacturing task. It is an important prerequisite for smooth operation of blockchain that how to design a reasonable crowdsourcing mechanism to facilitate SMRs to initiatively supply manufacturing service. The operation of blockchain-based manufacturing community is based on that SMRs cooperate to participate in PoW, a consensus algorithm, to accomplish interaction and transaction of manufacturing service. In general, SMRs contribute their own manufacturing resources to satisfy a service demand. When delivery is completed successfully by SMRs, the SMRs will be authorized to write data on new block and be awarded a certain amount of virtual economic incentives. After the new block is verified, it will be linked to the main blockchain in chronological order. PoW-based consensus algorithm is shown in Table 9.1.

The above algorithm can be described as follows: SMR nodes broadcast new manufacturing service demands to blockchain network and other SMR nodes stored manufacturing service data on a new block; every SMR node matches its manufacturing capacity with the manufacturing service demands and obtains PoW verified by the algorithms from contract layer; then, SMR nodes broadcast the new block to the blockchain network; other SMR nodes will admit that the new block is valid if all data of the new block is valid; other SMR nodes accept the new block and appends the hash value of the block to the hash value of original blockchain to form new extended chain.

The key of the algorithm is Step 3.3 that PoW is verified by the algorithms from contract layer. This chapter proposed a service capacity calculating based Pow, which avoid waste of computing resources and integrate supply and demand matching function at the same time. Taking 3D printing manufacturing service for example, service capacity calculation is as follows:

$$\mathrm{PoW} \propto k \times Q \times \mathrm{V} \times \frac{1}{P} \times \mathrm{M} \tag{9.3}$$

where k denotes cost marginal effect coefficient; Q refers to order quantity; V represents total geometric volume of parts; P denotes parts quality requirements coefficient, the higher the requirements, the larger the value; M represents parts printing difficulty coefficient, the more difficult, the larger the value.

Table 9.1 Process of PoW-based consensus algorithm

Algorithm name: PoW-based consensus algorithm

Input: manufacturing service demands
Output: manufacturing service self-organizing blockchain
Start
Step 1: Broadcast new manufacturing service demands to blockchain network;
Step 2: SMR nodes collect new manufacturing service demands and mark on blockchain;
Step 3: SMR nodes acquire PoW and permission to mark on blockchain by offering manufacturing service;
  Step 3.1: Real-time search for unacknowledged manufacturing service demands in blockchain network to form a demand set;
  Step 3.2: Match manufacturing service demands with manufacturing capacity and make an application to wait confirmation from demander;
  Step 3.3: Undertake and complete the manufacturing service demands, and obtain PoW;
Step 4: SMR nodes broadcast obtained PoW to blockchain network;
Step 5: SMR nodes acquire permission to mark on blockchain;
Step 6: Get validity verification from other SMR nodes, and other SMR nodes accept the block;
Step 7: SMR nodes calculate Merkle-root and records transaction data into block header;
Step 8: SMR nodes accept the block and appends the hash value of the block to the hash value of original blockchain;
End

9.4.2 *Smart Contract*

As shown in Fig. 9.8, smart contract system is a set of codified contractual clauses supported by the distribution blockchain technology. It is self-executing and self-enforcing, and composed of public smart contracts and customized smart contracts. The public smart contracts are fundamental and general computer proposals to maintain SMN running normally, such as admittance condition and incentive mechanism. The customized smart contracts enable MC autonomy that DMCs can customize their own voting mechanism for different purposes and CMCs draw enforceable reasonable interest distribution smart contract according to different demand orders. On the one hand, this system provides guidance for disorganized SMRs aggregating and regulates SMRs into autonomous SMN. On the other hand, it translates contract documentations into codified protocols, which greatly prevents SMRs from fraud and protects interests of SMRs from default because of its self-trigger mechanism.

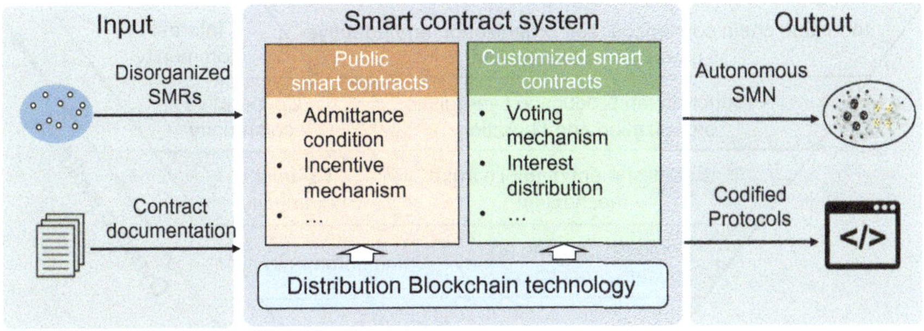

Fig. 9.8 Smart contract system

9.4.3 Evolution Mechanism of Blockchain-Driven Cyber-Credits

Social manufacturing is supported by manufacturing community around individualized demands to integrate manufacturing capacity from prosumers for personalized product development. Personalized product development requires cooperation of a group of prosumers and involves a series of coordination and negotiation of manufacturing service. The generation of manufacturing community is a dynamic evolution process.

Figure 9.9 shows that the generation and dynamic evolution process of a manufacturing community. They consist of five stages, namely, resource community, value community, co-trust community, cooperative community and interest community. The characteristics of manufacturing community are identified as follows: prosumer aggregates complementarily around personalized product to optimize configuration of manufacturing resources; product manufacturing process self-organizes and evolves with product lifecycle. In addition, manufacturing community enhances the trust among community members driven by economic interest. During adaptive process, manufacturing community achieves a blockchain-based consensus, and an ecosystem where production collaboration is accomplished by self-organizing is formed ultimately.

9.5 Examples

This chapter constructs an Ethereum-based decentralized application for self-organization of manufacturing community to support blockchain-based manufacturing community.

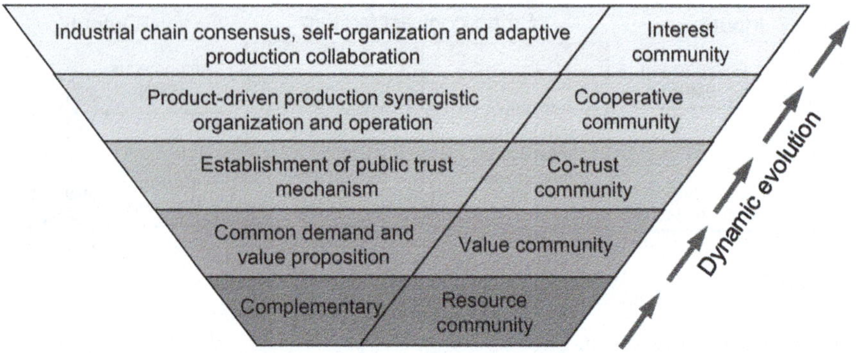

Fig. 9.9 Dynamic evolution process of manufacturing community

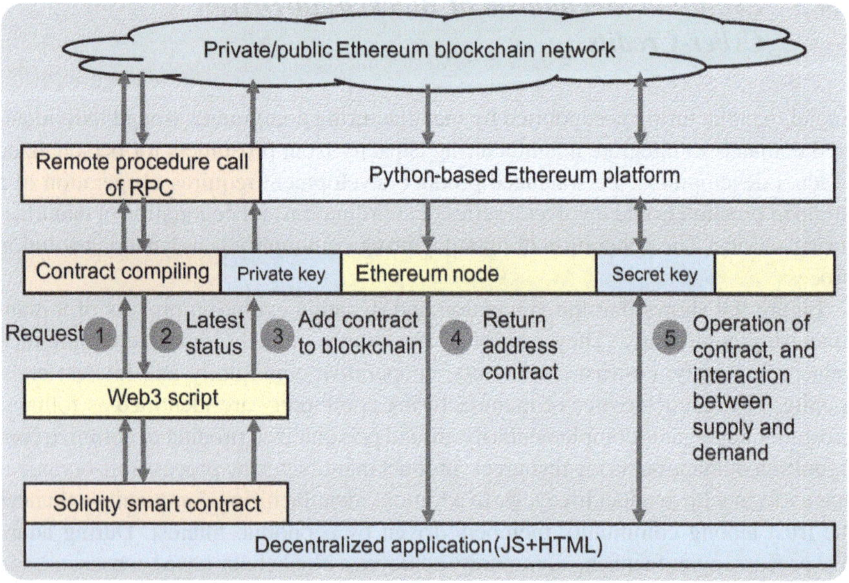

Fig. 9.10 Operation logic of DAPP for self-organization of community

9.5.1 Architecture and Runtime

Figure 9.10 shows the implementation logic of Ethereum-based decentralized application (DAPP). Based on Ethereum, Python programming language is applied to implement internet interaction of blockchain. Smart contracts are written by the popular and stable Solidity language.

Development of smart contract is based on MixEthereum IDE platform shown in Fig. 9.11. After Solidity contract is compiled and sent to test network, RPC (Remote Procedure Call) will create a private/public blockchain on the test network. Other

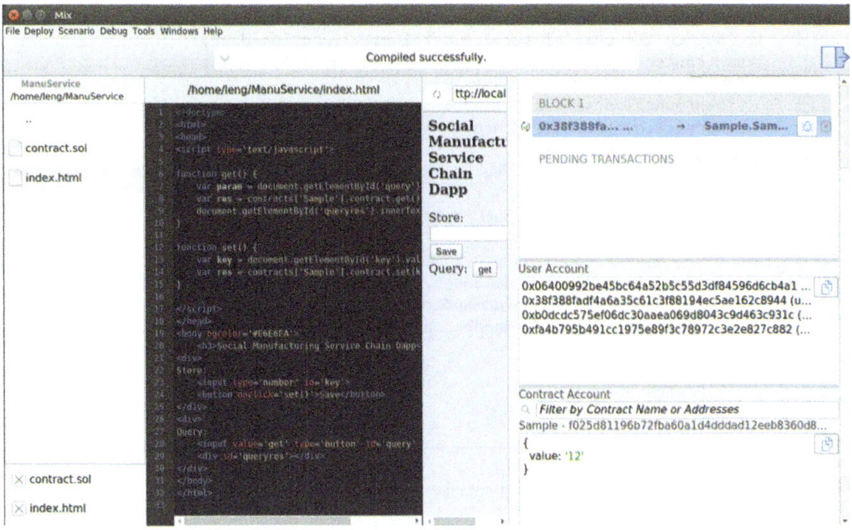

Fig. 9.11 Smart contract development based on MixEthereum IDE

nodes will synchronize the entire blockchain and are responsible for executing smart contract. Each SMR node has a public key and a private key. SMR node can create a digital signature with its private key, and other SMR nodes can use the public key of the SMR node to verify the digital signature. When an SMR creates an Ethereum node, the SMR will be assigned an address as its public key and the corresponding private key is kept by the SMR privately. In application layer, Web3 scripting language is used to support for human-computer interaction DAPP.

9.5.2 A Demonstrative Decentralized Application

As shown in Fig. 9.12, this chapter gives an example that a manufacturing service smart contract is developed based on Ethereum (a Turing complete smart contract scripting language), which defines a service demand issued by an outsourcer and requires that suppliers are up to three. Suppliers can apply for matching demands or cancel the application. All manufacturing service transaction rules are specified in smart contract. Smart contract is self-executing and programmable, which enables it to encapsulate interaction and transaction rules of manufacturing service, function as a software agent of blockchain-based manufacturing community, support diverse distributed applications, and self-organize decentralized manufacturing community. This chapter will provide some DAPP to support for blockchain-based manufacturing community.

```
contract(Service, function(accounts) {
  it("Notice the initial service settings", function(done) {
    var conference = Service.at(Service.deployed_address);
    Service.new({ from: accounts[0]  })
    .then(function(service) {
      service.quota.call().then(
          function(quota) {
            assert.equal(quota, 3, "The capabity doesn't match the demand!");
          }).then( function() {
            return service.numOrders.call();
          }).then( function(num) {
            assert.equal(num, 0, " Cannot generate more orders!");
            return service.demander.call();
          }).then( function(demander) {
            assert.equal(demander, accounts[0], "Demander doesn't match!");
            done();
        }).catch(done);
      }).catch(done);
    });
  });
```

Fig. 9.12 An example of manufacturing service smart contract

```
it("Outsource a part of the service", function(done) {
  var c = Service.at(Service.deployed_address);
  Service.new({ from: accounts[0] }).then(
    function(service) {
      var orderPrice = web3.toWei(.05, 'ether');
      var initialBalance = web3.eth.getBalance(service.address).toNumber();
      service.matchOrder({ from: accounts[1], value: orderPrice }).then(
        function() {
          var newBalance = web3.eth.getBalance(service.address).toNumber();
          var difference = newBalance - initialBalance;
          assert.equal(difference, orderPrice, "Difference should be what was sent");
          return service.numRegistrants.call();
      }).then(function(num) {
          assert.equal(num, 1, "there should be at least one registrant");
          return service.registrantsPaid.call(accounts[1]);
      }).then(function(amount) {
          assert.equal(amount.toNumber(), orderPrice, "Demander's paid but is not listed");
          done();
      }).catch(done);
  }).catch(done);
});
```

Fig. 9.13 A code fragment of Ethereum-based smart contract

Based on the manufacturing service smart contract shown in Fig. 9.12, the first step in implementing smart contract is to draft protocols for supply and demand matching. As shown in Fig. 9.13, this chapter establishes supply and demand matching process in the smart contract Service. The matching process is divided into two steps: outsourcer and supplier confirm the match between manufacturing capacity and service demands; check the number of suppliers whether to exceed the default maximum, and if more than the default, a mismatching will be returned. In terms of programming language, the Solidity language simplifies the traditional IF-ELSE syntax for the THEN-RETURN form, thereby reducing program nesting, flattening the code and allowing the call results to be returned asynchronously.

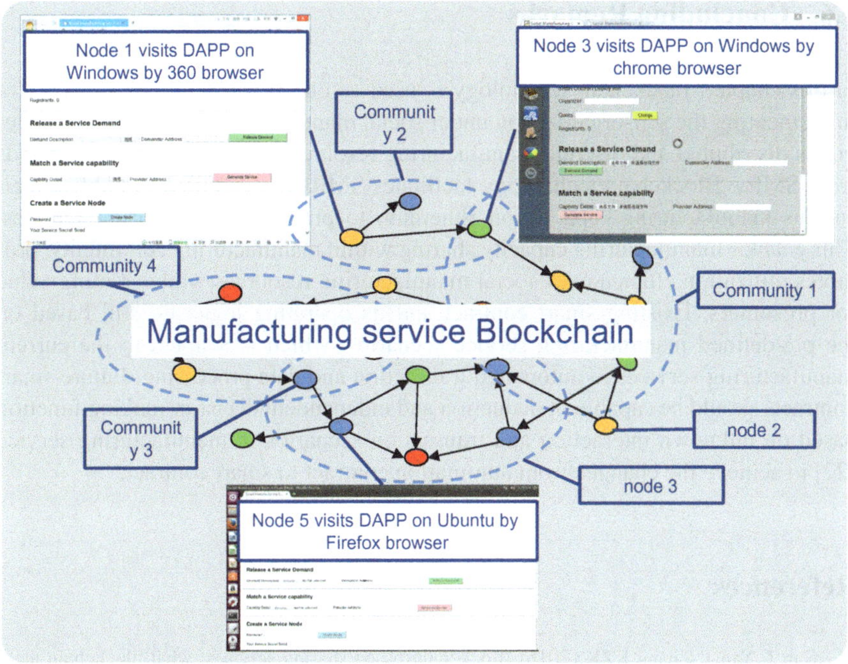

Fig. 9.14 Reliability of each system

The second step is to define the confirmation rules for service transaction. In smart contract Service, Fig. 9.13 describes that the SMR nodes (accounts[0]) and (accounts [1]) undertake partial manufacturing task at the price of order Price in the service transaction. After the service transaction is confirmed, smart contract will check the two SMR accounts and update the list of suppliers that take the manufacturing service demands published by outsourcers.

On the basis of the first two steps, next step is to develop DAPP based on smart contract. This chapter presents a sample code that reads transaction details of manufacturing service from the smart contract Service by Web3, as shown in Fig. 9.13. DAPP provides a user-friendly interface for manufacturing service smart contract, and is used to store and read transaction data from manufacturing service in decentralized blockchain network. As shown in Fig. 9.14, DAPP can be run on any SMR node to interact with Ethereum. Each node can submit transaction data from manufacturing service to blockchain and read the blockchain data from other SMR nodes.

9.6 Concluding Remarks

In this chapter, blockchain technology is used as one fundamental framework for implementing the self-organization under social manufacturing context, which integrates distributed and discrete manufacturing resources to satisfy service demands from SMEs. Blockchain technology facilitates SMRs to quantify and publish their surplus manufacturing capacity on Ethereum to provide manufacturing services. This enables manufacturing capacity sharing within manufacturing community, promotes utilization efficiency of social manufacturing resources while creating value for prosumers. Existing smart contract and its operating logic are still based on the pre-defined manufacturing service scenarios, which is adapted to the current manufacturing services in automated transaction and data processing. Future smart contracts should be capable of deduction and independent decision-making function based on unknown interaction and transaction scenarios in manufacturing service [22] to achieve the changes from automation contract to smart contract.

References

1. Sun J, Yan J, Zhang KZK (2016) Blockchain-based sharing services: what blockchain technology can contribute to smart cities. Financ Innov 2(1):26–34
2. Jiang PY, Leng JW, Ding K (2016) Social manufacturing: a survey of the state-of-the-art and future challenges. In: IEEE International Conference on Service Operations and Logistics, and Informatics, Beijing, China, 10–12 July 2016
3. Jiang PY, Ding K, Leng JW (2016) Towards a cyber-physical-social-connected and service-oriented manufacturing paradigm: Social Manufacturing. Manuf Lett 7:15–21
4. Jiang PY, Leng JW (2017) The configuration of social manufacturing a social intelligence way toward service-oriented manufacturing. Int J Manuf Res 12(1):4–19
5. Jiang PY, Leng JW, Ding K (2016) Social manufacturing: a survey of the state-of-the-art and future challenges. In: IEEE International Conference on Service Operations and Logistics, and Informatics, Beijing, China, 10–12 July 2016
6. Jiang PY, Leng JW, Ding K, Gu P, Koren Y (2016) Social Manufacturing as a sustainable paradigm for mass individualization. Proc Inst Mech Eng, Part B: J Eng Manuf 230(10):1961–1968
7. Leng JW, Jiang PY, Zhang F, Cao W (2013) Framework and key enabling technologies for social manufacturing. Appl Mech Mater 2382(312):498–501
8. George R, Paul D, Jurgenson N (2012) The coming of age of the prosumer. Am Behav Sci 56(4):379–398
9. Zhou Y, Xiong G, Nyberg T (2016) Social manufacturing realizing personalization production: A state-of-the-art review. In: IEEE International Conference on Service Operations and Logistics, and Informatics, Beijing, China, 10–12 July 2016
10. Aaron W, Primavera DF (2015) Decentralized blockchain technology and the rise of lex cryptographia. Web Pages, available at SSRN: https://ssrn.com/abstract=2580664
11. Apte S, Petrovsky N (2016) Will blockchain technology revolutionize excipient supply chain management? J Excipients Food Chem 7(3):76–78
12. Azaria A, Ekblaw A, Vieira T, Lippman A (2016) MedRec: using blockchain for medical data access and permission management. In: The 2nd International Conference on Open and Big Data, Vienna, Austria, 22–24 August 2016

13. Yli-Huumo J, Ko D, Choi S, Park S, Smolander K (2016) Where is current research on blockchain technology?—a systematic review. PLoS ONE 11(10):e163477
14. Satoshi N (2008) Bitcoin: a peer-to-peer electronic cash system. Web pages, available at: http://www.academia.edu/download/32413652/BitCoin_P2P_electronic_cash_system.pdf
15. Andrychowicz M, Dziembowski S, Malinowski D, Mazurek L (2014) Secure multiparty computations on bitcoin. In: Security and privacy. IEEE, pp 443–458
16. Richard D, Gareth O (2015) Rep on the block: a next generation reputation system based on the blockchain. In: International Conference for Internet Technology and Secured Transactions, London, United Kingdom, 14–16 December 2015
17. Leng JW, Jiang PY (2016) A deep learning approach for relationship extraction from interaction context in social manufacturing paradigm. Knowl-Based Syst 100:188–199
18. Leng JW, Jiang PY (2017) Mining and matching relationships from interaction contexts in a social manufacturing paradigm. IEEE Trans Syst Man Cybern: Syst 47(2):276–288
19. Leng JW, Jiang PY, Ding K (2014) Implementing of a three-phase integrated decision support model for parts machining outsourcing. Int J Prod Res 52(12):3614–3636
20. Leng JW, Jiang PY, Zheng M (2017) Outsourcer–supplier coordination for parts machining outsourcing under social manufacturing. Proc Inst Mech Eng, Part B: J Eng Manuf 231(6):1078–1090
21. Xu X, Du Z, Chen X (2015) Consensus model for multi-criteria large-group emergency decision making considering non-cooperative behaviors and minority opinions. Decis Support Syst 79:150–160
22. Huckle S, Bhattacharya R, White M, Beloff N (2016) Internet of things, blockchain and shared economy applications. Procedia Comput Sci 98:461–466

Chapter 10
Configuration of Social Manufacturing System

Jiewu Leng and Pingyu Jiang

10.1 Social Manufacturing Space and Communities

Social manufacturing (SocialM) paradigm involves crowd intelligences from cyber-physical-social space and social organizations (e.g., communities) to enable the social interactions among prosumers and the organic connections of socialized manufacturing resources (e.g., machine tools, design software, measurement equipment, and sensors) for co-creating individualized products.

10.1.1 Definitions and Descriptions

Definition 10.1 Social Manufacturing Space is a cyber-physical-social-based logically organized set of prosumers which have their own customized community spaces (CCSs), and can provide rapid product and manufacturing service retrieving and matchmaking, service evaluation, manufacturing activities sharing, information interaction for service demanders and providers [1–3].

Social manufacturing space includes one global space (GS), several local spaces (LSs), and various CCSs. GS is the space where all the prosumers share their public process dynamics and communicate with each other through decentralized social media or public social platform [4, 5], but LSs are the spaces where prosumers exchange cooperated manufacturing information with their on-going partners efficiently. Based on different rules of organization and classification granularity, social manufacturing space can further be sorted as follows: Core enterprises-based social manufacturing communities (ce-SMC), industrial chain-based social manufacturing communities (ic-SMC), and peer-to-peer type-based social manufacturing communities (pt-SMC). There are many similar crowdsourcing or community-based concepts including Maker Manufacturing [6], Peer-Production [7], Open Production [8],

P. Jiang, *Social Manufacturing: Fundamentals and Applications*, Springer Series in Advanced Manufacturing, https://doi.org/10.1007/978-3-319-72986-2_10

Crowd Manufacturing [9]. These definitions are different from other cloud platform-based social manufacturing concept.

Similar to the industrial park, ce-SMC is a traditional community where crowd of prosumers provide different production capabilities and manufacturing services for the core enterprise. However, the difference is that prosumers in ce-SMC are logically aggregated and geographically distributed. Contrary to the virtual enterprise and enterprise alliance concept, ic-SMC is a kind of vertical prosumers-centered community which focuses on the different value-added positions and changes the corporation mode especially for small and medium-sized enterprises. Pt-SMC is another horizontal prosumers-centered community where similar prosumers gather and provide similar manufacturing services for demanders [10, 11].

10.1.2 Framework and Operational Logic

Under the context of social manufacturing, customized community spaces (CCSs) of prosumers, where massive SMRs including both hardware and software are well self-organized.

Dynamic production capability of SMR should be configured firstly. According to different respective of organization rules, the CCSs can be embedded into different social manufacturing communities (SMCs), in which prosumers can interact with each other under the support of many social media and platforms. When an outsourcing or crowdsourcing task comes, a set of candidate prosumers from manufacturing community is gathered and matched proactively. After the matching is done and an outsourcing/crowdsourcing order is formed, the following-up cooperative manufacturing activities would be performed in the SMC [12, 13].

Social manufacturing space is identified by adopting two phases of data mining in social manufacturing paradigm, as shown in Fig. 10.1. In the first phase, a *semi-supervised learning algorithm* to automatically extract relations from the unstructured context data is developed. As a result, a social manufacturing space with the *heterogeneous social manufacturing network* manner for supporting the further demand-capability matchmaking is identified. In the second phase, for handling the complexity of networks and performing network search and inference, the system finds the potential relations from the established heterogeneous network by a *probabilistic graph model*-based network matching algorithm. It returns data sub-networks (potential relations) rank ordered by probability of the match, and assigns inference labels based on the semantics of matched relations. Finally, the system can enable the group level demand-capability matching in the cross-enterprise manufacturing outsourcing decision making progress.

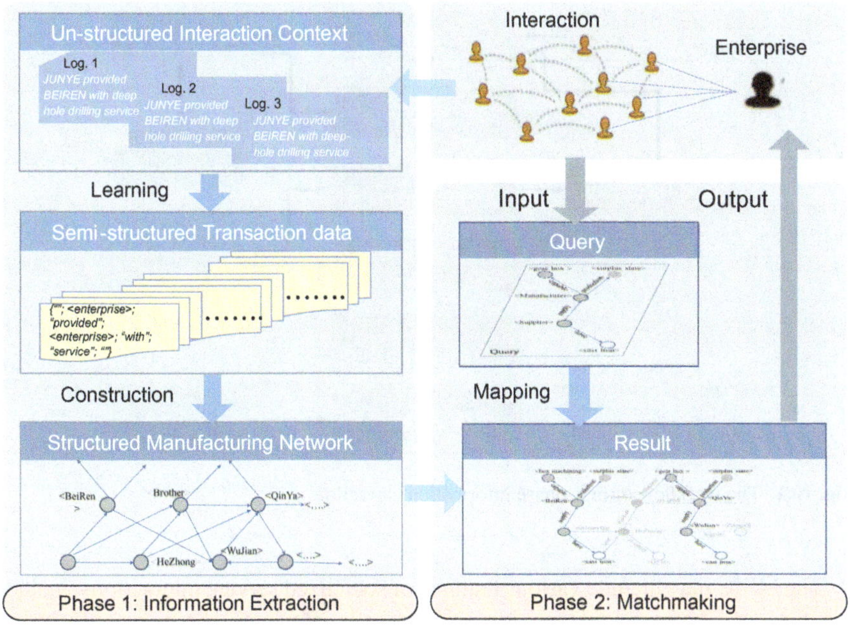

Fig. 10.1 The framework of social manufacturing space identification

10.1.3 Identification of Demand-Capability Candidates Through Interaction Contexts

The growing use of information and communication technologies, social media and platforms have generated explosion of massive interaction data forming Manufacturing Service Interaction Contexts (MSIC), which underlies a social manufacturing space comprised of highly flexible, multi-dimensional, and cooperative manufacturing relations among prosumers [14, 15]. This situation consequently leads two challenges for decision-makers.

The first one is the mining of various relations and interactions that occur across prosumers. This information can be used by decision-makers to understand how social manufacturing resources and capabilities are distributed and matched in the space, and identify potential partners to outsource or crowdsource manufacturing processes to them. Secondly, matchmaking of manufacturing service demand and capability goes beyond the old logic of "one enterprise produce one product"; a product goes through a crowd of prosumers involved in a number of coordinated manufacturing processes before finished. These crowd of prosumers or interaction relations can be identified as patterns from the network viewpoint. Because the relational and event data of the influences, interactions, and dependencies among prosumers is of highest significance, the decision-makers need automated decision-support to identify potential group-level manufacturing relations in the MSIC.

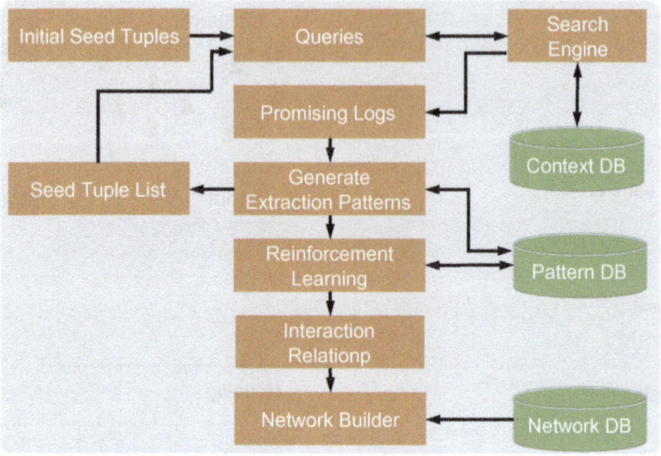

Fig. 10.2 The workflow of first-phase information extraction

The MSIC may include various features of socialized service interactions, including metadata, structural information, and semantic information (e.g., manufacturing specialization, supply connections, or service interactions). The key point is to extract the relevant information in related cross-enterprise manufacturing context instead of the entire document. The context record usually includes moments and events such that (1) each moment contains a manufacturing demand or capability, (2) each event represents a manufacturing service activity (i.e., a well-defined step in the machining process), (3) each moment implies an interaction instance (e.g., a manufacturing outsourcing order), and (4) each event includes two or more entities (e.g., the enterprise executing or initiating the activity). Many social media or decentralized platforms such as cloud ERP will produce information in these specific forms. An information extraction algorithm is proposed based on iterative tuple update and approximate pattern reasoning, as shown in Fig. 10.2.

Beginning with a few initial seed tuples, the algorithm generates patterns in an unsupervised manner and augments the seed list by capturing similar instances of the same category from unlabeled context data. The context database captures the interactions in social media and decentralized platform. The seed tuples consist of a number of entity-relations to initially extract related logs in context database which contains the relations in the seed tuples. Then, if a log does contain the seed tuples in one of sentences, it is labeled as a positive log. These labeled relations are further used to generate extraction patterns, which are stored in pattern database. The generated patterns are converted and added to seed tuples to retrieve promising logs. In the next iteration, the updated query list is derived from the generated patterns to further extract potential relations. Finally, the extracted interaction relations can be aggregated in a heterogeneous network database that contains organizational, temporal, informational, and social aspects in the social manufacturing space. The main flow of this algorithm is to perform the relations mining on the contexts to iteratively obtain extraction patterns and then use the patterns as query list for capturing new useful data in the social manufacturing space.

The critical point of data mining is a sustainable reinforcement technique that uses the extracted patterns to mine more tuples that belong to the same category context data. The patterns of the related entities are depicted in a flexible way. And *ELIZA* type pattern is introduced and improved to exploit limited syntactic and semantic information. For instance, a manufacturing enterprise interaction pattern is a tuple that comprises of two prosumers that correspond to a certain interaction, which could be defined as a 7-tuple:

$$T_i = < pref; e_1; inf_1; e_2; inf_2; rel\_tag; suf > \tag{10.1}$$

where $pref$, inf_1, inf_2, and suf are vectors associated weights. $pref$ is the prefix of sentence before e_1; inf_1 and inf_2 are the infix of sentences among e_1, e_2, and $rel\ tag$; and suf is the suffix of sentence after the relation tag (i.e., rel_tag). For each enterprise pair tuple <e_1, e_2>, it finds the sentences which connects e_1 and e_2 close to each other to generate patterns. For example, an expression of *"JUNYE provided BEIREN with deep-hole drilling service"* is characterized as a 7-tuple pattern *{""; <enterprise>; "provided"; <enterprise>; "with"; "service"; ""}*. These patterns can be used to depict those sentences in the text that contains various interactions.

Since there are a limited number of patterns as the initial seed tuples, the system retrievals the context sentences that match the patterns to generate more new seed tuples, and then starts the next iteration by using these new tuples to capture new promising patterns. Here, we use a *sentence alignment method* to match similar patterns, and then category them for patterns learning. The match between two 7-tuples (T_i, T_j) is defined as

$$match(T_i, T_j) = \sum_{s \in (pref, inf1, inf2, suf)} w_s * sim_s(i, j) \tag{10.2}$$

The similarity measurement between two *sentence segments* such as $pref_i$ and $pref_j$ is based on a *sentence alignment function* as shown in Eq. (10.3). The similarity $sim(i, j)$ between two sentences $X = (x_1, \ldots, x_i, \ldots, x_m)$ and $Y = (y_1, \ldots, y_j, \ldots, y_n)$ is defined as the score of the optimal alignment between the segment from x_1 to x_m of X and the segment from y_1 to y_n of Y.

$$sim(i, j) = \max \begin{Bmatrix} sim(i, j-1) + \log \frac{p(\text{"}-\text{"}, y_j)}{p(\text{"}-\text{"}) * p(y_j)} \\ sim(i-1, j-1) + \log \frac{p(x_i, y_j)}{p(x_i) * p(y_j)} \\ sim(i-1, j) + \log \frac{p(x_i, \text{"}-\text{"})}{p(x_i) * p(\text{"}-\text{"})} \end{Bmatrix} \tag{10.3}$$

where $p(x_i)$ refers to the presence probability of word x_i ("$-$" denotes black space), and $p(x_i, y_j)$ denotes the probability that x_i and y_j locate at the same position in two sentence segments. With respect to segment X of a length m and Y of a length n, the maximal local alignment from $(m+1)*(n+1)$ scores can be identified as the overall similarity score $sim(i, j)$; And thus, this sentence alignment method is flexible and can be easily implemented.

A new tuple T_{new} will be generated if there is a pattern P_i such that $match(T_{new}, P_i)$ is greater than the pre-specified threshold. Iteratively, more and more new tuples will result from different patterns associated with a match degree, which is the basis for deciding what new tuples can be added to the *Relation Table* that is being constructed.

The generated tuples and patterns should be evaluated for the feedback control in order to ensure their high quality. Only highly confident tuples can be used as seeds in the next iteration. The confidence of a tuple is based on the fitness as well as the number of the patterns that generate it, and thus it will be of confidence if it is inferred by many highly reliable patterns.

To get rid of unreliable patterns and tuples from further mining, a metric to evaluate reliable pattern P_i is introduced as

$$score(P_i) = \frac{U_i}{N_i} * \log(U_i) \tag{10.4}$$

where U_i is the number of unique tuples extracted by P_i and N_i is the total number of unique tuples extracted by P_i. Considering an extracted tuple T_j and the set of matching patterns $P = \{P_i, i \in (1, 2, \ldots, N)\}$ used to generate T_j, the confidence of T_j is evaluated as

$$conf(T_j) = 1 - \prod_{i=1}^{N}\left(1 - score(P_i) * match(P_i, T_j)\right) \tag{10.5}$$

The confidence calculation is crucial for the precision of tuple learning. Based on the confidence of the candidate tuples, we discard all tuples of low confidence in each iteration. This filtering strategy can significantly improve the quality of the learned patterns and tuples. By repeating above procedures, the Relation Table is dynamic constructed.

10.1.4 Constructing of Social Manufacturing Space

To exploit the extracted relation table more effectively, it must be further aggregated using a standardized semantic heterogeneous graph defined as *Heterogeneous Social Manufacturing Network* (HSMN). The HSMN involves multiple typed entities and multiple typed edges denoting different relations. Here, a semantic HSMN graph is formulated as

$$G = (V, E, vt, et) \tag{10.6}$$

where $V = \{v_1, v_2, \ldots, v_n\}$ denotes the vertex set and $E = \{e_{ij} | e_{ij} = v_i * v_j, 1 \leq i, j \leq n\}$ denotes the edge set. vt denotes the attributes for vertex v, such as activities, roles, properties, and features. Similarly, et denotes the attributes for edge e, encoding different type of semantic relations, temporal and spatial dependencies, interactions, and influences. HSMN is unique since it can conduct both semantic and structural abstraction.

10.1.5 Configuration of Social Manufacturing Space and Communities Based on Role Controlling

As the outsourcing tasks in CMC system are limited in a sharp scope, i.e. process machining outsourcing and workpieces machining outsourcing, the main roles involved in the system can be divided into three classes such as platform side, service provider side and service requestor side. In all, there are five kinds of roles in CMC belonging to the three classes like cloud platform providers, platform technical service providers, machining service providers, logistics service providers and service requestors. In addition, different from the operation mode of cloud computing that all services can be provided and employed through internet, actual logistics is essential for the cloud machining community because object flows (e.g. row material flows, WIP flows and finished workpieces flows) are inevitable. Thus, the logistics services provider is considered another important role in CMC. The five kinds of roles involved in CMC are listed and described in Table 10.1.

In addition, different roles have different jurisdictions in CMC. The corresponding jurisdictions of a specific role will be provided on a related cloud-desktop with a series of function and application forms. Here, the cloud-desktop is a customized desktop on which the corresponding applications will be listed when one login the system with a specific role. He/she can use these applications to fulfill some special functions very conveniently. Furthermore, as an individual may have more than one role in CMC, he/she can also re-login the system with another role.

10.2 Key Algorithms for Coordination Decision-Making of Social Manufacturing Service Relations

10.2.1 Concepts of Coordination Decision-Making

Coordination decision-making process comprises of partner selection and order coordination. The partner selection process is to identify prosumers with the highest potential for meeting a manufacturer's need consistently and at an acceptable overall

Table 10.1 Roles involved in CMC

Roles	Definitions
Cloud platform provider	A cloud platform provider is a company or an organization that provides the cloud machining platform for both machining services providers and services requestors. It takes charge of the overall plan of the platform, and outsources or crowdsources their IT issues to other technical service providers. In short, it is just responsible for the running issues of the platform
Platform technical provider	Some IT companies or individuals are employed to provide services for establishing and maintaining the cloud platform through outsourcing or crowdsourcing
Machining service provider	In order to make full use of the superfluous machining equipment and capabilities, manufacturing prosumers (or service-oriented machining workshops) that are considered as machining services providers in CMC can add their equipment into the system to provide services to who need them
Logistics services provider	Because object flows are inevitable in CMC, some specific logistics prosumers joint in the community to take charge of all the logistics issues involved
Service requestor	Limited by lack of equipment, technology, etc., some manufacturing prosumers choose to outsource part of their non-core machining tasks to cut down cost. They broadcast their tasks and demands through the cloud machining system and search for specific services

performance. The order coordination process is to timely achieve the most beneficial portfolio among prosumers. Making a coordination decision is a complicated task. Here, five distinctive characteristics resulting in the great difficulty are concluded as follows: multiple qualitative or quantitative criteria, multi-objective in the coordination of decisions, fuzzy and subjective assessment and judgment, competition and cooperation game between prosumers, and stochastic metric in customer demands and production capacities.

10.2.2 Partner Selection Based on VIKOR Group Decision Method

In this section, a technique based on fuzzy VIKOR method is applied to the ultimate partner selection, the goal of which is to evaluate prosumers from various aspects that can reflect their service level, enterprise status and other business performance [16–18]. The VIKOR methodology was proposed to deal with multi-criteria decision-making problems with conflicting and non-commensurable criteria. Firstly, linguistic variables are defined to capture the importance of each criterion and the ratings of prosumers with regard to these criteria. We

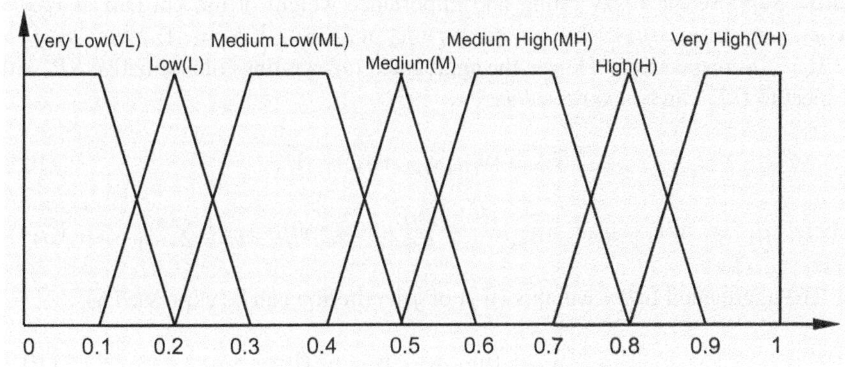

Fig. 10.3 Linguistic variables for importance weight of criteria and ratings

can use linear trapezoidal membership functions to embody the vagueness of these linguistic variables. The ultimate partner selection may be described by means of the following sets: a set of m candidate prosumers called as $SP = \{SP_1, SP_2, \ldots, SP_m\}$; a set of q criteria, $CT = \{CT_1, CT_2, \ldots, CT_q\}$; a set of k decision-makers called $DM = \{DM_1, DM_2, \ldots, DM_k\}$; a set of performance ratings of $SP_i(i = 1, 2, \ldots, m)$ with respect to criteria $CT_j(j = 1, 2, \ldots, q)$ called $V = \{v_{ij,}i = 1, 2, \ldots, m, j = 1, 2, \ldots, q\}$. The main steps of the selection process are:

Step 1: Arrange the decision-making group. A group of managers and experts from different functional areas within the enterprise have to be involved in the evaluation and selection. Since the DMs are different from each other in terms of experience and background, they must be organized according to the type of service tasks, enterprise's competitive situations and corporate strategies.

Step 2: Identify the appropriate linguistic variables. It is suggested that the DMs use the linguistic variables shown in Fig. 10.3 to evaluate the importance of each criterion and the ratings of alternatives with regard to these criteria. The linguistic variable can be represented in trapezoidal fuzzy numbers, for instance, "Medium Low (ML)" can be expressed as (0.2; 0.3; 0.4; 0.5). The membership function is

$$\mu_{ML}(x) = \begin{cases} 0, & x < 0.2 \\ \frac{x-0.2}{0.3-0.2}, & 0.3 \geq x \geq 0.2 \\ 1, & 0.4 \geq x \geq 0.3 \\ \frac{x-0.5}{0.4-0.5}, & 0.3 \geq x \geq 0.2 \\ 0, & x > 0.5 \end{cases} \quad (10.7)$$

Step 3: Aggregate the DMs' opinions. In this step, all the DMs' opinions to get the overall fuzzy weight of criteria and fuzzy rating of alternatives are aggregated. Suppose the fuzzy rating and importance weight of the kth DM as $\widetilde{v}_{ijk} = (v_{ijk1}, v_{ijk2}, v_{ijk3}, v_{ijk4})$, $\widetilde{w}_{jk} = (w_{jk1}, w_{jk2}, w_{jk3}, w_{jk4})$; $i = 1, 2, \ldots, m$, $j = 1, 2, \ldots, q$, respectively. Hence, the aggregated fuzzy ratings of alternative SP_i with respect to CT_j can be expressed as:

$$\widetilde{v}_{ij} = (v_{ij1}, v_{ij2}, v_{ij3}, v_{ij4}), \tag{10.8}$$

where $v_{ij1} = \min_k\{v_{ijk1}\}$, $v_{ij2} = \frac{1}{K}\sum_{k=1}^{K} v_{ijk2}$, $v_{ij3} = \frac{1}{K}\sum_{k=1}^{K} v_{ijk3}$, $v_{ij4} = \max_k\{v_{ijk4}\}$.

The aggregated fuzzy weights $(\widetilde{w}_j)$ of jth criterion can be expressed as:

$$\widetilde{w}_j = (w_{j1}, w_{j2}, w_{j3}, w_{j4}), \tag{10.9}$$

where $w_{j1} = \min_k\{w_{jk1}\}$, $w_{j2} = \frac{1}{K}\sum_{k=1}^{K} w_{jk2}$, $w_{j3} = \frac{1}{K}\sum_{k=1}^{K} w_{jk3}$, $w_{j4} = \max_k\{w_{jk4}\}$.

Finally, the aggregated decision information can be concisely represented in matrix format as follows:

$$\widetilde{V} = \begin{bmatrix} \widetilde{v}_{11} & \widetilde{v}_{12} & \cdots & \widetilde{v}_{1q} \\ \widetilde{v}_{21} & \widetilde{v}_{22} & \cdots & \widetilde{v}_{2q} \\ \cdots & \cdots & \cdots & \cdots \\ \widetilde{v}_{m1} & \widetilde{v}_{m2} & \cdots & \widetilde{v}_{mq} \end{bmatrix}, \quad \widetilde{W} = [\widetilde{w}_1, \widetilde{w}_2, \ldots, \widetilde{w}_q]^T. \tag{10.10}$$

Step 4: Defuzzify the DMs' opinions. Because $\widetilde{v}_{ijk}$, $\widetilde{w}_{jk}$, $\widetilde{v}_{ij}$ and $\widetilde{w}_j$ are linguistic variables represented by positive trapezoidal fuzzy numbers, a defuzzification of fuzzy decision matrix and fuzzy weight of each criterion into crisp values is needed here. The Center of Area (COA) method is applied to carry out the defuzzification process; the crisp value of a trapezoidal fuzzy number $\widetilde{B} = (b_1, b_2, b_3, b_4)$ can be calculated as:

$$\begin{aligned} defuzz(\widetilde{B}) = \frac{\int x \cdot \mu_B(x)dx}{\int \mu_B(x)dx} &= \frac{\int_{b_1}^{b_2}\left(\frac{x-b_1}{b_2-b_1}\right)\cdot xdx + \int_{b_2}^{b_3} xdx + \int_{b_3}^{b_4}\left(\frac{x-b_4}{b_3-b_4}\right)\cdot xdx}{\int_{b_1}^{b_2}\left(\frac{x-b_1}{b_2-b_1}\right)dx + \int_{b_2}^{b_3} dx + \int_{b_3}^{b_4}\left(\frac{x-b_4}{b_3-b_4}\right)dx} \\ &= \frac{-b_1 b_2 + \frac{1}{3}(b_3 - b_4)^2 + b_3 b_4 - \frac{1}{3}(b_1 - b_2)^2}{-b_1 - b_2 + b_3 + b_4} \end{aligned} \tag{10.11}$$

Then, we can obtain the crisp value of weight of criteria and rating of alternatives:

$$v_{ij} = defuzz(\widetilde{v}_{ij}), \quad w_j = defuzz(\widetilde{w}_j).$$

Step 5: Rank the alternatives by the values S_i, R_i and Q_i in ascending order. The compromise ranking in the VIKOR method is developed from the $Lp - metric$ used as an aggregating function in a compromise programming method:

$$L_{p,i} = \left\{ \sum_{j=1}^{q} \left[w_j (v_j^* - v_{ij}) / \left(v_j^* - v_j^- \right) \right]^p \right\}^{1/p}, \tag{10.12}$$

$$1 \leq p \leq \infty; i = 1, 2, \ldots, m;$$

where the best and worst value of jth criteria, also known as positive and negative ideal solutions, are $v_j^* = \max_i \{v_{ij}\}$, $v_j^- = \min_i \{v_{ij}\}$, respectively. $L_{1,i}$ [as S_i in Eq. (10.14)] and $L_{\infty,i}$ [as R_i in Eq. (10.15)] are used to formulate ranking measure.

$$S_i = \sum_{j=1}^{q} w_j (v_j^* - v_{ij}) / \left(v_j^* - v_j^- \right), \tag{10.13}$$

$$R_i = \max_j w_j (v_j^* - v_{ij}) / \left(v_j^* - v_j^- \right). \tag{10.14}$$

S_i is interpreted as concordance and provides DMs with information on the maximum group utility or majority. Similarly, R_i is interpreted as the minimum individual regret of the opponent. Then, index Q_i is obtained and based on the consideration of both the group utility and the individual regret of the opponent:

$$Q_i = \rho(S^* - S_i)/(S^* - S^-) + (1 - \rho)(R^* - R_i)/(R^* - R^-), \tag{10.15}$$

where $S^* = \min_i \{S_i\}$, $S^- = \max_i \{S_i\}$, $R^* = \min_i \{R_i\}$, $R^- = \max_i \{R_i\}$. ρ is the weight for the strategy of maximum group utility, whereas $1 - \rho$ is the weight of the individual regret.

Step 6: Determine the prosumers ultimately. Since a certain number n $(n < m)$ of best alternatives instead of only one are desired for the outsourced machining job, we modified the last step of original VIKOR method. The alternative prosumers $\left(SP^{(1)}, SP^{(2)}, \ldots, SP^{(n)}\right)$ who are the best ranked by the measure Q(minimum) are proposed as the compromise solution if the following two conditions, acceptance advantage and acceptance stability in decision making, are satisfied.

$C1$. Acceptable advantage:

$$Q\left(SP^{(n+1)}\right) - Q\left(SP^{(n)}\right) \geq 1/(m - 1),$$

where $SP^{(n)}$ and $SP^{(n+1)}$ are the alternatives with the n th and $(n + 1)$ th position in the ranking list by Q(minimum), respectively;

Table 10.2 Decision variables

Notation	Meaning
p_i	The price determined by The outsourcer to the prosumer i
m_i	Production capability (e.g. equipment number) devoted for the order determined by the prosumer
T	Common production interval determined by the outsourcer
D_i	Order quantity demand determined by the prosumer i
Q	Delivery batches determined by the outsourcer

$C2$. Acceptable stability in decision making:

The alternative prosumer $SP^{(n)}$ must also be better than anyone of $\left(SP^{(n+1)}, SP^{(n+2)}, \ldots, SP^{(N)}\right)$ ranked by S or R. This compromise solution is stable within a decision-making process, which could be the strategy of maximum group utility.

If one of the conditions is not satisfied, then a set of compromise solutions is proposed, which consists of

Alternative prosumers $\left(SP^{(1)}, SP^{(2)}, \ldots, SP^{(n+1)}\right)$ if only the condition $C2$ is not satisfied, or

Alternative prosumers $\left(SP^{(1)}, SP^{(2)}, \ldots, SP^{(n)}, \ldots, SP^{(N)}\right)$ if the condition $C1$ is not satisfied; $SP^{(N)}$ is determined by the relation $Q\left(SP^{(N)}\right) - Q\left(SP^{(n)}\right) < 1/(m-1)$ for maximum N.

10.2.3 Game and Coordination Decision of Manufacturing Service Relations

In this section, a kind of outsourcing relation is investigated: the manufacturer outsources a single type of parts machining task at different order quantity, different price, common production interval and common delivery batches to multiple prosumers who then perform the task at different devoted production capability levels. The outsourcer determines unit price for each prosumer, delivery batches, and the common production interval to minimize its cost. Prosumers in turn determine their optimal devoted production capability and order quantity to maximize their profits. This problem is modeled as a Stackelberg non-cooperative game [19–22] where the outsourcer is the leader and prosumers are followers, and formulated through a bi-level programming in which order information is exchanged among the followers. Due to the NP-hard nature of the bi-level model, a solution procedure based on a hierarchical imperialist competitive algorithm (HICA) is proposed to search the Stackelberg-Nash equilibrium solution. Tables 10.2 and 10.3 offers a review of related notations.

Table 10.3 Input parameters

Notation	Meaning
O	Total order amount
T_{per}	The most preferred production interval according to the production plan of outsourcer
h	Inventory cost per unit of outsourcer due to T is earlier to T_{per}
A_{si}	Machining setup cost per batch of prosumer i
A_s	Production setup cost per batch of outsourcer
A_l	Logistic setup cost per batch of outsourcer
C_i	Average machining cost per unit of prosumer i
k_i	Positive scaling parameter for average machining cost C_i
α_i	Average machining cost elasticity of prosumer with respect to the demand D_i
β_i	Average machining cost elasticity of prosumer with respect to the devoted production capability level m_i
l_i	Logistic cost per unit of prosumer i (in direct proportion to the distance)
h_{1i}	Holding cost of unfinished parts per unit of prosumer i
h_{2i}	Holding cost of finished parts per unit of prosumer i
P_{min}, P_{max}	Lower and upper limit of p_i decided by outsourcer
Q_{min}, Q_{max}	Lower and upper limit of Q decided by outsourcer
D_{imin}, D_{imax}	Lower and upper limit of D_i decided by outsourcer to sustain order elastic
m_{imin}, m_{imax}	Lower and upper limit of m_i due to the production capacity limitations of prosumers

Figure 10.4 represents a common form of the outsourcing interaction process. It is applicable for most outsourcing coordination case. Apparently, service is not a simple seller-buyer relation. In the interaction, there is a circulation of negotiation that outsourcer makes the first move by controlling the order prices, common delivery batches, and common production interval and prosumers react through optimizing their own order quantities and devoted production capabilities.

Bi-level problems are closely associated with Stackelberg games, which are characterized by two levels of optimization problems where the constraint region of the upper-level problem is implicitly determined by the lower-level optimization problem. The proposed Stackelberg non-cooperative game model can be formulated as a bi-level programming:

Upper-level problem: The outsourcer's cost equals to the whole charge by prosumers plus the logistic cost, setup and holding cost, given as follows:

$$\min U = \sum_{i=1}^{n} D_i \cdot (p_i + l_i) + Q \cdot (A_l + A_s) + \frac{h}{2} \cdot \left(2T_{per} - T\right) \cdot \sum_{i=1}^{n} D_i^{\circ} \quad (10.17)$$

subject to $T \in \left(0, T_{per}\right)$; $Q_i \in (Q_{min}, Q_{max})$; $p_i \in (P_{min}, P_{max})$.

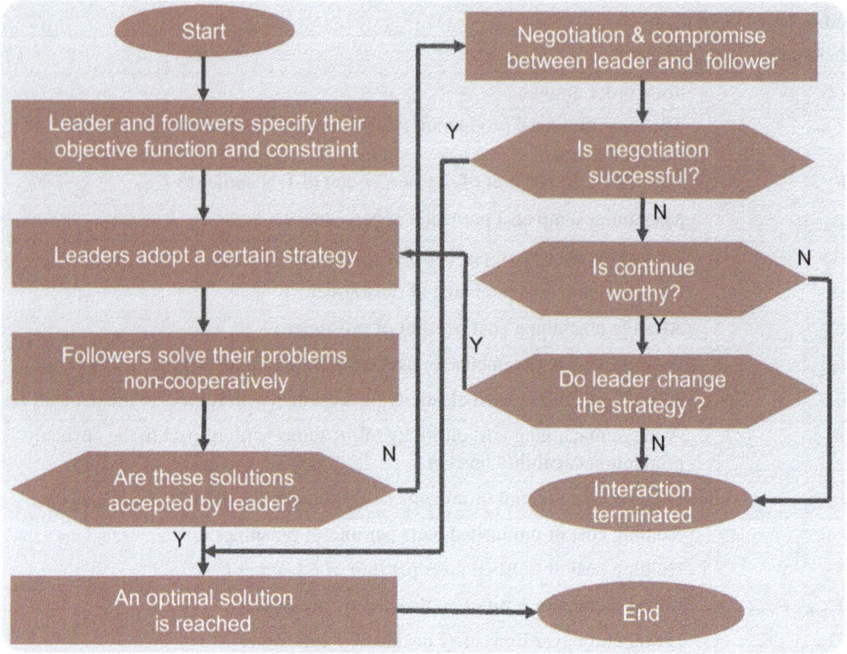

Fig. 10.4 The flowchart of the outsourcer–prosumers interaction

Lower-level problem: The prosumers' profit can be calculated as the revenue minus the machining cost, setup cost and holding cost of both unfinished and finished parts, given as follows:

$$\max L_i = D_i \cdot (p_i - C_i) - Q \cdot A_{si} \cdot m_i - \frac{h_{1i}}{2} \cdot \frac{D_i^2}{Q \cdot m_i} - \frac{h_{2i}}{2} \cdot \left(2T - \frac{D_i}{m_i \cdot Q}\right) \cdot D_i \tag{10.18}$$

subject to $\sum_{i=1}^{n} D_i = O; m_i \in (m_{imin}, m_{imax}); D_i \epsilon (D_{imin}, D_{imax})$.

10.2.4 Solving of Bi-Level Programing Based on HICA

To tackle NP-hard nature of the proposed bi-level programming, several solution procedures are proposed. The procedure initializes with a guess of the optimal upper-level decision values and moves this initial solution via a heuristic process to achieve a new solution. For each iteration, by solving the lower-level problem, the optimal reaction (Nash equilibrium) is obtained and returned to the upper-level.

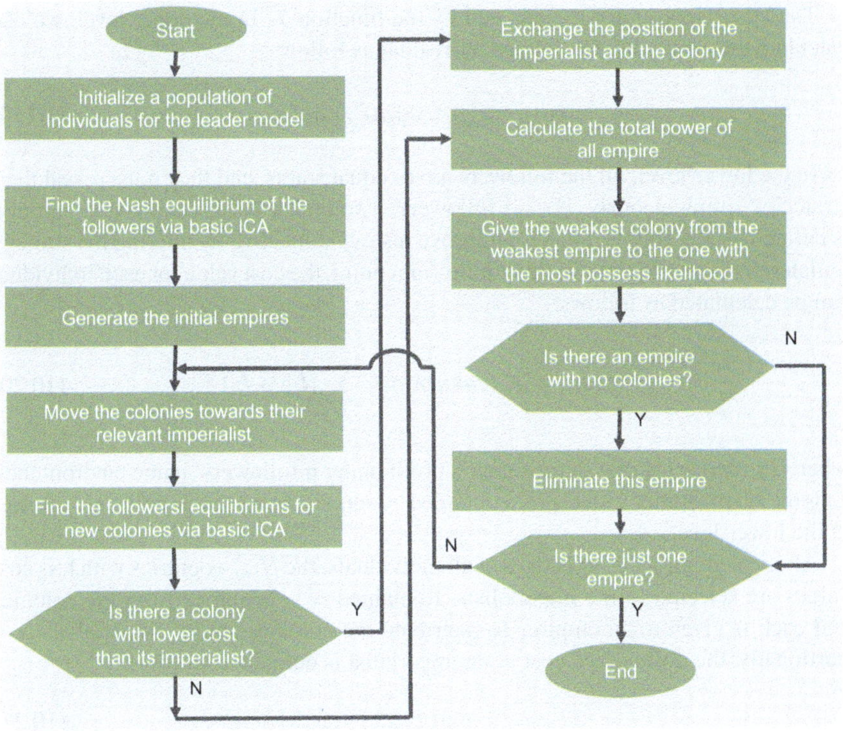

Fig. 10.5 The flowchart of HICA

This procedure continues until an optimal (Stackelberg-Nash equilibrium) or near-optimal solution is obtained for the upper-level problem. In this section, we develop an HICA based on the Imperialist Competitive Algorithm (ICA) by using some additional procedures and mechanisms to reach the high-quality solutions [23].

ICA starts with a randomly initial population of individuals, each called a country. Some of the most powerful countries are selected as imperialists and the rest form colonies which are then divided among imperialists based on imperialists' power. Imperialists with their colonies form empires, among which competition begins and colonies move towards their relevant imperialists. In the competition, weak empires collapse and powerful ones take possession of more colonies. At last, the most powerful empire will take the possession of other empires and wins the competition. In the following, we discuss some basic steps that will be called in HICA shown in Fig. 10.5.

Each individual is considered as an $3n + 2$ dimensions array of decision variables $country = [p_1, p_2, \ldots, p_n, D_1, D_2, \ldots, D_n, m_1, m_2, \ldots, m_n, T, Q]$. We initialize the population by generating N_{pop}. The N_{imp} best countries are then chosen as imperialists. We assume $N_{pop} = 50n$ and $N_{imp} = 10n$.

The cost of a country is evaluated by the function f. In the upper-level, we can calculate the cost value for a given individual as follows:

$$cost_U = f_U(country) = U^{\circ} \tag{10.19}$$

In the lower-level, all the followers are of equal status, and they must reveal their strategies simultaneously. For all followers, a so-called Nash equilibrium solution is defined as any follower cannot improve his own objective by altering his strategy unilaterally. Thus, to achieve this equilibrium point, the cost value for each individual can be calculated as follows:

$$cost_L = f_L(country) = \sum_{i=1}^{n} \left| L_i - \overline{L_\iota} \right| \tag{10.20}$$

where L_i represents the ith prosumer's profit under n followers' game environment, $\overline{L_\iota}$ is the ith prosumer's ideal profit obtained by supposing there was only one follower in the lower-level.

After calculating the objectives of all individuals, the N_{imp} countries with less cost values are selected as the imperialists. Remained N_{col} countries form the colonies and each is given to an empire. To distribute the colonies among imperialists proportionally, the normalized cost of an imperialist is defined as follows:

$$C_{imp(j)} = \max_i c_{imp(i)} - c_{imp(j)} \tag{10.21}$$

where, $c_{imp(j)}$ is the cost of jth imperialist and $C_{imp(j)}$ is its normalized cost. Having the normalized cost, the normalized power of the jth imperialist ($j = 1, 2, \ldots, N_{imp}$) is defined as

$$P_{imp(j)} = C_{imp(j)} / \sum_{i=1}^{N_{imp}} C_{imp(i)} \tag{10.22}$$

The number of colonies belongs to the jth imperialist are directly proportionate to its power:

$$NC_{imp(j)} = Round\left(P_{imp(j)} \times N_{col}\right) \tag{10.23}$$

The imperialists improve their colonies by moving all colonies towards them. Each colony moves by δ_i units along dimension i, while δ_i is a random variable with uniform distribution $\delta_i \sim U(0, \sigma_i \times \Delta_i)$, σ_i is an escalating parameter and Δ_i is the distance between the imperialist and the colony along dimension i. The parameter σ_i is set as follows:

$$\sigma_i = \begin{cases} 2 & i = 1, 2, \ldots, 3n \\ 1.2 & i = 3n + 1, 3n + 2 \end{cases}$$

This step plays a crucial role in both convergence rate of the algorithm and the probability of falling into local optima. Usually, to explore different points around the imperialist, a random amount of deviation θ_i is added to the direction of colony movement towards the imperialist. This deflection angle θ_i is chosen randomly with a uniform distribution $\theta_i \sim U(-\gamma, \gamma)$ and $\gamma = \pi/4$ Rad.

Here, to further improve the performance of the algorithm for solving Stackelberg games, an adaptive controller (AC) is adopted to adapt the movement deflection angle θ_i based on the cost distribution of each iteration. This adaption mechanism suggests that the deflection vector should be increased if the country density is high, and it should be decreased if the country density is low. It can be expressed as follows:

$$\varphi_i^{t+1} = \begin{cases} \theta_i^{t+1} \times \frac{1}{1-\frac{cost_{\min}}{cost_{\max}}} & \frac{cost_{\text{ave}}^t}{cost_{\max}^t} > a, \frac{cost_{\min}^t}{cost_{\max}^t} > b \\ \theta_i^{t+1} \times \left(1 - \frac{cost_{\min}}{cost_{\max}}\right) & \frac{cost_{\text{ave}}^t}{cost_{\max}^t} < a, \frac{cost_{\min}^t}{cost_{\max}^t} < b \\ \theta_i^{t+1} & else \end{cases}, \tag{10.24}$$

where $cost_{\min}^t$, $cost_{\text{ave}}^t$, $cost_{\max}^t$ define the minimum, average, maximum cost of countries at iteration t, respectively, φ_i^{t+1} denotes the movement deflection angle at iteration t + 1. Adaptive assimilation can enhance the ability of escaping from local optima and fast convergence rate.

The other basic calculation and steps such as exchanging the positions of the imperialist and one colony, total power of an empire and imperialistic competition are the same with original ICA and not discussed detailed here for the concise reason. Bringing together the above discussion, Fig. 10.5 represents the flow chart of the HICA adopted to solve the Stackelberg game.

10.3 Generation of Order-Driven Manufacturing Service Community for Mass Individualization

According to the discussion mentioned above, an ontology-based algorithm is introduced to generate order-driven Manufacturing Service Community (MSC). As shown in Fig. 10.6, the specific implementing procedure is correspondent with several steps such as description and release, retrieval and matching, selection and evaluation. Furthermore, as the core of algorithm, the retrieval and matching process includes: (1) decomposing the order task into sub-task at a manufacturing feature level through an ontology-based inference engine; (2) matching the most similar SMRs in the community to perform each sub-task based on inference engine(This step outputs the candidate SMRs sets); (3) adapting the ant colony optimization (ACO) to compose the SMRs which belong the corresponding sub-tasks under manufacturing constraints given by the service requestor; and (4) generating and exporting the candidate services for service requestor to be concerned.

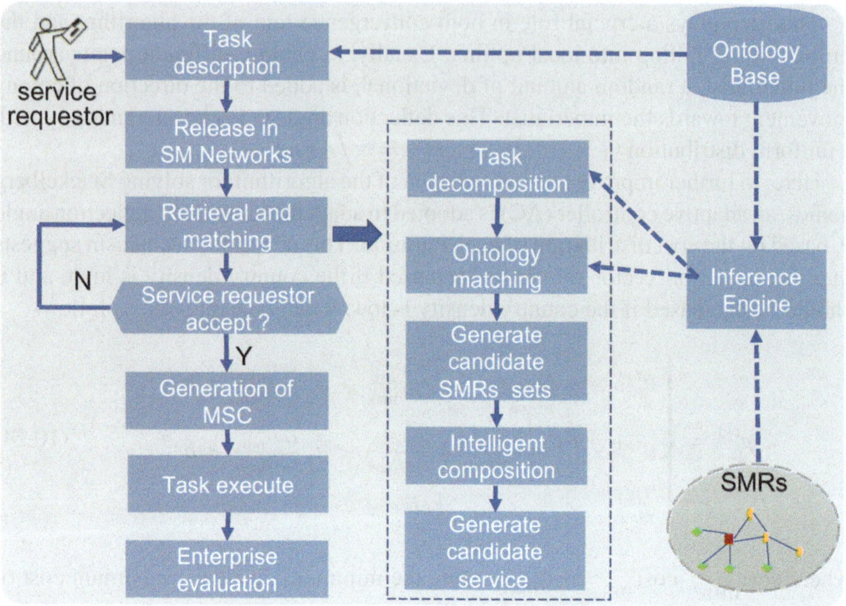

Fig. 10.6 Generation of order-driven dynamic community

There are various resources providers in MSC, which have short-term or long-term collaboration relations with the community. Trust mechanism plays an important role in the collaboration among different prosumers. Considering this point, there are many professional groups established in the community. To be specific, the resource providers with good reputation and similar machining capability are clustered into a group to complete the machining tasks of similar parts or processes. For example, through analyzing the historical data collected from machining tasks of large gear, the excellent resource providers are selected to establish a group. When a new machining tasks of gear released, the resources providers in this group could be selected first. Therefore, the resources providers within the group have higher trust and the resources providers out of the group have lower trust. The acquaintance model can be used to describe the trust of resources providers, which is formulated as AM $= (A, B, C)$, where A denotes the resources providers set in the group; B denotes the trust relation set; C denotes operational symbol of trust relation. The trust degree between resources provider and could be calculated using the above model too.

10.3.1 Modeling of Manufacturing Services for Mass Individualization

There are many definitions of manufacturing service from different perspectives, e.g., value creation, manufacturing capability and demand-supply interaction. To more accurately describe the process flow details of the mass individualized or personalized manufacturing services, this study focuses on the process flow perspective. A manufacturing service is a set of collaborative manufacturing activities governed by business logic and rules through a group of prosumers, using equipment to operate on raw materials transforming to a product. A service process model is defined as follows:

Definition 10.2 (*Service process*) A service process model is a graph

$$x_i = (V_i, E_i, vt_i) \tag{10.25}$$

where $V_i = \{v_{i,1}, v_{i,2}, \ldots, v_{i,n}\}$ denotes the service activities vertex set and $E_i = \{e_{i,jk} | e_{i,jk} = v_{i,j} * v_{i,k}, 1 \leq i, j \leq n\}$ denotes the relational edge set. $vt_i = \{vt_{i,1}, vt_{i,2}, \ldots, vt_{i,n}\}$ denotes the attribute sets for V_i. This definition is also in accordance with flow charting disciplines.

Definition 10.3 (*Service activity*) From the viewpoint of manufacturing capability, the service activities includes four properties: activity type, manufacturing material, service quality and manufacturing feature. An accurate descriptive model of service activity $vt_{i,j}$ is built as follows:

$$vt_{i,j} = \{ID, pro, name, type, mat, fea, qua\} \tag{10.26}$$

The elements stand for ID, provider, name, activity type, material, manufacturing feature(s) and service quality, respectively. As shown in Fig. 10.7, the activity type and manufacture feature properties are abstract classes, so each represents a hierarchy of subclasses which are classified according to their metrics.

Definition 10.4 (*Service process reference model*) A process reference model contains the major similarities among the existing processes, and serves as a sustainable blueprint for the development of a new service process. This definition is coined with business and software process reference model.

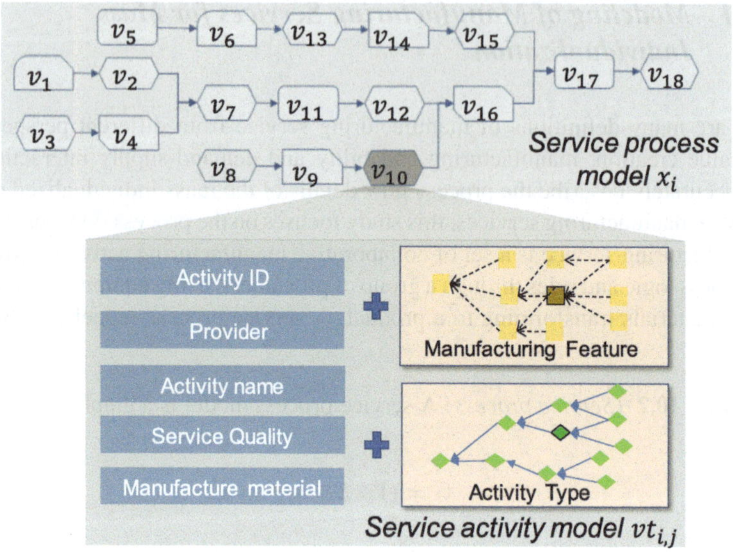

Fig. 10.7 An illustration of service process

10.3.2 Construction of Service Community Based on Reference Models

The developing of process reference models based on Granular computing (GrC) comprises of three phases. The first phase takes both the services activity and process sequence similarities as a measuring tool for services process similarity. This phase adopts the ontology subsumption and semantic similarity calculating techniques. The second phase introduces GrC theory into the clustering analysis of service processes, and constructs a quotient space family with hierarchical structure based on the theory of fuzzy quotient space. In this phase, the more representative and more accurate process granules are acquired from in an optimal information granular layer. The third phase concerns: analyzing the frequent common sequence among different services processes in every information granule under the best granularity layer, subsequently discovering the abstraction of corresponding activities in the sequences, and finally compositing into the process reference models.

In the determined optimal granular layer, the typical process sequence for each information granule can be acquired. To achieve a higher completeness of the reference model, we adopt the longest common subsequence (LCS) rather than the shortest common subsequence algorithm.

Given that $x_{A,i} = \left(v_{A,1}, v_{A,2}, \ldots, v_{A,i}\right)$ and $x_{B,j} = \left(v_{B,1}, v_{B,2}, \ldots, v_{B,j}\right)$ are subsequences of two service processes $x_A = \left(v_{A,1}, v_{A,2}, \ldots, v_{A,i}, \ldots, v_{A,p}\right)$ and $x_B = \left(v_{B,1}, v_{B,2}, \ldots, v_{B,j}, \ldots, v_{B,q}\right)$, respectively, where $1 \le i \le p$ and $1 \le j \le q$. A longest common subsequence $LCS_{(i.j)}$ of x_A and x_B is defined as a common subsequence with the maximum length. Here, considering the similarity between service activities, an improved dynamic programming approach is proposed for calculating $LCS_{(i.j)}$:

Step 1: start from the last node in the (or the rest of) service process.

Step 2: match backward one by one based on the recursive formula as shown in Eq. (10.27):

$$LCS_{(i,j)} = \begin{cases} \max\{LCS_{(i-1.j)}, LCS_{(i.j-1)}\} & sim_{act}(v_{A,i}, v_{B,j}) < s_t, \\ LCS_{(i-1.j-1)} \cup \{C_{super}(v_{A,i}, v_{B,j})\} & sim_{act}(v_{A,i}, v_{B,j}) \geq s_t, \\ 0 & i = 0 \| j = 0 \end{cases} \tag{10.27}$$

where s_t denotes the similarity threshold that separates the similar activities and un-similar activities, which is pre-specified by the developer. $C_{super}(v_{A,i}, v_{B,j})$ denotes the abstraction (i.e., super-class of properties) of two activity nodes, which will be detailed in the following sub-section.

Step 3: recursively repeat step 1 and 2 to get the final LCS result.

Step 4: if more than two individuals are included inside one granule, there will be repeated dynamic programming between the rest individuals with the obtained $LCS_{(i.j)}$ of two initial selected individuals.

During the automated programming of typical process sequence for a new reference model, there may exist conflicts/contradictions, namely, a relation cannot be aggregated into a uniquely defined process sequence when two pairs of activities having different sequential relations in two individuals. Generally, conflicts can be automatically resolved by pre-defined rules. For example, in the above programming, the conflicting relations are omitted in the reference model, so as they can be ordered either way in the practical implementations for a new service process. However, it suffers from the probability of oversimplifying the process sequence, and consequently leading to a less representative or over generic reference model. Moreover, many conflicts appear in implementation processes, whereas it is essential to leave these decisions to the designers. Not every process sequence refers to a direct causality, and these interpretations can be depended on human interaction rather than an algorithm.

In the obtained LCS of each granule, the activity nodes of services in each information granule may be different from each other, since they only have similarities greater than the pre-specified threshold in the LCS dynamic programming process. Therefore, it needs a step of abstraction of service activity nodes in LCS for achieving a higher generality and flexibility of unified process reference model, which can be calculated based on ontology theory. Suppose that there has i services in a granule, the abstraction of the jth activity nodes in the LCS is the super-class, which is in the lowest-level in the domain ontology, subsuming all jth activity nodes of services within granule, which can be formulated as follows:

$$vtr_{j,j\in LCS} = C_{super}\left(vt_{1,j}, vt_{2,j}, \ldots, vt_{i,j}\right) \tag{10.28}$$

The last step is the further assembly of the identified typical process sequences and abstracted activities into a real new and syntactically connected process reference model. We define a process reference model PRM_i of the ith grule as a graph

$$PRM_i = (Vr_i, Er_i, vtr_i) \tag{10.29}$$

where $Vr_i = \{vr_{i,1}, vr_{i,2}, \ldots, vr_{i,n}\}$ denotes the abstracted service activities vertex set and $Er_i = \{er_{i,j,k} | er_{i,j,k} = vr_{i,j} * vr_{i,k}, 1 \leq j, k \leq n_i\}$ denotes the identified typical service process set. vtr_i denotes the lowest super-class attributes for abstracted activities Vr_i.

In this step, this synthesis can be easily fulfilled automatic by stacking operations on the three datasets Vr_i, Er_i, and vtr_i. Finally, the frequent subgraphs are iteratively merged into PRM_i based on their mutual relations. After the graphs of process reference models are generated, the developers can manually change and reconfigure the node and edge elements of the model to meet the specific requirements.

10.3.3 Polymorphous Manufacturing Service Community Configuration

To understand the scope of manufacturing service community configuration, the framework is depicted in Fig. 10.8. Three important modules are included in the framework, that is, Customized Community Space (CCS) configuration, Social Manufacturing Community(SMC) organization, and manufacturing service discovery and matching. These modules form the generic logic to implement social manufacturing.

CCS configuration. As the core component of manufacturing community, CCS provides an independent space for prosumers to configure their manufacturing resources, release their manufacturing capabilities, and personalize industrial web applications they actually required. Besides, prosumers can also be the acquire supply/demand information from their partners, and manage their own commercial activities as well. To support the implementation of CCS, technologies such as manufacturing resources description and dynamic production modelling should be developed.

SMC organization. Geographically dispersed CCS could be organized into different kinds of SMCs according to different rules, such as core-manufacturers based community, industry chain-based community, process type-based community, etc. SMC is a bridge for prosumers to communicate efficiently and timely with their business partners by using many social network platforms. In the SMC, prosumers could release their order demands, production capabilities and commercial activities, which are available for all the prosumers in the SMC.

Manufacturing service discovery and matching. After CCS configuration, manufacturing resources in the SMC are properly classified using granular computing

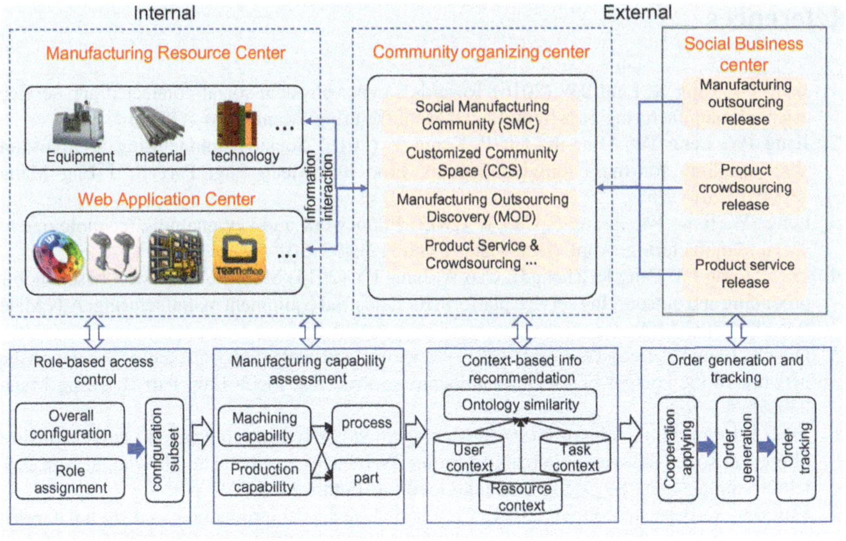

Fig. 10.8 Polymorphous manufacturing service community configuration

based multi-perspective classification. Simultaneously, resources of manufacturing service provider are evaluated according to their credit to support searching and matching between production capabilities (from manufacturing service providers) and order demands (from manufacturing service demanders), which are resolved by a hybrid hierarchical algorithm. An order relation between the demander and provider is generated in SMC after game theory-based bargaining and transaction.

10.4 Concluding Remarks

Considering that outsourcing and crowdsourcing order tasks related to specialized machining processes and parts are becoming one of the most significant manufacturing service and value-added activities during the collaborative production procedures, a set of decision-support tools are proposed to select potential prosumers and SMRs with the capability of monitoring and assessing the performance of their partners. The concept of order-driven Manufacturing Service Community (MSC) is introduced to provide a systematic solution for users to integrate the manufacturing with services so as to realize the added-value of production activities. An MSC performs order tasks related to machining processes and parts through the intelligent integration of SMRs, service matching and finding, service running and monitoring in a certain community, in which the prosumers have a better order relation and creditworthiness to the service requestor.

References

1. Jiang PY, Ding K, Leng JW (2016) Towards a cyber-physical-social-connected and service-oriented manufacturing paradigm: Social Manufacturing. Manuf Lett 7:15–21
2. Jiang PY, Leng JW, Ding K, Gu P, Koren Y (2016) Social manufacturing as a sustainable paradigm for mass individualization. Proc Inst Mech Eng, Part B: J Eng Manuf 230(10):1961–1968
3. Leng JW, Jiang PY, Zhang F, Cao W (2013) Framework and key enabling technologies for social manufacturing. Appl Mech Mater 2382(312):498–501
4. Liu C, Leng JW, Ding K, Zhang C, Cao W, Jiang PY (2014) Mobi-POSP: a web-based mobile producing and outsourcing service platform for industrial equipment manufacturing. Adv Mater Res 889:1306–1309
5. Ding K, Jiang PY, Leng JW, Cao W (2016) Modeling and analyzing of an enterprise relationship network in the context of social manufacturing. Proc Inst Mech Eng, Part B: J Eng Manuf 230(4):752–769
6. Johar I, Lipparini F, Addarii F (2015) Making good our future exploring the new boundaries of open & social innovation in manufacturing. Web Pages, available at: http://uk.ukwon.eu/File%20Storage/4970285_7_SIE-Making-Good-our-Future-May-2015.pdf
7. Kostakis V, Papachristou M (2014) Commons-based peer production and digital fabrication: the case of a RepRap-based, Lego-built 3D printing-milling machine. Telematics Inform 31(3):434–443
8. Wulfsberg JP, Redlich T, Bruhns FL (2011) Open production: scientific foundation for co-creative product realization. Prod Eng Res Dev 5(2):127–139
9. Bonvoisin J, Boujut J (2015) Open design platforms for open source product development: current state and requirements. In: Proceedings of the 20th International Conference on Engineering Design, Politecnico Di Milano, Italy, 27–30 July 2015
10. Dahlgren E, Goecmen C, Lackner K, Van Ryzin G (2013) Small modular infrastructure. Eng Econ 58(4):231–264
11. Panchal JH, Le Q (2014) Product development by self-organized virtual communities. In: Advances in computers and information in engineering research. ASME Press
12. Jiang PY, Leng JW (2017) The configuration of social manufacturing a social intelligence way toward service-oriented manufacturing. Int J Manuf Res 12(1):4–19
13. Jiang PY, Leng JW, Ding K (2016) Social manufacturing: a survey of the state-of-the-art and future challenges. In: IEEE International Conference on Service Operations and Logistics, and Informatics, Beijing, China, 10–12 July 2016
14. Leng JW, Jiang PY (2016) A deep learning approach for relationship extraction from interaction context in social manufacturing paradigm. Knowl-Based Syst 100:188–199
15. Leng JW, Jiang PY (2017) Mining and matching relationships from interaction contexts in a social manufacturing paradigm. IEEE Trans Syst Man Cybern Syst 47(2):276–288
16. Opricovic S, Tzeng GH (2007) Extended VIKOR method in comparison with outranking methods. Eur J Oper Res 178(2):514–529
17. Brans JP, Vincke P, Mareschal B (1986) How to select and how to rank projects—the PROMETHEE method. Eur J Oper Res 24(2):228–238
18. Aksoy A, Ozturk N (2011) Supplier selection and performance evaluation in just-in-time production environments. Expert Syst Appl 38(5):6351–6359
19. Amir R, Grilo I (1999) Stackelberg versus Cournot equilibrium. Games Econ Behav 26(1):1–21
20. McKelvey RD, McLennan A (1996) Computation of equilibria in finite games. Handb Comput Econ 1:87–142
21. Moya S, Poznyak AS (2009) Extraproximal method application for a StackelbergNash equilibrium calculation in static hierarchical games. IEEE Trans Syst Man Cybern Part B Cybern 39(6):1493–1504

22. Yugang Y, Liang L, Huang GQ (2006) Leader–follower game in vendor-managed inventory system with limited production capacity considering wholesale and retail prices. Int J Logistics Res Appl 9(4):335–350
23. Atashpaz-Gargari E, Lucas C (2007) Imperialist competitive algorithm: an algorithm for optimization inspired by imperialistic competition. In: IEEE Congress on Evolutionary Computation, Singapore, 25–28 September 2007

Chapter 11
Execution of Social Manufacturing

Pulin Li, Jiajun Liu and Pingyu Jiang

11.1 Shaping of Order-Driven Dynamic Manufacturing Communities

Under the context of social manufacturing, a manufacturing enterprise can encapsulate its own various facilities and be presented as a socialized manufacturing resource (SMR). Accordingly, it cooperates with other SMRs to complete product orders according to the tasks all the SMRs undertake respectively. In this process, all the available manufacturing enterprises and their social business relationships which are concerned with the above product orders can be abstracted as a social manufacturing network or system, where manufacturing enterprises are considered as nodes and their social business relationships as edges. In this chapter, we define these encapsulated enterprises in the form of SMRs as social manufacturing nodes (SMNs).

The SMNs compete and cooperate with each other so as to undertake product orders with the support of social sensors. Then, some cooperative SMNs gather together and gradually evolve into an order-driven dynamic manufacturing community (DMC), which plays an important role in enabling the social manufacturing system that is also presented as a kind of social manufacturing network from the angle of interconnection. In these different evolved DMCs, SMNs are used for completing product orders autonomously and collaboratively. This section will describe how to shape the DMCs [1].

In order to support the shaping of order-driven DMCs, it is necessary to figure out how to attach extended cyber-physical systems (ECPSs) to correspondent SMNs, how to realize the interconnection among SMNs, and how to deal with the cyber-credit problems in a DMC. Furthermore, a product order needs to be decomposed into detailed manufacturing service tasks and a task tree is constructed to express the handling sequence of the manufacturing service tasks. Around the task tree of a product order, SMNs autonomously undertake manufacturing service tasks and are organized into order-driven DMCs. Some performance indexes are established to analyze the performances of these DMCs.

P. Jiang, *Social Manufacturing: Fundamentals and Applications*, Springer Series in Advanced Manufacturing, https://doi.org/10.1007/978-3-319-72986-2_11

11.1.1 Attaching of a Series of Extended CPS Nodes to an SMN

To support the shaping process of order-driven DMCs, it is necessary to attach a series of ECPSs to correspondent facilities inside an SMN. Here, an ECPS node stands for the combination of existing CPS, radio frequency identification (RFID) and social sensors, and is attached to a facility in an SMN, referring to Chap. 6. In fact, it is a social factory model.

Inside DMCs, an SMN is configured with a series of ECPS nodes so as to realize the connection between physical space and cyber space. So, SMNs powered with correspondent ECPS nodes are capable of collecting original data, executing instructions, and sharing real-time information [2]. Furthermore, manufacturing service tasks can be autonomously executed through the coordination and share among ECPS nodes.

In fact, a product order can be decomposed into several sub-orders, including production orders related to manufacturing service tasks, according to the BOM of the product. For example, when a production order is obtained by an SMN, the engineer related to this node writes the manufacturing data concerning process planning, production instructions, etc., into a smart workpiece (SW) via social sensors. The ECPS nodes inside the SMN communicate with SWs and obtain the schedules and instructions from them by gaming and matching mechanism. Then the SMN performs the production operations according to the schedules and instructions. With the help of correspondent ECPS nodes, meanwhile, this SMN can also sense its own status and share it with other SMNs.

ECPS nodes are highly autonomous, not all their decisions are made by the central control unit. This means that all the exceptions in social manufacturing will be handled in time. For example, if one ECPS node breaks down, another alternative node will be elected for substitution [3]. The specific working principle and process of ECPS nodes are expressed in Fig. 11.1.

Here, the manufacturing engineers determine the process plans, schedules and production instructions of SWs and writes such data into their attached RFID tags to make sure that SWs can follow the proper manufacturing routes. An SW is delivered to an SMN and then the machining requirements are broadcasted to each ECPS node in the SMN.

Idle ECPS nodes with suitable manufacturing capabilities will accept the machining request according to the "*demandability*" matching mechanism. These ECPS nodes need to be evaluated and then the evaluation results are sent back to the SW to determine if the optimal ECPS node is chosen. The SW will send the detailed machining instructions to the selected ECPS node and the physical facilities such as machining equipment, sensors, actuators, and fixtures will be activated according to the production instructions.

When a production disturbance, including that a downtime in the machining equipment occurs, the ECPS node will detect the exception according to the real-time status data collected by its sensors. Whether the exception can self-recover will be deter-

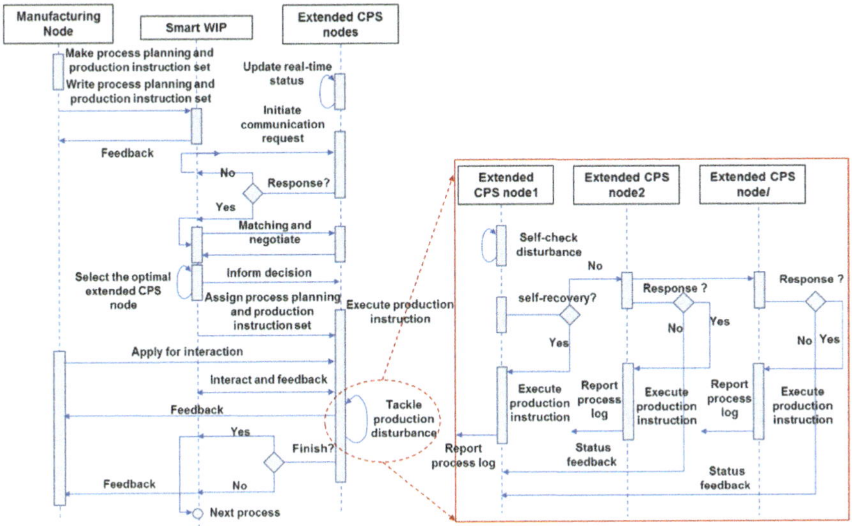

Fig. 11.1 Working principle of extended CPS in an SMN

mined with the rules/knowledge stored in its local database. If it can self-recover, some of the corresponding solutions will be offered, and if not, it will communicate with other ECPS nodes nearby for assistance. According to the status feedbacks from other ECPS nodes, the best alternative node is selected to execute the following-up operations. In addition, all the adjust information will be sent to the upper information system in the SMN level for recording. As soon as the first production process is completed, the SW will communicate with other ECPS nodes for the next one, and another ECPS node will execute the next operation until all the production processes are finished.

11.1.2 Interconnecting Between SMNs with Social Sensors

In a social manufacturing network, manufacturing services are provided by mass collaborations of SMNs to satisfy the needs of product orders. In order to ensure that the process of mass collaboration runs smoothly, it is necessary for SMNs to build seamless interconnection with each other. The interconnection among SMNs is implemented by using social sensors that are combined respectively into different ECPS nodes in SMNs. The functions of social sensors include data sensing, data verifying, data encrypting, data transmitting and role relationships managing.

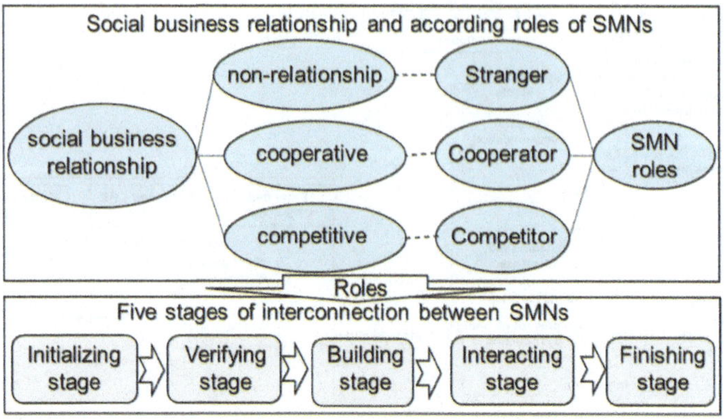

Fig. 11.2 Five stages of the interconnection between SMNs

The interconnection among SMNs can be divided into three categories based on their social business relationships, which are non-relationship, cooperative relationship, and competitive relationship. Accordingly, the roles of SMNs can be defined as "*stranger*", "*cooperator*", and "*competitor*" respectively. "*Stranger*" means that SMNs do not have interconnection history and know nothing about each other. In this situation, the level of interconnection is very low, and the social sensor will only interact a small amount of basic information referring to enterprise profiles. During working processes, SMNs gradually know more about each other, and the roles of these SMNs would transform into "*cooperator*" or "*competitor*". The cooperative activities among SMNs include product orders sharing, manufacturing service planning, real-time production information sharing, etc. And all these activities are implemented and ensured through using social sensors. As for the competitive relationship, the interaction data should be treated carefully for security issues.

The interconnection among SMNs can be separated into five stages, i.e., initializing stage, verifying stage, building stage, interacting stage, and finishing stage, as shown in Fig. 11.2. The detailed description can be explained with the following example [4].

Assume that there are two SMNs, that is, SMN_1 and SMN_2, and their attached social sensors are $sensor_1$ and $sensor_2$ respectively. Then, the interaction process between SMN_1 and SMN_2 can be described as follows.

- SMN_1 searches for target SMN_2 and sends the connecting requests labeled with the corresponding role to SMN_2 through its social $sensor_1$ in the initializing stage.
- When social $sensor_2$ of SMN_2 receives the connecting request, it will check the identity of SMN_1 in the verifying stage. If valid, it will respond to SMN_1 and send a feedback message to SMN_2. If not, social $sensor_2$ won't respond to the connecting request.
- After social $sensor_1$ receipts the response, it will build an encrypted connection channel between social $sensor_1$ and $sensor_2$ in the building stage.

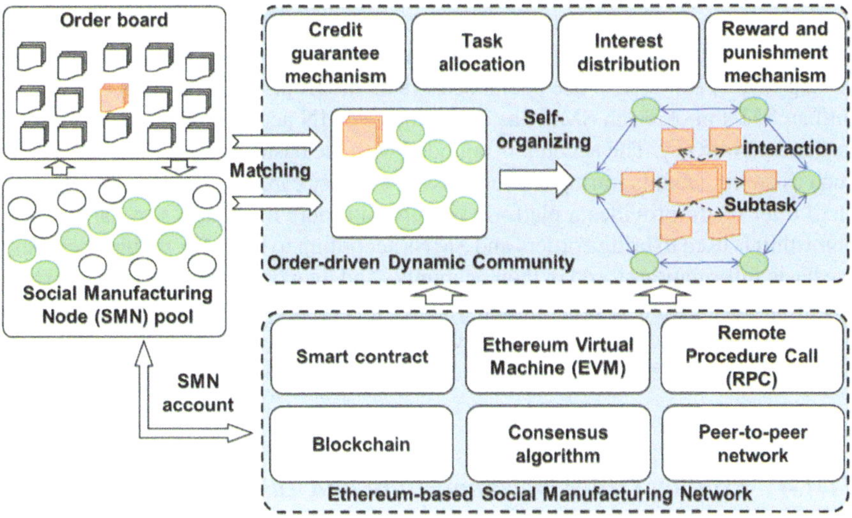

Fig. 11.3 Blockchain shell for order-driven dynamic communities

- In the interacting stage, these two SMNs implement real-time interconnection through the encrypted connection channel. During this process, an amount of information is transmitted to each other, including social business context, real-time production information sharing, product quality data, etc. It must be noted that the information is handled and encrypted by using social sensors based on the roles of SMNs.
- When the interconnection between these two SMNs has finished, their social sensors will terminate the connection in the finishing stage.

11.1.3 Constructing of Blockchain Shell for Dynamic Manufacturing Communities

SMNs cooperate with each other to undertake product orders by using order-driven DMCs in the form of self-organization. In the DMCs, there are a lot of matching, gaming, cooperating, competing and interacting processes. These processes generate a lot of data which might be easily disturbed if there is no assurance with a trust and security mechanism. As shown in Fig. 11.3, a blockchain shell for order-driven DMCs is constructed to support the above processes [5] based on the blockchain principles for cyber-credits in the context of social manufacturing described in Chap. 9.

Blockchain shell for DMCs is composed of *Ethereum*-based social manufacturing network, order board, SMN pool and order-driven DMCs.

Here, *Ethereum*-based social manufacturing network is the basis, including blockchain, consensus algorithm, peer-to-peer network, smart contract, *Ethereum* virtual machine (EVM), and remote procedure call (RPC). It enables the self-organization and autonomy of order-driven DMCs and facilitates the interactions among SMNs [6]. SMNs use social sensors as the front-end interfaces and are put into an SMN pool. Each SMN has an exclusive SMN account with a pair of public key and private key. The account is bounded with the basic information of the SMN, such as credit, manufacturing capacity, asset, delivery capacity, customer evaluation, etc. Order board provides a platform to publish orders for SMNs. And a matching algorithm is used to bridge orders and SMNs according to the order requirements and the basic information stored in their accounts. And then order-driven DMCs are built with four customized smart contract systems, i.e., credit guarantee mechanism, task allocation, interest distribution, and reward and punishment mechanism. These four smart contract systems can regulate the executing process of order-driven DMCs.

11.1.4 Product Order Decomposition and Task Tree Construction

Based on the description of the first three sections, the interconnection among SMNs can be realized and SMNs have chances now to get and accomplish product orders. But a product order always requires various and massive manufacturing services and an SMN can only provide limited manufacturing services during the progress because of its limited manufacturing service capabilities. So it is necessary to decompose manufacturing service tasks [7], which is also a kind of production order. The correspondent decomposition strategies deal with:

- determining the granularity of decomposition based on product quantity and bill of material (BOM) of the product order,
- decomposing the product order into specific manufacturing service tasks according to the granularity,
- appending the manufacturing service requirements to the decomposed tasks, including machining accuracy, roughness, and other extra requirements,
- evaluating the manufacturing time and executed cost of each decomposed manufacturing service task,
- planning the manufacturing service sequence of the decomposed tasks based on the part relationships in the BOM, and
- building the task tree based on the manufacturing service sequence.

The above strategies are demonstrated with a task tree in Fig. 11.4. In the task tree, task nodes are ordered based on their manufacturing service sequence. It is seen that the task nodes from the lower layer are supposed to be processed before the ones in the upper layer. In another word, when executing the manufacturing service tasks in the upper layer, task nodes in the lower layer have been finished already. In addition, each task node also contains some attributes of corresponding manufacturing service task, including delivery time, machining requirements, payment and other additional information.

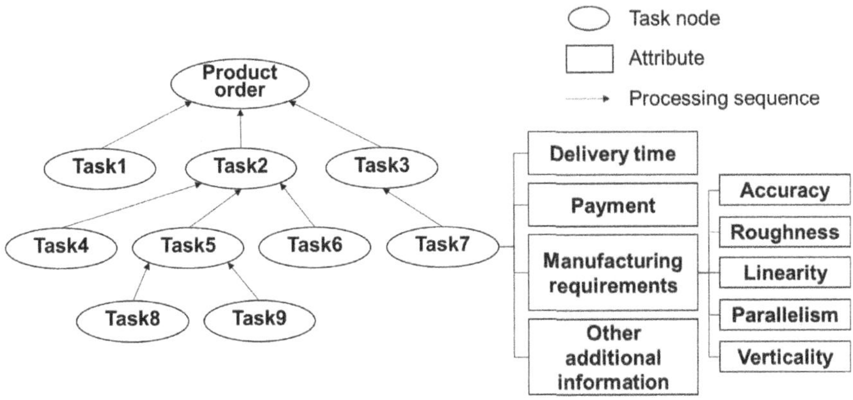

Fig. 11.4 An example of task tree of a product order

11.1.5 Formation of Order-Driven Dynamic Manufacturing Communities

Around the task tree of a product order, SMNs autonomously take manufacturing service tasks based on their current service capacities in the form of order-driven DMCs. The self-organization of such DMCs can just be analyzed by using the evolutionary dynamics theory [8].

From the viewpoint of evolutionary dynamics theory, the order-driven DMCs can be regarded as a population, called as community population, which is composed of two sub-populations: manufacturing service task sub-population and SMN sub-population, as shown in Fig. 11.5. The self-organization of order-driven DMCs can be regarded as an evolutionary process between manufacturing service task sub-population and SMN sub-population. The purpose of the evolutionary process is to achieve global symbiosis relationship between the manufacturing service tasks sub-population and SMN sub-population. Furthermore, the global symbiosis relationship is composed of local symbiosis relationships where an SMN take one or more manufacturing service tasks. Assume that manufacturing service tasks are independent and the evolution of local symbiosis relationships are in the similar process, the evolutionary process of the order-driven DMCs can be simplified to analyze how a manufacturing service task and an SMN are respectively evolved into local symbiosis.

According to the evolutionary dynamics theory, if a population containing two individuals achieves symbiosis, the fitness degrees of the two individuals are equal. It suggests that when an SMN undertakes one manufacturing service task, the fitness degree of the SMN is equal to the fitness degree of the manufacturing service task. The fitness degree of manufacturing service task is a function of task payment, task handling requirements and task delivery time. The fitness degree function of the SMN is just related to the quoted price, the manufacturing service capabilities and

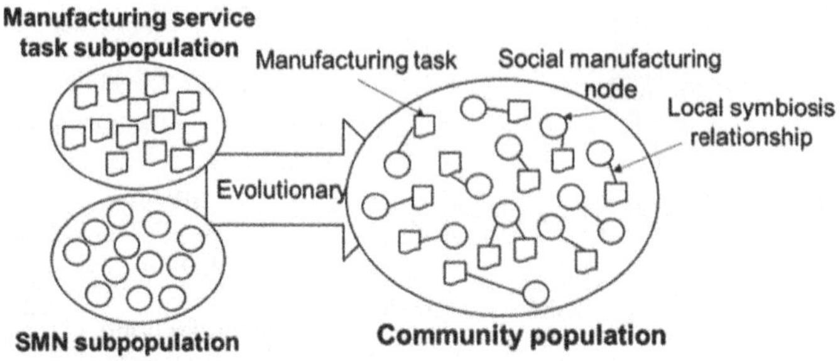

Fig. 11.5 The community population of a dynamic community

the delivery time. Based on the fitness degree function, the evolutionary process can be described as that the SMN and the manufacturing service task adjust their fitness degree constantly to reach an agreement on the premise of maximizing their own profits. If symbiosis cannot be achieved, it means that the SMN cannot provide manufacturing services for the task.

SMNs take product orders autonomously and are self-organized into DMCs under the context of social manufacturing. They will gather together to complete product orders by participating in manufacturing service tasks in the form of DMCs.

11.1.6 Building of Performance Indexes Related to Performances of Order-Driven Dynamic Manufacturing Communities

After constructing order-driven DMCs, the structure of the DMCs is temporarily in a stable status. Here, such a temporary DMC can be abstracted as a directed weighted complex network, where the nodes of the network stand for SMNs, and the edges of the network denote cooperative relationships among SMNs. The directions of the network's edges indicate the manufacturing service sequence executed by SMNs. The cooperation among SMNs may take place in different time, hence there may exist many connections among SMNs. It is worth mention that the weights of edges indicate the number of cooperation among SMNs, hence the weights should be integer values.

The topology of the directed weighted network indicates the composition of the order-driven DMCs. Some performance indexes are used for deeply analyzing the performances of such order-driven DMCs. The indexes can be divided into three main categories, that is, order indexes, node indexes and community indexes. To some extent, these indexes reflect the essential performances hidden in product orders, SMNs and DMCs. The performance indexes and their meanings can be seen in Table 11.1.

Table 11.1 Performance indexes and their meanings

Indexes	Sub-indexes	Meanings
Order indexes	Task load	The total amount of tasks in the dynamic community
	Difficult coefficient	The difficulty degree to complete the product order
Node indexes	Node betweenness	Measures of node centrality, and usually represents the importance of an SMN in the dynamic community
	Node depth	The order of an SMN in the task planning
Community indexes	Service capability	The maximum amount of services could be provided by the dynamic community
	Correlation coefficient	The strength of cooperation among SMNs in the dynamic community
	Community robustness	The ability of the dynamic community to resist destruction (usually caused by SMN leaving). It can be measured by the extent of damage to which the community can withstand

11.2 Planning and Real-Time Scheduling for Social Manufacturing Execution

Section 11.1 describes the shaping process of order-driven DMCs. In this sub-section, the planning and real-time scheduling of social manufacturing system are presented.

Manufacturing execution planning is also one of the vital problems in the context of social manufacturing. It is about how to organize the various manufacturing service tasks from different SMNs to fulfill product orders. In the social manufacturing system, order-driven task planning is considered from two aspects, that is, among SMNs and inside an SMN. In addition, initial configuration and production preparation related to SWs, ECPS nodes, social factory models and task planning results, must be done before production activities start.

As soon as starting production activities, SMNs complete their manufacturing service tasks by using real-time scheduling that is actually the execution procedure of planning results [7]. Here, the real-time scheduling of manufacturing service tasks also involves two levels, that is, inside an SMN and among SMNs. Within an SMN, the real-time scheduling is executed autonomously according to the social factory model, which is concurrently supported by data sampling and cyber-credit recording. Among SMNs, a blockchain-based strategy is introduced to implement data sampling and cyber-credit recording so as to ensure the integrity, confidentiality, and credibility of the production data. In this sub-section, a smart-contract-based decentralized application (DAPP) is deployed to enable real-time tracking of manufacturing service tasks among SMNs.

11.2.1 Order-Driven Task Planning Among SMNs

When SMNs undertake manufacturing service tasks from the task tree of a product order, order-driven DMCs are respectively formed at the same time. After that, task planning among SMNs should be implemented to make sure that the product order could be smoothly carried out.

Task planning among SMNs is a gaming process where SMNs negotiate with each other to reach an agreement on the sequence of manufacturing service tasks. Under the context of social manufacturing, SMNs are in peer-to-peer relationship in the correspondent order-driven DMCs in the role-driven manner. It is necessary to develop a decentralized strategy to support the task planning. Firstly, SMNs elect a leader SMN through online voting. Afterwards, SMNs put forward their proposals on the task planning based on their real-time service capacities and send them to the leader SMN. Then the leader SMN coordinates all the proposals from SMNs to form a draft schedule of manufacturing service tasks, and broadcasts the draft schedule to the correspondent order-driven DMCs. SMNs check and modify the draft schedule until all the SMNs agree with it.

The scheduling agreement by all the SMNs is determined as the final scheme. Here, the leader SMN will protocol a smart contract called as task planning contract, which is a codified final scheme with corresponding default clauses. When the task planning contract becomes effective, each SMN will get a copy of it. Referring to the description in Chap. 9, we know that the task planning contract will execute the breach clauses automatically when an SMN breaks a contract.

It must be pointed out that the approved schedule among SMNs is manufacturing-service-task-level task-schedule. Such a schedule specifies the sequence of running manufacturing service tasks and the correspondent SMNs that will undertake the tasks.

11.2.2 Order-Driven Task Planning Inside an SMN

On the basis of description in the previous sub-section, we understand that the task planning among SMNs concerning order-driven DMCs is completed and the manufacturing service tasks of SMNs are identified. Afterwards, SMNs can autonomously decompose their manufacturing service tasks to be undertaken into process-flow-level workpiece-schedules related to production process plans and facility-level process-schedules related to specific production processes. Such schedule and its sub-schedules are presented in the form of "*Gantt*" chart. Here, social factory model is used for plotting the above task planning concerning an SMN. As shown in Fig. 11.6, the task planning inside an SMN can just be divided into two stages, i.e., the initialization stage and the planning stage.

In the initialization stage, when an SMN undertakes manufacturing service tasks, its social sensors will send the information about the manufacturing service tasks to

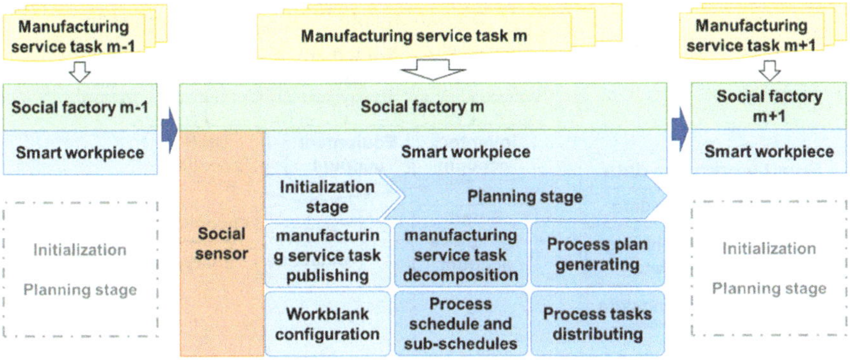

Fig. 11.6 Task planning inside an SMN

its social factory model. When the social factory model receives the information, it will keep the information and publish the manufacturing service tasks in the task list. Then the social factory model appoints its crew(s) to carry out the manufacturing service task. As a start point, all the workblanks are configured as SWs one by one with the writable RFID tag and initial manufacturing information is respectively written into the RFID tag of each SW. This ensures that SW is capable of interacting with the social factory model. This stage aims to prepare connection between the SW and the various ECPS nodes inside this social factory model.

In the planning stage, the social factory model will decompose the manufacturing service tasks into process-flow-concerned production process plans related to workpieces to be fabricated and generate process-flow-level workpiece-schedules. Furthermore, it will also use specific production processes related to a specific facility to generate the facility-level process-schedules based on possible running status of the facility. In this way, the task planning results will be presented as a "*Gantt*" chart in which the workpiece-schedules followed by the facility-level process-schedules are illustrated and depend on correspondent production process plans.

11.2.3 Data Sampling and Cyber-Credit Recording Inside an SMN

After the task planning inside an SMN, we start to run its social factory model to carry out the above three kinds of planned schedules by enabling various ECPS nodes to sample production data. At the same time, cyber-credit recording systems also need to run so as to solve the problem concerning trust and security during operating the SMN. Here, the purpose of the data sampling is to acquire original operating data of ECPS nodes to support self-decision-making, self-diagnosis, visualization, etc. Cyber-credit recording is evaluated based on the historical work performance. It provides an effective tool to enable transparent control and management in this

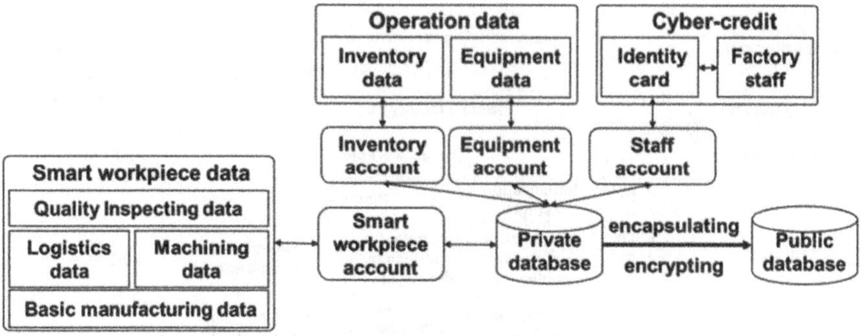

Fig. 11.7 Data sampling and cyber-credit recording inside an SMN

social factory model. Data sampling and cyber-credit recording inside an SMN are presented in Fig. 11.7. Here, the data sampling deals with collecting SW data and operation data.

(1) ***The sampling of SW data***

SW data is composed of basic manufacturing data, logistics data, production data, quality inspecting data, etc. When the SMN undertakes a manufacturing service task, the basic manufacturing data of the task will be sent to it firstly. Afterwards, the SMN will create all the SWs accounts for the manufacturing service task based on the above basic manufacturing data. In fact, the SW account is uniquely bound to the SW according to the manufacturing service task. After that, the SW has right to write and read logistics data, production data and quality inspecting data by using its SW account according to step-by-step progress on running the task. At the same time, the SW can interact with correlated ECPS nodes in real time via interaction interfaces of social sensors.

In the machining procedure of a SW, for example, logistics data, machining data, and quality inspecting data are closely related to the machining processes of the SW. Taking a machining process as an example, in the beginning, a logistics equipment transports the SW to a machining equipment, and the SW gets the logistics data and writes them into its SW account. When the SW is handled with the machining equipment, the SW interacts with the machining equipment to obtain machining data and transmits them to its SW account simultaneously. After finishing the current machining process, the SW would be transported once again to the quality inspection station. Similarly, the logistics data are stored. The quality inspector then performs an examination and updates the inspecting data to the SW.

It must be declared that written-in data to the SW are either indexes of these data or themselves depending on the storage capability of an RFID tag attached to the SW. Detailed production data can be obtained by using the indexes to enquire backend manufacturing database. Here, the RFID tag attached the SW looks like a very tiny temporary database.

(2) ***The sampling of operation data***

Operation data can be divided into two classes, i.e., inventory data and equipment data.

By using unique SW accounts, a warehouse configured with interactive interfaces is able to exchange inventory data with SWs. SWs will be stored in different allocations according to their sizes when they enter the warehouse. The sensor network deployed in the warehouse captures inventory data and updates them. In contrast, if SWs leaves the warehouse, inventory data are removed from the SW account and saved as historical data.

Equipment data are collected by various physical sensors, such as temperature sensors, energy sensors, acceleration sensors, force sensors, etc. Generally speaking, equipments refer to machining tools, assembling workstations, logistics facilities, quality inspecting workstations and other auxiliary tools. Each equipment is installed with plenty of add-on sensors that act as an ECPS node to satisfy its practical needs in data sampling. These sensors through the ECPS node monitor the real-time running status of the equipment at a specific sampling frequency, and the original data collected by the sensors are directly saved in a backend manufacturing database for proper analysis because of their nature of big data. In addition, these data can also be found through the index data stored in correspondent SWs.

It should be noted that all the sampling data need to be stored in a private manufacturing database that belongs to a specific SMN and can be not accessed by other SMNs except some inventory data from public warehouses. Under this case, a public database is used to save such public data, including encapsulated and encrypted data from the private manufacturing database and social business context data generated through the social interactions with other SMNs.

(3) ***Cyber-credit recording***

Cyber-credit is a comprehensive evaluation of SMN staffs, which reflects in everyday work. Each staff is offered with a unique identity card that is linked to their staff account. This enables the SMN to monitor the staff in real time. At the same time, with the help of the identity card, the staff can interact with the SMN within his jurisdiction. The identification card is responsible for the collection of interactive data and authorized to write the data into the staff account. When a staff executes a task, the SMN monitors and evaluates the staff performance via the identity card, and the evaluation result is sent to the staff account. The staff can't intervene in the whole process, but he can check the information in his staff account. On the basis of staff's cyber-credit records, the cyber-credit records of an SMN can also be created. Such records can be integrated into the above blockchain shell.

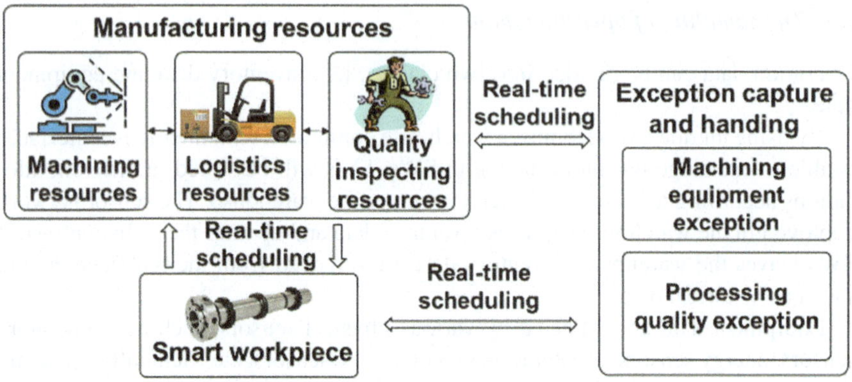

Fig. 11.8 The real-time scheduling of manufacturing services

11.2.4 Real-Time Scheduling of Manufacturing Services Inside an SMN

In the real production cases, it is impossible for an SMN to plan all manufacturing services for smart WIPs in advance due to unexpected disturbances. In this situation, the SMN needs to run its planned schedules in three levels mentioned above according to its real-time running status. Especially, running status from facilities or equipments inside an SMN, which deal with machining, assembling, logistics, quality inspecting, etc., would influence in depth how to execute the above planned schedules. The purpose of real-time scheduling is to coordinate the facilities to fit with dynamic changes from unexpected disturbances which either happen in such facilities or are oriented from both manufacturing service tasks, correspondent planned production process flows, and specific production processes related to a special facility. It means that the planned schedules must be modified to a new changeable "*Gantt*" chart if one of the above unexpected disturbances happens. Figure 11.8 describes the real-time scheduling of manufacturing services.

When a machining equipment is in an idle state, for example, it sends its state data to SWs. When an SW chooses the machining equipment on the basis of the state data, it interacts with logistics equipment to transport itself to the machining equipment in an optimal route. And then, the machining equipment starts the machining service task. After finishing the machining process, the SW interacts with logistics equipment once again to deliver itself to the quality inspecting station and further examines whether or not its machining quality satisfies the quality requirements. If the SW is qualified, it will be delivered to a workpiece buffer waiting for the next machining process. This process won't stop until all the processes are finished. Finally, the SWs are delivered to the customers or following SMNs. Actually, this is a normal way without unexpected disturbances and means that planned schedules run correctly and no revisions of "*Gantt*" chart happen.

In fact, exception capturing and handing are essential for the real-time scheduling. On one hand, when some production processes of the SW are unqualified, the SMN assesses the situation of the process and makes decisions for next step. If the SW can be re-handled to satisfy the processing requirements, a new production process would be added to the production process flow. Otherwise, the SW would be marked as a waste part and distributed to waste part area in the inventory. On the other hand, it is inevitable that some machining equipment may break down. In this situation, the rest of machining tasks will be rescheduled to generate a new "*Gantt*" chart, and the trouble with the machining equipment must be tackled immediately.

11.2.5 Data Sampling and Cyber-Credit Recording Among SMNs

In the order-driven DMCs, SMNs cooperate with each other to fulfill the product orders by integrating their various facilities and production process flows during running the planned schedules. In order to enable cooperative activities, it is necessary to apply data sampling and cyber-credit inside an SMN not only for obtaining the running status of the SMN but also for bridging different SMNs. It is also the basic premise that the real-time scheduling mechanism works. To ensure data integrity, confidentiality and credibility, the data sampling and cyber-credit recording among SMNs are implemented using a blockchain-based strategy, as shown in Fig. 11.9. Here, each SMN has a unique node account with a pair of private key and public key. The data and the cyber-credit of the SMN are stored respectively in the private memory and the public memory related to their node accounts. With the private key and the public key, the SMN can write the data of manufacturing service task into the private memory. The cyber-credit can be updated by the DMC with the public key and a valid dynamic password. The dynamic password is generated randomly by the node account when an SMN undertakes manufacturing service task and participates in the DMC. The usage of the valid dynamic password depends on one of two conditions. The first is that the password is used only during the valid period and the second is that the password is available only when more than 67% DMC members agree. In addition, an order account is created to store and update the status data of product orders. The order account also owns a pair of private key and public key, but the private key has a limitation that its validity is similar to the validity of dynamic password in the node account. It is worth mention that all the data related to node or order accounts are recorded in the blockchain.

During running the planned schedules, data sampling among SMNs is closely related to manufacturing service tasks. Firstly, SMNs collect all the raw data of the manufacturing service tasks and store them in private database. Afterwards, raw data about product orders are encapsulated as public data, which are further classified into processing data, quality data and logistics data. These three kinds of public data are autonomously transmitted to the private memory of SMN node accounts

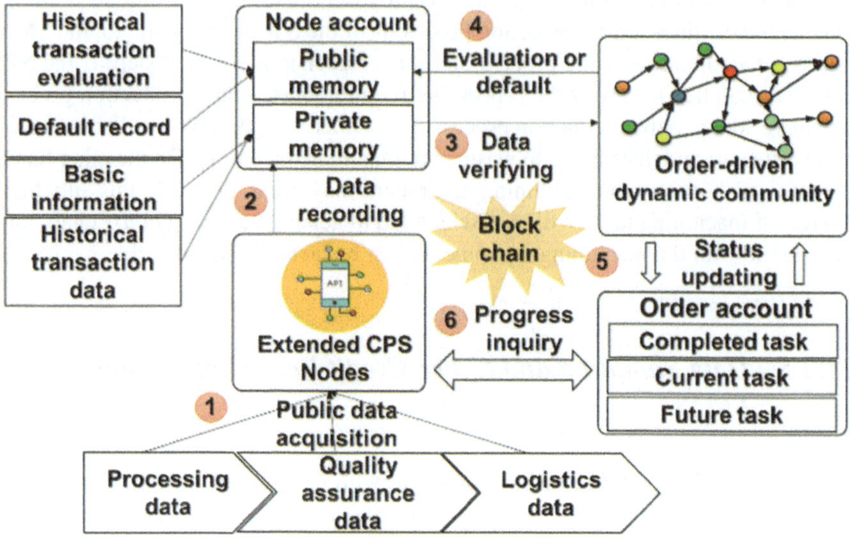

Fig. 11.9 Data sampling and cyber-credit recording among SMNs

via ECPS nodes. Data in the private memory are readable for other SMNs in the same DMC. When an SMN completes its manufacturing service task, the DMC reads and verifies the data of the manufacturing service task. If the data are valid, they are recorded in blockchain by the DMC and then a transaction evaluation of the manufacturing service task is written into the public memory of the node account with the valid dynamic password. At the same time, the DMC updates the status to the order account with its corresponding valid private key. All the SMNs have authority to access the progress on the product orders. In contrast, if an SMN tries to tamper the data, it will be punished economically and disqualified for the manufacturing service task. In addition, a default is recorded in the node account.

Based on the procedure mentioned above, the sampled data and cyber-credit records related to node accounts increase gradually as time goes on and are kept in blockchain permanently. In this way, the blockchain plays a significant role in guaranteeing the inter-enterprise cooperation among SMNs.

11.2.6 Real-Time Tracking of Manufacturing Services Among SMNs

To support the mass cooperation among SMNs in the order-driven DMC, during running the planned schedules, it is necessary to enable the real-time tracking of manufacturing services. On the one hand, such a real-time tracking mechanism pro-

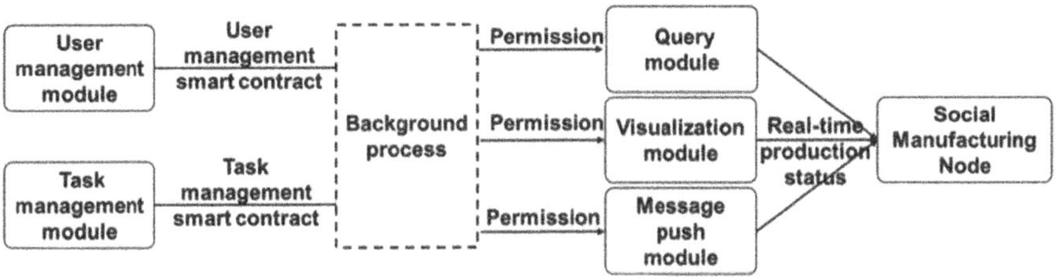

Fig. 11.10 Smart contract-based decentralized application for real-time tracking

vides SMNs with the current running status of product orders and enables SMNs to prepare something for up-coming manufacturing service tasks. On the other hand, it makes customers know the "*footprint*" their product orders go to.

As shown in Fig. 11.10, the real-time tracking mechanism works by means of using a smart-contract-based DAPP, which is composed of five functional modules, i.e., user management module, task management module, visualization module, query module and message push module. The first two functional modules are customizable for the DMC based on smart contract. With the help of the user management module, here, a user management smart contract is developed to assign permissions for SMNs, which makes real-time production data shareable only among the SMC members. The task management module is used to satisfy different tracking requirements from different product orders, where the DMC manages the tracking requirements of each SMN via a smart contract related to task management. When smart contracts related respectively to user management and task management are established by the DMC, they are compiled as a background procedure to support the operation of the DAPP.

The left three functional modules are common for all the product orders. The visualization module functions as a "*Kanban*" to vividly present the real-time production running status of manufacturing service tasks, which provides an interface for SMNs to update the real-time running status of their own manufacturing service tasks. The query module assists SMNs to acquire the real-time tracking data that they care. The message push module can push the real-time production data to SMNs autonomously when production activities are in progress.

11.3 Evaluation of Manufacturing Services in Social Manufacturing System

Social manufacturing system [9], powered with DMCs and SMNs, offers many possibilities for enabling different manufacturing service capabilities. The self-organizing mechanism, planning and operating principles, and key enabled technologies concerning manufacturing services are also discussed in depth in the above sections. Here, we will briefly describe how to evaluate the manufacturing services in the context of social manufacturing.

The distributed blockchain technology and smart contracts can be used to build a credit and security mechanism for social manufacturing [10]. On the one hand, the credit and security mechanism settles credit problems in social manufacturing network by modeling cyber-credit recording procedure of SMNs or SMRs. On the other hand, distributed blockchain technology ensures that the original context of social business interactions and the production data of SMNs are credible. These data can be used to evaluate manufacturing services in social manufacturing system. In order to successfully carry out the evaluation, we need to solve the following three problems.

The first problem is how to build quantitative indicators for evaluating manufacturing services. In fact, evaluating manufacturing services is closely related to two aspects, that is, original social business interaction context and SMN production data. It is obvious that recorded data are unstructured. It means that it is hard to use these data for evaluation. Therefore, how to uniform the hybrid data into quantitative indicators is the key to evaluate manufacturing services in the context of social manufacturing.

The second problem is how to build a multi-level evaluation system aiming at three targets related to manufacturing services respectively in the levels of SMNs, DMCs and social manufacturing system. This evaluation is carried out based on different datasets from the above levels [11]. Here, evaluating manufacturing services in each level is a start point to combine three targets together. In addition, ranking of SMNs and DMCs can be also implemented as a "by-product" in terms of using this multi-level evaluation system.

The third problem is how to build quantitative approach for value calculation of cyber-credits. Although the cyber-credit mechanism can be constructed, a practical methodology for calculating the credit scores of various SMNs still need to be developed. On the one hand, it settles trust and security issues in the mass cooperation among unfamiliar SMNs. On the other hand, it is concerned with solving the problem of manufacturing services evaluation and provides a real-time visible "*Kanban*" of credit rating for production decisions.

On the basis of analysis mentioned above, an evaluation system for multi-level manufacturing services (ES-MMS), together with the cyber-credits space (CCS) configuration, is proposed to solve the three problems mentioned above.

The evaluation is carried out respectively from node-level, community-level, and system-level. All of the multi-level evaluations are based on the historical social business contexts, original production data, manufacturing service task data, product order data, etc. To illustrate the operational logic of ES-MMS and CCS, some clarifications should be made first.

Clarification 1: *Evaluation System for Multi-level Manufacturing Services (ES-MMS)* is a system which adopts systematic engineering thinking to establish a multi-level framework for evaluating manufacturing services in the context of social manufacturing.

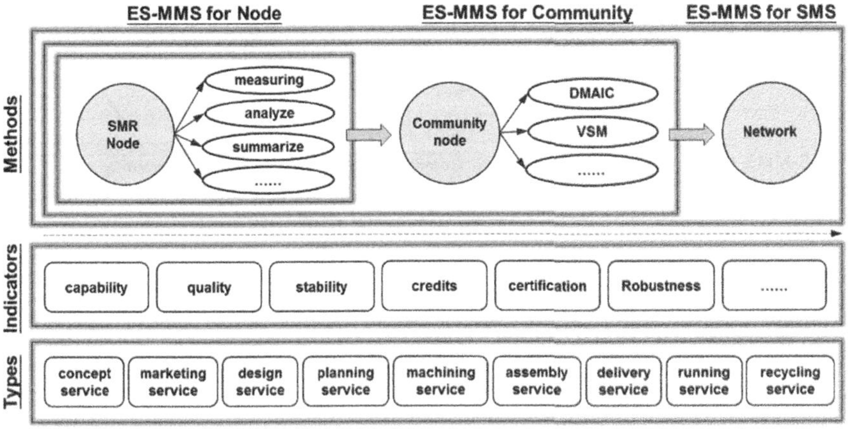

Fig. 11.11 Multi-level ES-MMS in the social manufacturing system

ES-MMS deals with product manufacturing activities in the whole stages of a product life cycle in social manufacturing. It needs to consider various issues concerning quality, capability, stability, responsiveness, tracing/tracking, credits, certification, etc., as shown in Fig. 11.11. ES-MMS will not only focus on the performance of a single indicator but also adapt to multiple indicators related to different application issues. For example, when evaluating a logistic-service SMN or DMC, the delivery capability and cost controlling capability might be the main indicators, but other issues should also be considered.

Clarification 2: *Cyber-Credits Space* (CCS) is a kind of "*space*" where all the possibilities related to cyber-credit recording and scoring issues in the levels of SMNs and DMCs are described formally in the context of social manufacturing.

CCS covers all the product manufacturing activities in the whole stages of a product life cycle. In the stage of product requirement analysis and design, CCS is main influencing factors of creating social relationships. In the stages of production, delivery, running and recycling, CCS becomes passive and is affected and modified by the ES-MMS. CCS also has two layers, that is, SMN layer and DMC layer. The relationship between ES-MMS and CCS can be seen in Fig. 11.12.

11.3.1 Recording of Manufacturing Services Contexts as Historical Data

As shown in Fig. 11.12, the recording of various original data is the basis of the ES-MMS and CCS. It is important to extract the valuable information from a large number of original data including social business contexts and manufacturing data. Since these data are usually distributed and unstructured, they need to be further handled

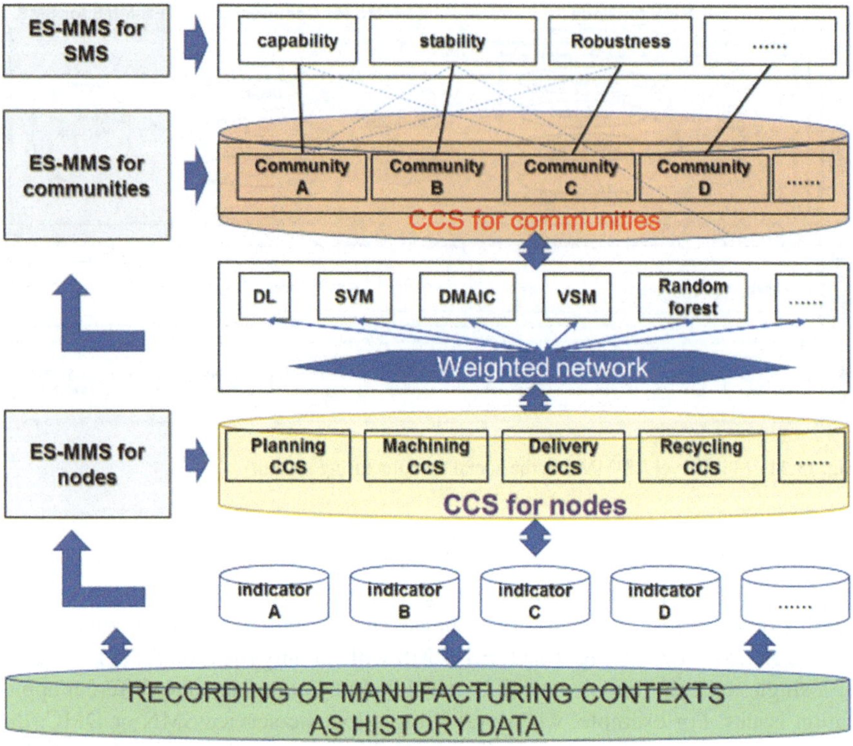

Fig. 11.12 Relationship between ES-MMS and CCS

and then are sent to a backend database. The purpose to do as this is to prepare data rich enough for realizing the multi-level evaluation of social manufacturing services. Figure 11.13 just illustrates how the procedure concerning ES-MMS and CCS works.

As shown in Fig. 11.14, furthermore, the CCSs of both SMNs and DMCs can be regarded as a physical three-dimensional space, where "*x*-*y*" plane stands for manufacturing space and z-axis indicates credit scores. Nodes in different colors indicate different credit scores. For instance, the green nodes which actually are SMNs mean to have a higher credit score. Some SMNs which have common benefits collaboratively for doing manufacturing service tasks will gather together as a DMC shown in the dotted circle in the Figure. The next sub-section will discuss how to calculate the value of the credit.

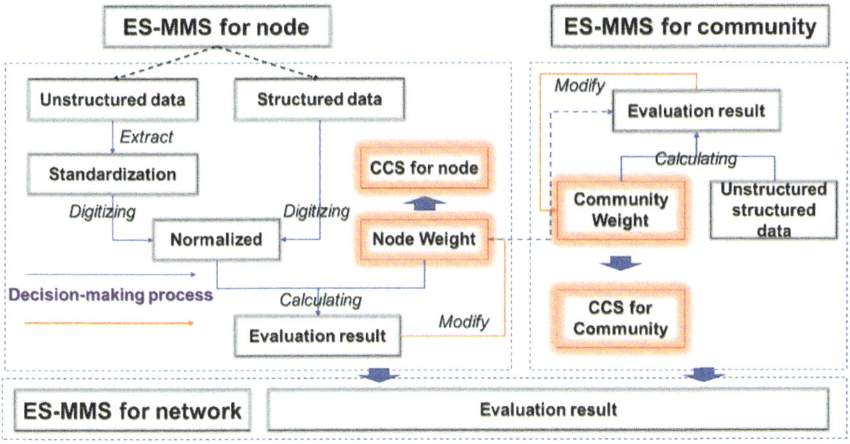

Fig. 11.13 Contexts/data analytics on SMN, community, and network based on ES-MMS and CCS

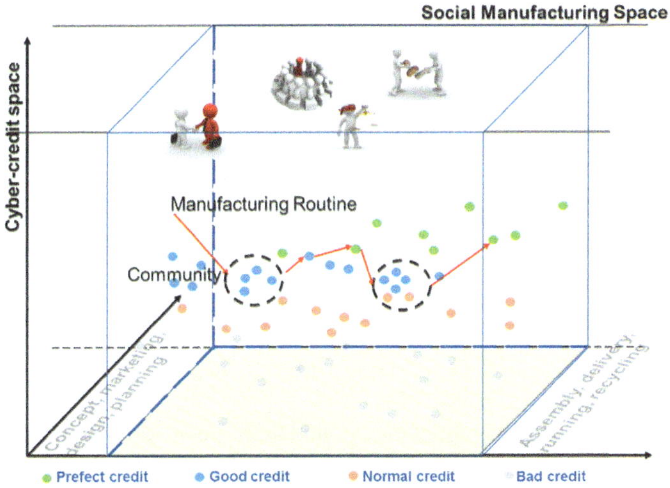

Fig. 11.14 Manufacturing space and CCS

11.3.2 Evaluation of Manufacturing Service of an SMN

Original data in an SMN has mainly two forms, that is, unstructured and structured data.

On one hand, the unstructured data, such as the voice of customer (VOC) should be analyzed for extracting useful information. These data might exist in pre-processing, in-processing and after-processing stages. And some labels can be used to distinguish the satisfaction degree of customers. Taking the VOC as an example, some quantized labels can be attach to the comments, and a mathematical architecture will be introduced to calculate and evaluate the manufacturing services. All these labels are the attributes of the nodes, which can form a set as

$$UD_i = \{ud_{i1}, ud_{i2}, \ldots, ud_{ij}, \ldots, ud_{in}\} \tag{11.1}$$

where ud_{ij} indicates the j-th unstructured index of the node i. There are n indexes in the node i totally. The corresponding weights of the indexes can be formalized as follows.

$$WUD_i = \{wud_{i1}, wud_{i2}, \ldots, wud_{in}\} \tag{11.2}$$

where wud_{ij} indicates the weight of the ud_{ij}, and there are also n corresponding weights in the node i.

Note that the weights are dynamic for the different manufacturing service scenarios. For instance, when evaluating a logistical service, the weight of delivery punctuality should be higher. These weights are dynamically adjusted while the manufacturing services are executed.

On the other hand, structured data from original contexts or data, such as machining quality performance, should be measured and normalized. These indexes can be obtained via measuring tools or statistical tools. These data can be written as

$$D_i = \{d_{i1}, d_{i2}, \ldots, d_{ij}, \ldots, d_{im}\} \tag{11.3}$$

where d_{ij} is the j-th structured index of the node i. There are m indexes in the node i totally. Similar to the unstructured data, there are also weights for the structured ones.

The corresponding weights set can be expressed as

$$WD_i = \{wd_{i1}, wd_{i2}, \ldots, wd_{im}\} \tag{11.4}$$

where wd_{ij} indicates the weight of the d_{ij}, and there are m corresponding weights in the node i.

Then the evaluating result of manufacturing service of the node i can be presented as

$$MMS_{node} = \sum (UD_i * WUD_i) + \sum (D_i * WD_i) \tag{11.5}$$

Note that weights of the indexes vary with different conditions. In addition, the original weights are defined as one. The weights need to be modified according to the following formula.

$$w_i = \frac{\xi_c}{\xi_s} w_i' \tag{11.6}$$

Table 11.2 Different manufacturing services with their evaluation indexes

Service name	Main evaluation indexes
Concept	Practicality Cost Novelty …
Marketing	Time Volume staff in …
Design	Structural novelty Structural machinability Design qualification …
Planning	Scheduling redundancy Executable Robustness …
Machining	Process quality Process stability Process Capability …
Assembly	Interchangeability Assembly coordination …
Delivery	Package integrity Punctuality …
Running	Anti-disturbance Monitorability …
Recycling	Convenience Sustainability …

where ξ_c denotes the complaint rate for the index; ξ_s stands for the satisfaction rate of the index; w_i is the adjusted weight and w_i' is the unadjusted one.

A detailed explanation of the evaluation indexes for manufacturing services is presented in Table 11.2.

11.3.3 Evaluation of a Dynamic Manufacturing Service Community

Corresponding to the SMN's evaluating, the evaluation of a dynamic manufacturing service community can be regarded as a combination evaluation of different SMNs. And an SMN may contain one or more types of manufacturing service, so the evaluation of a community may transfer to the evaluation of the multiple services.

The evaluation process of a community can be illustrated in Fig. 11.15.

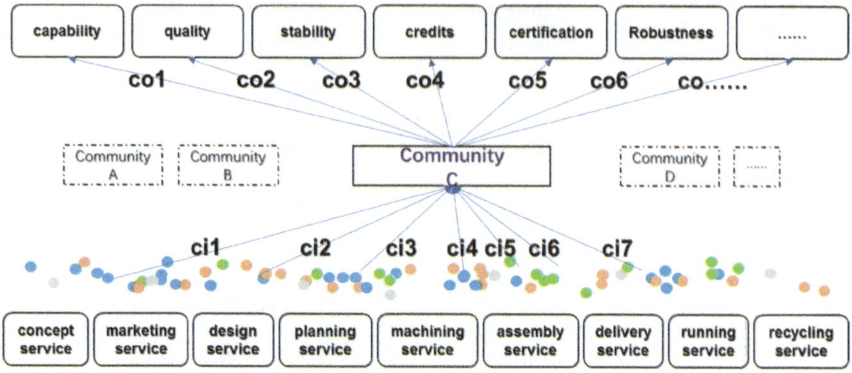

Fig. 11.15 Evaluation of manufacturing service community

The evaluation indexes include capability index, quality index, stability index, credits index, etc. These indexes can be formalized as follows

$$E_c = \{c_{o1}, c_{o2}, \ldots\ldots c_{on}\} \tag{11.7}$$

where E_c is a set of all the evaluable indexes and c_{oi} stands for a specific index. For example, c_{o1} stands for the capability of a community C and all the inputs' indexes, i.e., the manufacturing services, in E_c may have influences on its evaluation value.

So, a sensitivity matrix approach is used to express the relationship between the manufacturing services inputs and the evaluable indexes outputs of the community C.

$$\begin{bmatrix} c_{ik1} & c_{ik2} & \cdots & c_{ikn} \end{bmatrix} \mathbf{M}_c = \begin{bmatrix} c_{ok1} & c_{ok2} & \cdots & c_{okm} \end{bmatrix} \tag{11.8}$$

where c_{ikj} is the j-th input evaluating value and c_{okj} is the j-th output one of the SMN in the k-th sampling in the community. And

$$\mathbf{M}_c = \begin{pmatrix} \frac{\partial c_{ok1}}{\partial c_{ik1}} & \frac{\partial c_{ok2}}{\partial c_{ik1}} & \cdots & \frac{\partial c_{okm}}{\partial c_{ik1}} \\ \frac{\partial c_{ok1}}{\partial c_{ik2}} & \frac{\partial c_{ok2}}{\partial c_{ik2}} & \cdots & \frac{\partial c_{okm}}{\partial c_{ik2}} \\ \vdots & \vdots & \ddots & \vdots \\ \frac{\partial c_{ok1}}{\partial c_{ikn}} & \frac{\partial c_{ok2}}{\partial c_{ikn}} & \cdots & \frac{\partial c_{okm}}{\partial c_{ikn}} \end{pmatrix} \tag{11.9}$$

is the sensitivity matrix for the relationship of n inputs and m outputs of community C. And k is the number of samples, which changes along with the evolution of community C. In another word, $\mathbf{M}_c$ is based on the historical contexts/data and affected by the new contexts/data. Once a new community has been formatted, $\mathbf{M}_c$ can be used for the evaluation of historical contexts/data.

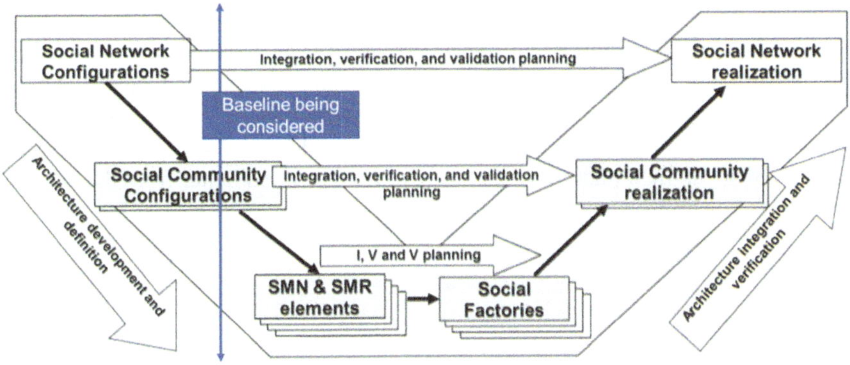

Fig. 11.16 Vee model of SMS

11.3.4 Evaluation of Social Manufacturing System

It is very difficult to evaluate a social manufacturing system or network because it has too many dynamically changeable characteristics as a socio-technical system. In addition, it is also too hard to clearly explain whether a system is good or not in detail.

From systems engineering perspective, getting a better performance of social manufacturing system obeys to an iterative procedure concerning the top-down synthesis, development, and operations of modeling its real-world in a near optimal manner.

Under the consideration of the above factors, a Vee model [12] is used for illustrating how to model a social manufacturing system in the whole stages of a product life cycle, as shown in Fig. 11.16. In the model, time and system maturity are described from left side to right side. The core of the Vee model depicts the evolving baseline from user requirements agreement that is helpful to identify a system concept to definitions and development of system components that will comprise of final social manufacturing system. For more complicated consideration, the dual-Vee model is under using for describing social manufacturing system.

11.4 Concluding Remarks

The execution of social manufacturing system is dependent on its key work principles, which include three aspects, that is, the shaping of order-driven DMCs, the planning and real-time scheduling of manufacturing service tasks, and the evaluation of manufacturing services. In fact, shaping an order-driven DMC is a basis of running a social manufacturing system. The interconnection inside an SMN and among SMNs are two key procedures of shaping a DMC. Realizing the efficient and effective manufacturing service task planning and real-time scheduling means that the social manufacturing system can run very well respectively in the levels of SMNs, DMCs, and system. And the use of a multi-level evaluation system declares such a fact that the social manufacturing system is dynamically changeable and can be evolved to a higher level depending on evaluation-result-driven continuous improvement. Finally, it is predictable that the use of dual-Vee model would make the configuration and runtime of social manufacturing system become better and better from the angle of systems engineering.

References

1. Ding K (2017) Cyber-physical-social interconnection-driven social manufacturing system and its key operating technologies. Doctor of Engineering, Xi'an Jiaotong University, Xi'an
2. Jiang PY, Ding K, Leng JW (2016) Towards a cyber-physical-social-connected and service-oriented manufacturing paradigm: Social Manufacturing. Manuf Lett 7:15–21
3. Tao F, Cheng Y, Zhang L, Nee AYC (2017) Advanced manufacturing systems: socialization characteristics and trends. J Intell Manuf 28(5):1079–1094
4. Ding K, Jiang PY (2017) Social sensors (S2ensors): a kind of hardware-software-integrated mediators for social manufacturing systems under mass individualization. Chin J Mech Eng 5(30):1150–1161
5. Nakamoto S (2008) Bitcoin: a peer-to-peer electronic cash system. Web Pages, available at: http://bitcoin.org/bitcoin.pdf
6. Swan M (2015) Blockchain: blueprint for a new economy. O'Reilly Media Inc., CA, USA
7. Leng JW, Chen QX, Mao N, Jiang PY (2018) Combining granular computing technique with deep learning for service planning under social manufacturing contexts. Knowl-Based Syst 143:295–306
8. Nowak MA, Sigmund K (2004) Evolutionary dynamics of biological games. Science 303(5659):793
9. Jiang PY, Leng JW, Ding K, Gu P, Koren Y (2016) Social manufacturing as a sustainable paradigm for mass individualization. Proc Inst Mech Eng, Part B: J Eng Manuf 230(10):1961–1968
10. Liu JJ, Jiang PY, Leng JW (2017) A framework of credit assurance mechanism for manufacturing services under social manufacturing context. In: 2017 13th IEEE Conference on Automation Science and Engineering (CASE), Xi'an, China, 20–23 August 2017
11. Li PL, Jiang PY (2015) Research on quality-oriented outsourcing decision architecture for small-batch parts of multistage machining processes. In: Proceedings of the 22nd International Conference on Industrial Engineering and Engineering Management, Guangzhou, China, 6–9 September 2015
12. Walden DD, Roedler GJ, Forsberg KJ, Hamelin RD, Shortell TM (2015) INCOSE systems engineering handbook, 4th edn. Wiley

Chapter 12
Industrial Cases Concerning Social Manufacturing

Pingyu Jiang

12.1 Starting Points

One of the most important key points for social manufacturing paradigm is for a group of manufacturing enterprises to grow up an Internet-based sustainably ecological enterprise circle and dynamically implement Internet-based connecting and communicating behaviors in business under the context of social manufacturing. At the same time, this sustainably ecological enterprise circle would present the natures listed as follows:

- microlization and minimalization of manufacturing resources related to a big-and-middle-scale manufacturing enterprise,
- self-organization of socialized manufacturing resources (SMRs),
- virus-like propagation of organizational structure,
- share and competition of both capabilities and business benefits for SMRs inside a manufacturing community or among different manufacturing communities,
- dynamically distributive infrastructure,
- big-data-driven decision-making and performance optimization, and
- novel industrial software model to be used.

On the top level, specially, social manufacturing paradigm works in the form of social manufacturing network which consists of dynamically changeable manufacturing communities. While any socialized manufacturing resource, which either comes from the decomposition of big-and-middle-scale manufacturing enterprise or is originally an independent micro-and-small-scale manufacturing enterprise, is able to act as a node in different communities. This social manufacturing network runs under a distributive environment, enables the above natures like self-organization and virus-like propagations, is very huge, shape-always-changeable and shape-unknown, and looks like a field-sensitive Internet-based ecological enterprise circle.

P. Jiang, *Social Manufacturing: Fundamentals and Applications*, Springer Series in Advanced Manufacturing, https://doi.org/10.1007/978-3-319-72986-2_12

In fact, there are two types of phenomena in this circle. The first type of phenomenon is that it is possible to exist one or several leading manufacturing enterprises either in the level of manufacturing communities or in the level of social manufacturing network. Here, the leading enterprises don't act the role of SMRs but look like a kind of service provider in the above two levels so as to gather correspondent SMRs around them. The service providers for the level of manufacturing communities sometimes use "*platforms*" as the front-ends of correspondent manufacturing communities to respectively face to their SMRs. They seem to be a kind of "*platforms*" owner. The service provider for the level of a social manufacturing network is just concerned with running a sustainably ecological enterprise circle and needs to integrate different service providers from the level of manufacturing communities. The second type of phenomenon is that all the SMRs connected to both manufacturing communities and a social manufacturing network relatively have the equal power and are without leading ones. Here, it is also possible that some SMRs can work together in the above two levels in the natures of self-organization, virus-like propagation, and share and competition without the control of one or several leading enterprises. The key point we need to declare here is that such SMRs only utilize some public platforms and social media like *Taobao*'s e-commerce platform, *Wechat*, *Facebook*, etc., as their business interaction "*blackboards*" and "*places*" so as to create and enable collaborative relationships. At the same time, these resources aren't controlled by the public platforms and social media. It is the best way to develop a new industrial software model instead of the above public platforms and social media so as to support this kind of amazing working mode. Extremely thinking about this phenomenon, "*everyone is a manufacturer*" may be realized by means of using this new social-media-like industrial software model together with personal 3D printers.

This also means that we can use three types of control mechanisms to enable the distributive implementations of social manufacturing paradigm, which include completely-centralized-control, partially-centralized-control and centerless-self-control mechanisms. Here, the controllable domain of completely-centralized-control mechanism, which is operated by leading manufacturing enterprises, is on the level of different manufacturing communities. Its presenting mode is one or several "*platforms*" as the front-end of the correspondent manufacturing communities. The controllable domain of partially-centralized-control mechanism, which is also operated by leading manufacturing enterprises, is on the level of a social manufacturing network. Similar to the above situation, its presenting mode is concerned with a group of interconnected "*platforms*" respectively as the front-ends of different manufacturing communities and sometimes works in the form of sustainably ecological enterprise circle. The above "*platforms*" also include public platforms and social media mentioned before.

On the basis of descriptions mentioned above, we can find many industrial cases in manufacturing industry, which fit with the first type of phenomenon, for example, *Haier*'s ecological enterprise circle innovation in the aspects of organization and runtime logic, distributive implementation related to X-part manufacturing, etc. Here, the leading manufacturing enterprises in the ecological enterprise circle either try to partially-centralized-control this circle in a wider product and service domain or completely-centralized-control manufacturing communities in this circle in a more narrow product and service field. While SMRs just join the circle as nodes which play two different roles, either working collaboratively and coequally, or under the management of another key manufacturing enterprise.

We can also find some industrial cases in manufacturing industry, which are correspondent with the second type of phenomenon, for example, *RepRap* open-source 3D printer manufacturing enterprise circle on *Alibaba*'s *Taobao* e-commerce platform. This is a typical centerless-self-control mechanism for a distributive implementation of social manufacturing network. The credit and security identifications are guaranteed by the *Alibaba*'s *Taobao* e-commerce platform through a pre-payment system. We believe blockchain model would be the best way to enable the credit and security identifications not only for the working results but also for product manufacturing activities in the whole stages of a product life cycle.

From the angle of how to use SMRs, how to shape an organizational structure and how to realize a runtime logic, three industrial cases which respectively represent the above three types of control mechanisms of the distributive implementations are described as follows in order to demonstrate how to preliminarily realize a social manufacturing paradigm in reality.

12.2 Industrial Case: *Haier*'s Distributive Implementation with Partially-Centralized-Control Mechanism for Social Manufacturing

Haier is originally a big manufacturing enterprise group in *Tsingdao*, China and produces both home-used and industrial electronic products like refrigerators, washing machines, air conditioners, etc. [1]. Currently, *Haier* is doing a pioneer industrial practice in which *Haier* wants to become a leader of world-class Internet-based and sustainably ecological enterprise circle mainly focusing on intelligent home and industrial solutions of using electronic products and correspondent services. Such an industrial practice can be classified into the catalogue related to a kind of initially distributive implementation for the social manufacturing paradigm with the help of partially-centralized-control mechanism. Figure 12.1 just shows what *Haier* wants to do right now.

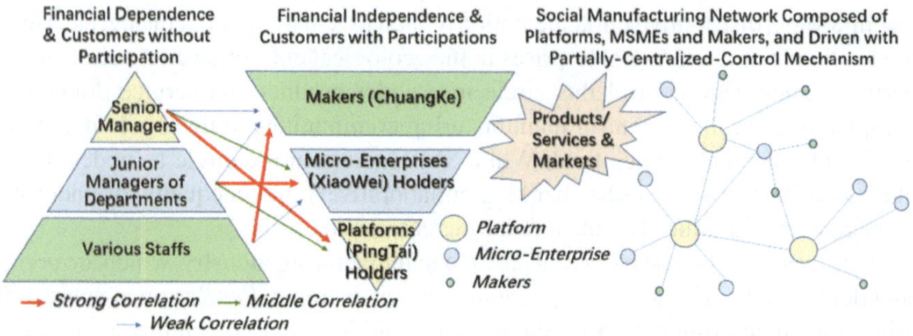

Fig. 12.1 Haier's initial implementation of social manufacturing paradigm driven with partially-centralized-control mechanism

12.2.1 Socialization of Manufacturing Resources

The first key issue of enabling *Haier*'s initial social manufacturing paradigm is how to use SMRs. *Haier*'s fundamental SMRs are from the inside, which focus on [2]:

- decomposing original manufacturing enterprise group either into several controllable and dependent platforms, also called as "*PingTai*" in Chinese, or into a lot of half-independent micro-and-small-scale manufacturing enterprises (MSMEs), also called as "*XiaoWei*" in Chinese, and
- promoting some staffs to become half-or-completed independent makers, called as "*ChuangKe*" in Chinese.

The purpose of *Haier*'s microlization, minimalization and socialization of its manufacturing resources is to create a new sustainably ecological enterprise circle in which *Haier* hopes its manufacturing resources, together with other SMRs that don't belong to it, can work collaboratively, and takes a leading place inside the circle. So it is necessary for *Haier* to develop and control the platforms which are actually the web portals or front-ends of correspondent manufacturing communities in a social manufacturing network, and create a new platform for financial investments to potential makers and MSMEs so as to incubate them growing-up, enhance *Haier*'s own core competitive power and enlarge the number of controllable MSMEs and makers. At the same time, *Haier* also hopes to control most of core MSMEs related to product design, assembly, supply chain and marketing functions, which either are actually decomposed from the inside or invested by it. Figure 12.2 illustrates current platforms *Haier* has built.

Haier's extended SMRs are from the outside, which deal with:

- integrating or investing MSMEs from the outside of Haier under the consideration of "*win-win*" benefits, and
- integrating or investigating potential makers from the outside of *Haier*.

Here, SMRs form the outside of *Haier* are also integrated into correspondent platforms mentioned above. In addition, makers developed well would be transferred

Fig. 12.2 Haier's partial platforms as front-ends in manufacturing communities

to MSMEs. In this way, *Haier* has created a large pool of SMRs and is using them to build a field-sensitive, Internet-based and sustainably ecological enterprise circle for electronic products and services.

12.2.2 Organizational Architecture

The second key issue of enabling *Haier*'s initial social manufacturing paradigm is how to organize SMRs into manufacturing communities and further into a social manufacturing network. Actually, *Haier* changes its original organizational structure of the manufacturing enterprise group into the three-layer architecture, which deals with platforms, MSMEs, and makers.

Different from the social manufacturing paradigm driven with the centerless-self-control mechanism, as shown in Fig. 12.3, *Haier*'s social manufacturing network has a backbone sub-network which consists of all the platforms (called as "*PingTai*" in Chinese) and core MSMEs (called as "*XiaoWei*" in Chinese), which are related to product design, assembly, supply chain and marketing functions, and decomposed from the inside. As an important extension, MSMEs and makers from the outside of *Haier* construct an extended sub-network. Such an extended sub-network makes *Haier*'s initial social manufacturing paradigm, driven with partially-centralized-control mechanism, much more active. It is obvious that this sustainably ecological enterprise circle has the basic characteristics of self-organization, dynamically extendable nodes, virus-like propagation, product order share and competition, etc.

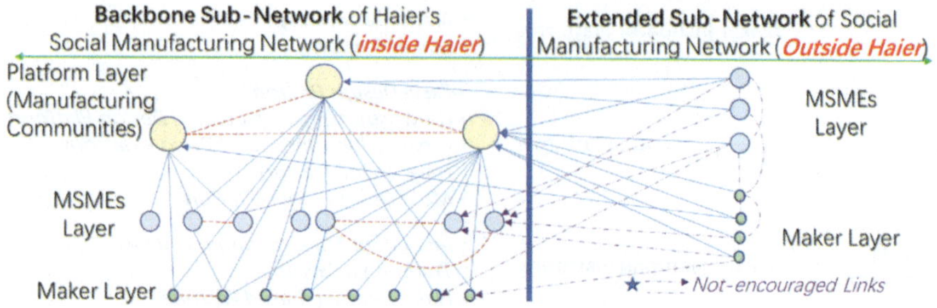

Fig. 12.3 Organizational architecture of Haier's social manufacturing network driven with partially-centralized-control mechanism

It must be pointed out that MSMEs and makers from the outside of *Haier* do not have their own manufacturing communities or platforms because they have to use *Haier*'s manufacturing resources. Of course, it means that *Haier* holds the right of designing rules and would cause the problem of benefit conflict sometimes.

12.2.3 Runtime Logic

The third key issue of enabling *Haier*'s initial social manufacturing paradigm is how to implement a runtime logic by means of using both the manufacturing communities and the social manufacturing network in the form of product-order-driven schemes which depend mainly on the partially-centralized-control mechanism and cover all the product manufacturing activities in the whole stages of a product life-cycle.

It is very clear that a key point to build a runtime logic is where to get and how to finish a product order referring to its type. Generally speaking, a product order essentially consists of the type and the number of products to be fabricated, lead time, specific design requirements if needed, total prices, etc. While a product is composed of a series of parts and assemblies. The parts and the assemblies of a product can be divided further into three catalogues, that is, self-made, bought and outsourced parts and assemblies. At the same time, the number of products to be fabricated decides that the production mode deals with fabricating one-of-a-kind product, small batch of products, or middle and mass volume of products.

These characteristics of the product order imply such two facts that the shape of supply chain depends on outsourcing and crowdsourcing mechanisms, and the shape of design and production flows is just concerned with the correspondent production mode which is individualized production, mass customization, or mass production. Here, value creations of the outsourcing mechanism are mostly from the manufacturing enterprise or the socialized manufacturing resource which holds the product order. This means that the outsourcing mechanism relies mainly on "*acquaintance model*". But value creations of the crowdsourcing mechanism are just concerned

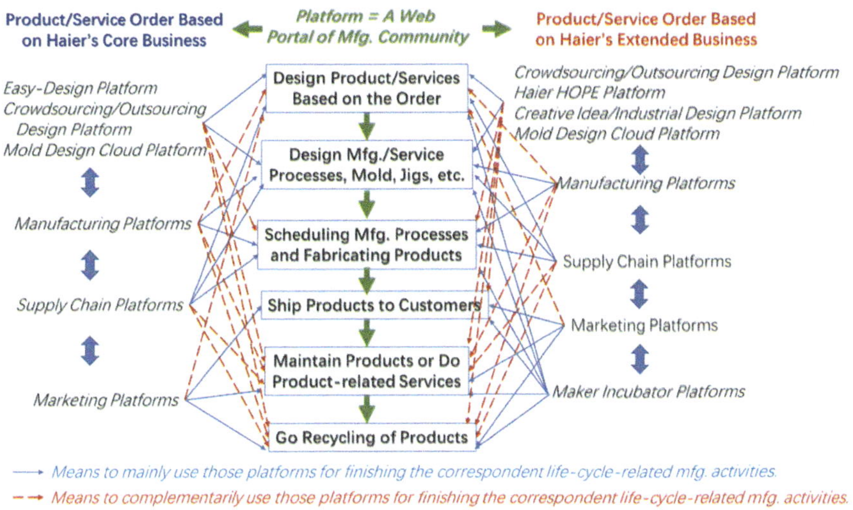

Fig. 12.4 Workflow of using product/service-order-driven platforms for product manufacturing activities during product life cycle

with all the participants. It means that the crowdsourcing mechanism depends essentially on "*stranger model*". In the moment, *Haier*'s product orders include all the characteristics mentioned above.

Depending on either *Haier*'s core business mainly related to SMRs inside *Haier* or extended business mainly concerned with ones outside *Haier*, for Ha*ier*'s case, different runtime strategies which deal with shaping a flow of accessing different platforms are used for reaching different goals. Here, shaping the flow of using different platforms is related to product-order-driven manufacturing activities in the whole stages of a product life cycle. Platforms are web portals which are also the front-ends of correspondent manufacturing communities respectively. Figure 12.4 just shows the situations for both *Haier*'s core business and extended business.

It must be emphasized that one of the biggest differences locates how to reach a designing goal. For *Haier*'s core business, design tasks concerning a product/service order are mainly done with "*Easy-Design*" platform which is inner-used as a web portal of design community. Design teams which are also called as design-type MSMEs or "*XiaoWei*" inside *Haier* will compete with each other to get the whole or partial design tasks. This kind of competition is on the basis of either outsourcing or crowdsourcing mechanisms. At the same time, some of design tasks are also put into an open crowdsourcing and outsourcing design platform for designing competition in a much larger range outside *Haier*. For *Haier*'s extended business, design tasks related to a product/service order are mainly done with a series of platforms which have the characteristics of crowdsourcing and outsourcing mechanisms. This means that product and service spectrums can be extended to an unimaginable range large enough if *Haier* wants. Of course, design teams inside Haier can also join the competition to obtain design tasks they want. Here, what we want to emphasized on is the role of makers. Some very new ideas for new product/service development are form makers. They also use different platforms to transfer their new ideas to realizable product/service order. It is obvious that design teams and makers will use different platforms if necessary. This means that they finish design tasks through interacting among different communities.

Another key point we have to declare here is to share *Haier*'s SMRs under the context of social manufacturing network, which is also an inter-connection among manufacturing communities. Those SMRs include *Haier*'s own resources and resources outside *Haier* which are integrated into correspondent platforms and managed by *Haier*. Here, *Haier*'s own resources are a backbone to implement design, production, supply chain and marketing capabilities. Not only for *Haier*'s core business but also for its extended business, both of them share all the resources integrated to manufacturing, supply chain and marketing platforms. It makes it possible for makers or any MSMEs to transfer their product/service designs into reality even if they have no factories, no mature supply chain and marketing ways, etc. It must also be pointed out that crowdsourcing and outsourcing mechanisms play an important role in assigning tasks concerning design, production, supply chain and marketing.

12.2.4 Analysis of Haier*'s Mode and Its Trends in the Future*

It is clear that *Haier*'s mode is a kind of initial social manufacturing paradigm driven with partially-centralized-control mechanism. The purpose that *Haier* creates the mode is to shape a sustainably ecological enterprise circle and rule it to an extreme.

Viewing from the angle of integrating and using SMRs, *Haier* decomposes its own manufacturing resources and staffs into platforms, MSMEs and makers. SMRs from the outside of *Haier* can be only integrated into *Haier*'s correspondent platforms as a supplement. Under this case, the following enabled technologies would be useful [3, 4]:

- classifying or clustering SMRs by using deep learning algorithms, VSM method, etc., and analyzing and visualizing them by using statistics and big data theory,
- building a credit and security mechanism to guarantee such resources on the Internet by using the block chain method, and
- constructing maturity-degree models by using fuzzy multi-objective and analytic target cascading (ATC) methods to evaluate the capabilities of using those resources.

Viewing from the angle of shaping an organizational structure, SMRs from the inside of *Haier* are firstly organized into a three-layer architecture and linked with each other to shape a backbone of social manufacturing network which consists of a lot of manufacturing communities. SMRs from the outside of *Haier* are only classified into two-layer structure and then connected individually with *Haier*'s platforms,

MSMEs and makers respectively in each layer. It is obvious that SMRs from the outside of *Haier* are only used as a supplement of *Haier*'s social manufacturing network. It is also not found that outside resources are firstly constructed into manufacturing communities and then either integrated directly with *Haier*'s platforms or set concurrently up a new "*platform*" and become simply a member of *Haier*'s platforms. *Haier*'s power will be cut down if the outside resources do as this. Accordingly, the following enabled technologies would be useful [5]:

- shaping the social manufacturing network and analyzing its static and dynamic performances in two aspects of both the network and nodes by using complex network theory,
- finding the new mechanism of value co-creations by using organizational improvement methods supported with lean and 6-Sigma technologies, and
- exploring the mechanism of collaborative management and power decomposition for *Haier*'s sustainably ecological enterprise circle.

Viewing from the angle of implementing runtime logic, *Haier*'s core business and extended business use different runtime strategies although they all use crowdsourcing and outsourcing mechanisms. The reason is that *Haier* wants to govern its sustainably ecological enterprise circle to an extreme. The key of reaching this goal is to control product/service design schemes and find new design ideas as many as possible. Concurrently, *Haier* wants to provide its production, supply chain and marketing capabilities to make such design ideas into reality under its own control. In addition, the role of makers is put up to a new height. It must be declared that some analysis details only in the high level are discussed in this section except some details in the lower level like tracking and tracing product/service quality, planning and scheduling production activities, constructing a social factory, etc. Under this case related to both high-level and low-level details, the following enabled technologies would be useful [4, 6]:

- improving product/service-order-driven runtime procedures by using lean and 6-Sigma methods,
- using a block-chain-based credit and security mechanism to monitor runtime procedures,
- visualizing every product/service-order-driven runtime procedure by using statistics and big data theory,
- developing a social factory model as the reference of constructing new production nodes by using systems engineering, CPS/IoT, artificial intelligence, socio-technical system theory, etc.,
- improving production planning and scheduling models to fit with the requirements of social manufacturing network, and
- tracking and tracing product/service quality in a product/service-order-driven runtime procedure.

Haier has actually done much more leading work in constructing a sustainably ecological enterprise circle and using the concepts of social manufacturing paradigm driven with partially-centralized-control mechanism. It is also clear that *Haier* can do more according to using correspondent enabled technologies mentioned above. Such enabled technologies either can solve some problems *Haier* exists now or make *Haier*'s mode much better. In addition, what we would like to emphasize here is to build and run a block-chain-based credit and security mechanism for using SMRs efficiently and effectively on Internet. At the same time, how to balance and share the power, duties and benefits of this sustainably ecological enterprise circle and realize the value co-creations would be the first thing that needs to be taken into consideration.

Depending on the product and service spectrums *Haier* wants to extend, there exists also a risk that *Haier* would be out of control for its partial SMRs, especially in the aspect of platforms development and evolution. It is because these partial SMRs work for the correspondent products and services either *Haier* is not familiar with or wants to give up. Under this case, such resources would shape new manufacturing communities and develop new platforms. These new platforms would run gradually in the form of independence. It is also the reason that *Haier* is doing a distributive implementation of social manufacturing paradigm driven with partially-centralized-control mechanism.

12.2.5 An Example from Mold Manufacturing

Haier is constructing a very large of sustainably ecological enterprise circle with a kind of initial social manufacturing paradigm in which *Haier* tries to control and manage partial or key product lines, services and SMRs inside this circle. To demonstrate its core contents, we will describe a typical example from *Haier*'s mold manufacturing which deals with designing, machining and assembling and delivering tasks, and depends on decomposing, organizing and running methods of correspondent mold manufacturing resources.

The systematic changes of *Haier*'s mold company also follow the idea shown in Fig. 12.2 although it is only a very small "*combining block*" for the circle mentioned above. It means that *Haier*'s mold company firstly has to change its mold manufacturing resources into:

- cloud platform for mold manufacturing, which is actually a web portal as the frond-end of mold manufacturing community,
- four types of MSMEs concerning mold designing, mold machining and assembling, CNC machining specially for makers, 3D printing specially for makers, and
- makers as individual mold designers.

Here, four-type MSMEs and makers are the members of the mold manufacturing community. In addition, a national engineering center for molding technologies, as a supporting base, would help makers, MSMEs and platform to solve various problems concerning the basic molding theory and techniques.

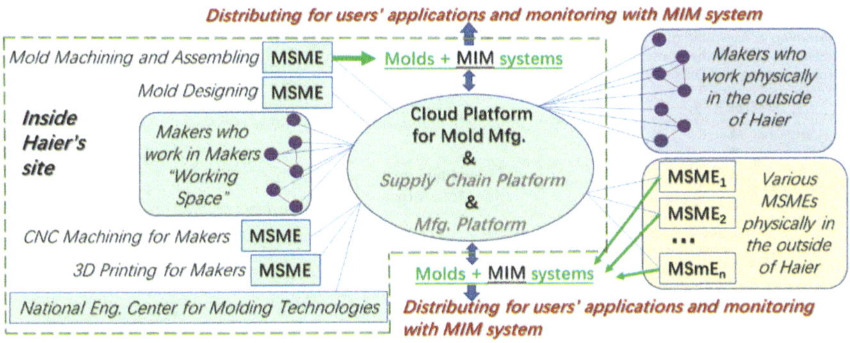

Fig. 12.5 Organizational structure of current Haier mold company

As the same as *Haier*'s policies related to socialization, self-organization and limited virus-like propagation of manufacturing resources, MSMEs and makers from the outside of *Haier* are also collected by means of using the above cloud platform.

Figure 12.5 just shows this new organizational structure around the current *Haier* mold company. Here, *Haier*'s other existing platforms that don't belong to the mold company, e.g., supply chain platform, manufacturing platform, are also integrated into the company's organizational structure so as to use them as a supplement. Another important plug-and-play embedded device is Mold Information Management (MIM) system, which is attached to a mold and used for reporting running-status of the mold to the cloud platform. In this way, the running conditions of the mold can be controlled and monitored remotely.

It must be mentioned here that the company is also running a makers' "*working space*" where makers not only are good at mold design tasks but also at any design tasks they want. In the following discussion, we simply ignore the makers' capabilities except mold design ones.

As soon as gathering different mold manufacturing resources to shape an organizational structure, we can discuss the runtime logic issue. Accordingly, mold orders are a driver to enable the runtime logic, which depends on both the method of getting the orders and the characteristics of presenting either one-of-a-kind production mode or small-volume-of-a-kind production mode. In addition, mold orders related to producing various sheet metal parts, plastic parts, etc., deal with correspondent molds for producing:

- *Haier*'s appliances and other products, and
- various products not belonging to *Haier*.

As soon as a mold order arrives, as shown in Fig. 12.6, the correspondent information of the order would be announced in the cloud platform. Depending on the type of the order and available competing methods to get the order, *Haier*'s mold company can select different running paths to finish mold designing, machining and assembling, and delivering tasks on the basis of outsourcing or/and crowdsourcing mechanisms.

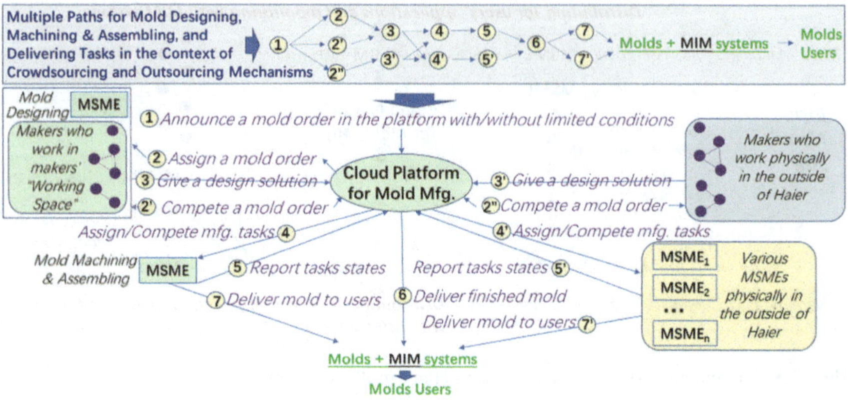

Fig. 12.6 Runtime logic to finish mold orders in different ways driven by outsourcing and crowdsourcing mechanisms

It can be seen in Fig. 12.6 that there are three paths to do a mold designing task related to a mold order.

A mold order, which deals with *Haier*'s appliances and other products, is announced by an MSME or maker from the inside of *Haier*, and is set up with attached limitations, there are two paths to do the mold designing task:

- assigning the mold designing task directly to either *Haier*'s mold designing MSME or specific makers working at makers' "*working space*" under the control of outsourcing mechanism, and
- competing the mold designing task among *Haier*'s mold designing MSME and makers working at makers' "*working space*" under the control of both crowdsourcing and outsourcing mechanisms,

For a mold order which deals with either *Haier*'s appliances and other products without attached limitations, or various products not belonging to *Haier*, a typical path to get the mold designing task besides the paths mention above is:

- to compete the mold designing task among all the mold designing MSMEs and makers inside and outside *Haier* by means of using the cloud platform under the control of crowdsourcing mechanism.

After finishing the above mold designing task, also seeing in Fig. 12.6, we can use two paths for doing the correspondent mold machining and assembling task:

- assigning the mold machining and assembling task directly to a or several specific mold machining and assembling MSMEs inside or outside *Haier*, depending on its/their production capabilities, and
- competing the mold machining and assembling task among all the mold machining and assembling MSMEs inside and outside *Haier* by means of using the cloud platform under the control of crowdsourcing and outsourcing mechanisms.

During the mold machining and assembling process, the MSMEs where undertake the machining and assembling tasks may report correspondent production states to the cloud platform. In this way, the online monitoring of the production process can be reached. As soon as finishing the machining and assembling task, as an option, an MIM system can be installed to the mold in the form of a tiny embedded web server. And then all the produced molds would be delivered to their users.

The mold attached with the MIM system, as the tangible "*product*" plus with the intangible "*pressing services*", can be used to construct the product service system. It is one of the most important functions of the MIM system. In terms of selling "*pressing services*" instead of selling molds, *Haier* mold company extends its capability through acting as a "*press services*" provider and can get much more benefits. Here, the MIM system records the times of pressing. Users can pay their costs of using the molds according to the pressing times.

It must be declared again that *Haier*'s mold company is only a very small "*combining block*" in *Haier*'s ecological enterprise circle. Its capabilities in aspects of integrating SMRs, extending product and service lines, etc., present a kind of huge energy and power, which come from its characteristics such as self-organization, virus-like propagation, capability share, etc. This small case also demonstrates the fascination and the perspective of using the social manufacturing paradigm.

12.3 Industrial Case: Distributive Implementation with Centerless-Self-control Mechanism for Social Manufacturing of *RepRap* Open-Source 3D Printers

RepRap 3D printer is a kind of desktop one firstly invented by *Adrian Bowyer* and then was further developed into different types and iterations in an open-source manner. Today many MSMEs, individuals and even makers are designing and producing their own 3D printer products based on the *RepRap* open-source specifications [7]. Different from the *Haier*'s case, which is a typical industrial case of social manufacturing paradigm driven with the partially-centralized-control mechanism, *RepRap* open-source 3D printers are designed, produced, and distributed in a distributive implementation driven with the centerless-self-control mechanism on the basis of some e-commerce and crowdfunding platforms like *Jingdong*'s crowdfunding platform [8], *kickstarter.com* [9], *Alibaba*'s *Taobao* network [10] and *Amazon* network [11]. Similar to the above section, we will discuss such a distributive implementation of the social manufacturing paradigm from the angle of how to use SMRs, how to shape an organizational structure and how to realize a runtime logic.

12.3.1 Use of Socialized Manufacturing Resources

Normally, SMRs related to fabricating *RepRap* open-source 3D printers can be used under the control of a kind of centerless-self-control mechanism and are closely concerned with correspondent product manufacturing activities covering the whole stages of a product life cycle. There isn't any leading socialized manufacturing resource or manufacturing enterprise to govern others. They are gathered dynamically around different "*physical*" manufacturing communities (PMC or PMCs), which are shaped potentially with the help of either e-commerce and crowdfunding platforms like *Jingdong*'s crowdfunding platform, *kickstarter.com*, *Alibaba*'s *Taobao* network, *Amazon* network, etc., or BBS like *RepRap forums* [7], in willingness by themselves. They also work not only competitively but also collaboratively in order to reach a balance through sharing benefits related to manufacturing tasks. Here, the term "*manufacturing*" covers the whole stages of a product life cycle and is not limited in the stage of product production.

Furthermore, a very big "*physical*" social manufacturing network can also be shaped in terms of interconnecting the above PMCs. In fact, SMRs to construct either a PMC or a "*physical*" social manufacturing network in a much larger range deal mainly with several types of participants who are representatives or owners of the SMRs and are also designers, producers, crowd-funders, consumers, and owners of other auxiliary-type resources respectively. Around either BBS or e-commerce and crowdfunding platforms which play a role in acting as the front-ends of the correspondent PMCs and searching for designers, producers, crowd-funders and consumers respectively, such participants can join one or more e-commerce platforms and BBS. It means that any participant can be a member of different PMCs. In other words, we can also classify "*logical*" manufacturing communities (LMC or LMCs) as designer communities, producer communities, crowd-funder communities, consumer communities, etc., according to roles of participants. Actually, the above "*logical*" manufacturing communities are participant-role-based. It can be viewed that, as shown in Fig. 12.7, these participators' role-related manufacturing communities in a logical meanings may establish a "*logical*" social manufacturing network so as to support correspondent participant-centered business interactions in all the product manufacturing activities for producing *RepRap* open-source 3D printers. It is also clear that public social media like *WeChat*, *Facebook* or *QQ* are the front-ends of such "*logical*" manufacturing communities, together with using limited interaction tools provided by PMCs. Here, actuated SMRs, including actuated MSMEs, actuated individuals, etc., only mean that they have already become the members of some LMCs. While unactuated SMRs imply such a fact that they are potential SMRs to join the LMCs (also in Fig. 12.7).

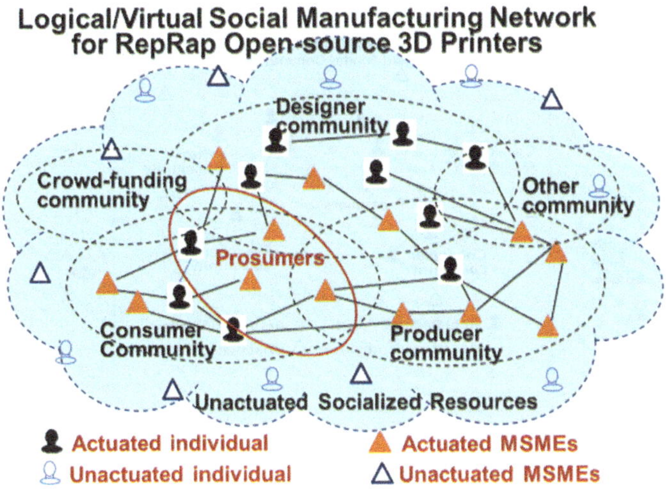

Fig. 12.7 Logic Social Manufacturing Network for RepRap Open-source 3D printer Manufacturing Activities in Product Life Cycle

12.3.2 Organization Principles

In the "*physical*" social manufacturing network for producing *RepRap* open-source 3D printers, the SMRs, including designer resources, producer resources, consumer sources, crowd-funder resources, and other auxiliary-type resources, involve deeply in the manufacturing processes of *RepRap* open-source 3D printers, and register physically in either BBS or e-commerce and crowd-funding platforms which are actually the front-ends of PMCs in the form of self-driven mode. Here, such PMCs are connected with each other through either the above participants who concurrently join the different PMCs and the correspondent LMCs mentioned above. Generally speaking, the SMRs in these PMCs are homogeneous and are dynamically self-organized and connected with each other. Accordingly, participants share ideas and information, compete for product orders, collaborate with one another even for the same manufacturing tasks, etc.

It must be pointed out, as shown in Fig. 12.8, all these product manufacturing activities are carried out through using LMCs attached to PMCs. Here, PMCs depend on BBS, e-commerce and crowd-funding platforms, etc., as their front-ends. LMCs just use both public social media like *WeChat* and *QQ* and limited interaction tools provided by PMCs as their front-ends. In this way, the organizational architecture is built in the form of embedding LMCs into PMCs and presented as a two-layer community architecture. It is also worth to be emphasized that this architecture is very suitable to enable the crowdsourcing and the crowdfunding mechanisms, together with the outsourcing mechanism for reaching a distributive implementation of the social manufacturing paradigm driven with centerless-self-control mechanism.

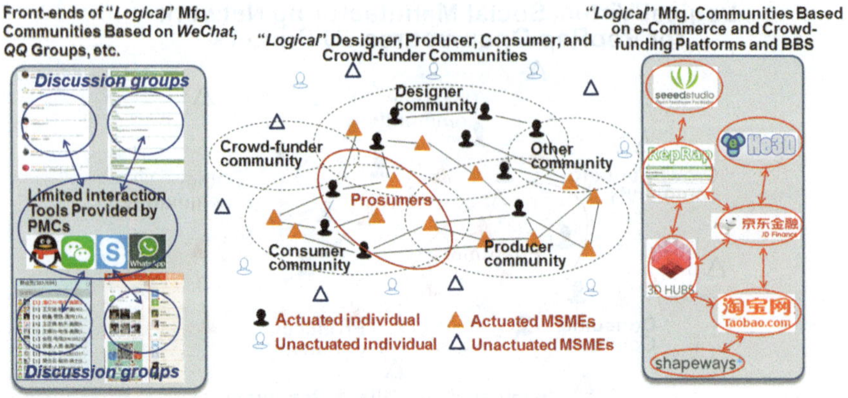

Fig. 12.8 Organizational architecture related to a social manufacturing implementation for producing RepRap open-source 3D printers

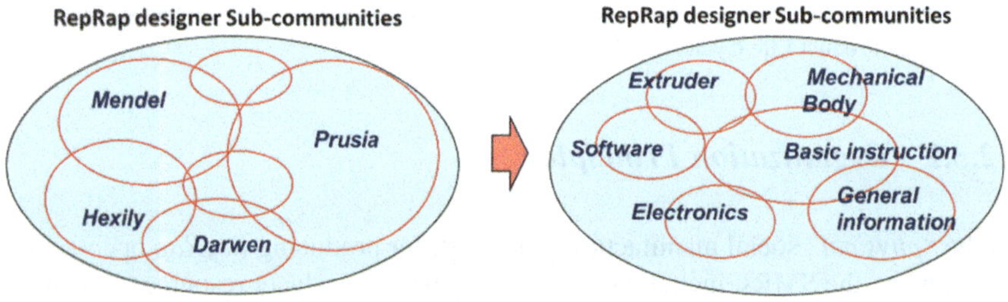

Fig. 12.9 RepRap "logical" designer sub-communities from different perspectives

Because of the importance of LMCs, sometimes we will decompose further them into more detailed sub-communities from different perspectives. For the purpose of competitions, a "*logical*" designer community connected with the "*physical*" *RepRap* official forum, for example, can be divided further into either product-type-related sub-communities shown in the left-side of Fig. 12.9 or component-type-of-a-product-related sub-communities shown in the right-side of Fig. 12.9. Here, the product-type-related sub-communities deal with designing the different types of open-source 3D printers such as *Mendel*-type, *Prusa*-type, *Hexily*-type, *Darwen*-type, etc. The component-type-of-a-product-related sub-communities are concerned respectively with designing extruder, mechanical body, software, electronics, etc.

12.3.3 Runtime Logic

There are at least two types of manufacturing ways for producing *RepRap* open-source 3D printers under the context of social manufacturing driven with centerless-

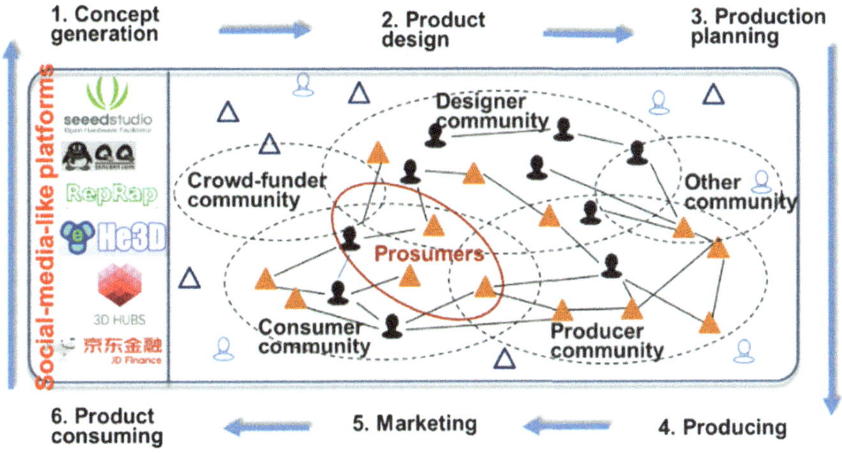

Fig. 12.10 General runtime logic for RepRap open-source 3D printer productions under the context of social manufacturing

self-control mechanism, that is, open-source-community-based volume and individualized production, and crowdfunding-based customized production. While the credit and security mechanism to guarantee the above manufacturing ways depends on the online-paying controls of correspondent PMCs which use e-commerce and crowdfunding platforms as the front-ends like *Jingdong*'s crowdfunding platform, *kickstarter.com*, *Alibaba*'s *Taobao* network, and *Amazon* network.

These manufacturing ways are all carried out, depending on SMRs clustered into both PMCs and LMCs, inside an organizational architecture of social manufacturing. Figure 12.10 just shows a generic six-step workflow which is suitable for enabling the above two types of manufacturing ways and includes the steps of concept generation, product design, production planning, producing, marketing, and product consuming.

The first type of manufacturing way is on the basis of open-source-community-based volume and individualized production. Under this manufacturing way, there exist at least a designer community around a BBS and several producer communities attached to an e-commerce platform. Here, a mature and open-source design scheme of *RepRap* open-source 3D printers must have existed and have also been known by most of the producers who belong to different producer communities. At the same time, the designer community has enough capability to modify the original design scheme and makes the difference so as to fit with the needs of individualized production. For the volume production, as soon as a producer gets a big product order on an e-commerce platform, he/she can organize different producer communities through freestyle competitions or collaborations to accomplish this product order in the natures of share, virus-like propagation and self-organization, because he/she has no capability to do all the tasks. For the individualized production, we need to consider the efficiency of the production besides finishing individualized product orders in the form of one by one. Under this case, the first thing is to revise the design scheme which is related to an individualized product order to satisfy the requirements of the specific consumers. The second thing is to cluster and combine all the current individualized product orders into a unified virtual product order so as to be suitable for organizing production activities in the form of mass individualization. And then this kind of virtual product order is gotten by the competitions and shares among different producer communities so as to finish the correspondent production tasks. Here, both outsourcing and crowdsourcing mechanisms will play a role in reaching the goal of finishing the product order including such a virtual one.

The second type of manufacturing way is concerned with crowdfunding-based customized production. In this case, an initiator who is a representative of either some MSMEs or individuals including makers possesses a good and special design scheme which attracts specific consumers' attentions and satisfies their requirements. The correspondent designer and producer communities related to this initiator are just connected with a crowdfunding platform which is the front-end of a PMC. The specific consumers will prepay the money on the crowdfunding platform in order that the correlative producers can get enough money to start the production procedures. Depending on the volume of the products requested by such consumers who have prepaid the money, the initiator can gather enough designer and producer communities to finish such a product order in terms of using outsourcing and crowdsourcing mechanisms in the natures of share, virus-like propagation and self-organization. It's obvious that the initiator needs to get enough consumers' prepay before starting this type of manufacturing way. The more times the initiator uses this way, the less willingness the consumers response it.

12.3.4 Analysis and Discussions

Product manufacturing activities in the whole stages of a product life cycle, related to *RepRap* open-source 3D printers, is actually driven with centerless-self-control mechanism under the context of social manufacturing. The workflow is completely different from the *Haier*'s one and has the natures of self-organization, virus-like propagation, share, etc., because no any leading socialized manufacturing resource or manufacturing enterprise can govern the above two-layer community architecture which consists of LMCs and PMCs, and the correspondent social manufacturing network.

In fact, the PMCs are implicitly embedded into either some e-commerce and crowdfunding platforms or BBS forums, and respectively use them as their front-ends. The interconnections inside or among these PMCs just depend on the correspondent role types of participants who are also the members of different LMCs. In addition, such platforms and forums run under the control of the socialized third-party, which look just like a kind of social media to support the above two-layer community architecture and make sure that the credit and security mechanism for product manufacturing activities works on the basis of both the online-prepay control of the platforms and open-source protocols. Of course, the running logic of these resources deals with crowdfunding, crowdsourcing and outsourcing mechanisms

From the angle of both using SMRs and shaping an organizational architecture, there are no "*master-slave*" relationship among them during product manufacturing activities related to *RepRap* open-source 3D printers. These SMRs are freely gathered into a two-layer community architecture which consists of LMCs and PMCs by using self-organized and virus-like propagation manners, are used if needed, and work not only competitively but also collaboratively in order to reach a balance through sharing benefits oriented from product manufacturing activities. We would like to point out again that e-commerce and crowdfunding platforms and BBS forums act as the front-ends of PMCs and enable the credit and security mechanism based respectively on online-repay controls and open-source protocols. In order to use and self-organize the SMRs more efficiently and effectively, at least the following key enabled technologies are necessary accordingly:

- to develop a series of new independent social-media-like platforms, as a replace of both e-commerce and crowdfunding platforms and BBS forums,
- to study new enabled technology for aggregating SMRs, and
- to look for blockchain-driven credit and security mechanism to replace current online-prepay control manner.

From the angle of runtime logic, the above two types of manufacturing ways are completely different from the traditional one. Here, mass individualization concerning the context of social manufacturing is one of the final production goals we would like to pursue. To an extreme situation, at the same time, it is also possible that everyone can become a manufacturer if he/she wants. It means that both crowdfunding and crowdsourcing mechanisms, together with outsourcing mechanism, are very strong drivers for a distributive implementation of social manufacturing driven with the centerless-self-control mechanism. In order to enable a runtime more efficiently and effectively, at least the following key enabled technologies are necessary accordingly:

- to study an enabled technology to generate a virtual product order which fit with the production needs of mass individualization, and
- to develop a distributive production technology to gather available SMRs in the form of self-organization and virus-like propagation, and let them work both competitively and collaboratively so as to take a balance through sharing benefits related to production tasks.

Compared with *Haier*'s industrial case, this case focuses on product manufacturing activities related to open-source products working on both e-commerce and crowdfunding platforms and BBS forums. The centerless-self-control mechanism of SMRs is the key point. In addition, the dynamic use of these resources makes product manufacturing activities able to be extended to an unlimited range we can't imagine. It can also be said that the hung energy in a society would be released to create a new era of manufacturing.

12.3.5 *A Crowdfunding Example Concerning* RepRap *Open-Source 3D Printer*

In the context of social manufacturing, the biggest gap to be overcome for most individual designers is the financial problem that would obstruct a good product design scheme to become useful products. Crowdfunding is just the good way to find the financial support before someone can start production activities. Here, an example entitled as *Athorbot* project [12], which deals with producing a kind of *RepRap* open-source 3D printer, is utilized to demonstrate the entire product manufacturing process. This process exactly demonstrates how the project takes advantage of the crowdfunding method based on *Jingdong*'s crowdfunding platform so as to acquire the financial support for its production [8].

Athorbot project starts with an idea that most of the current *RepRap* open-source 3D printers only have one extruder. It means these printers cannot construct objects with different colors. In order to solve this problem, the project firstly establishes a small team and makes a preliminary design related to a kind of new *RepRap* open-source 3D printer for multi-color object generations. However, the team is without enough money to produce the 3D printers and announces its "*call for consumers*" proposal on *Jingdong*'s crowdfunding platform. The online proposal sets up the minimal number of the new *RepRap* open-source 3D printers to be produced and the price per the 3D printer. Consumers who are interested in buying this kind of 3D printer need to prepay the money on the *Jingdong*'s crowdfunding platform. The deal goes to success when the number of the 3D printers ordered is more than the minimal number to be set up. And then the production stage runs under the consultation of the design team. After finishing the production procedure, most of these 3D printers are delivered to the consumers who have repaid the money. The rest of the 3D printers are sold on *Jingdong* or *Taobao* e-commerce platform. Figure 12.11 just shows the principle how to do as this.

It must be pointed out that three PMCs are involved into this distributive implementation of social manufacturing paradigm driven with the centerless-self-control mechanism. These three PMCs utilize the following "*platforms*" respectively as their front-end:

- *RepRap* BBS forum as the front-end of design-driven PMC,
- *Jingdong*'s crowdfunding platform as the front-end of both consumer-based and crowd-funder-driven PMC, and
- *Taobao* e-commerce platform as the front-end of both producer-driven and consumer-based PMC.

Here, the interactions among different "*platforms*" can also be realized with the help of "*WeChat*" social media tool among correspondent LMCs which respectively consist of designers, consumers, crowd-funders, producers, etc. According to the crowdfunding principle, crowd-funders can be either consumers or retail traders.

The designer team works mainly on a "*logical*" designer community which is connected with the *RepRap* BBS forum and deals with all the steps of the manufacturing

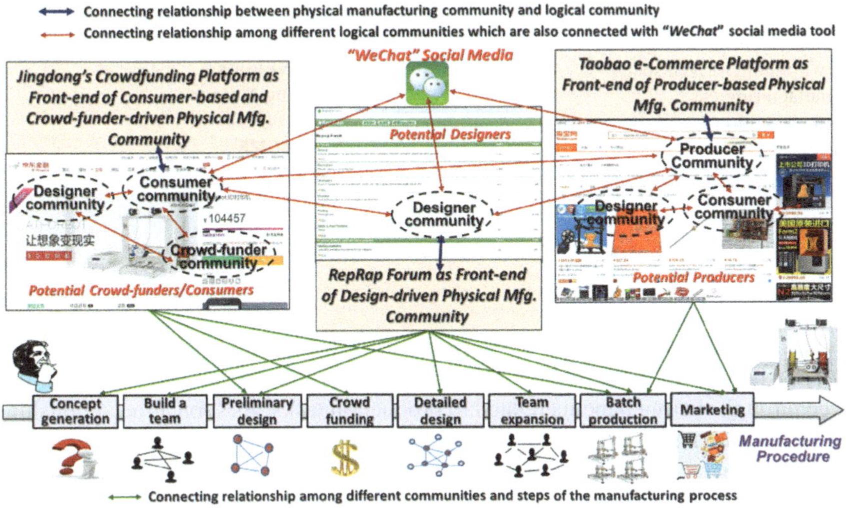

Fig. 12.11 Manufacturing process for Athorbot 3D printers

procedure for the 3D printers. In addition, designers including the members of the above design team would also consist of different "*logical*" designer communities inside both *Jingdong*'s crowdfunding platform and *Taobao* e-commerce platform. But the functions of these two "*logical*" designer communities are completely different. The "*logical*" designer community in *Jingdong*'s crowdfunding platform is used for making the design-related consultation and searching for the crowdfunding opportunities. The "*logical*" designer community in *Taobao* e-commerce platform is just concerned with the functions of explaining and revising the design details for easy-manufacturing.

In this distributive implementation, consumers also act the "*crowd-funders*" role of the financial support through prepaying manner on *Jingdong*'s crowdfunding platform. It is obvious that there exist other two LMCs, that is, consumer community and crowd-funder community, besides the designer community mentioned above on this platform. Here, the platform looks like an "*agent*" which acts as a credit and security mechanism to ensure the success of the deal. The correspondent PMC and LMCs which uses *Jingdong*'s crowdfunding platform are related to three steps of the manufacturing procedure shown in Fig. 12.11, which respectively are "*Preliminary Design*", "*Crowdfunding*" and "*Batch Production*".

Since the design team is not capable of producing this batch of 3D printers, it is necessary for it to call potential producers through *Taobao* e-commerce platform. In this way, the "*logical*" producer community, together with the correspondent production-driven PMC which uses *Taobao* e-commerce platform as the front-end, is created. In addition, designer community mentioned above is used for explaining and revising the design details. Consumer community probably including both consumers who have prepaid the money and are without payments is built concurrently.

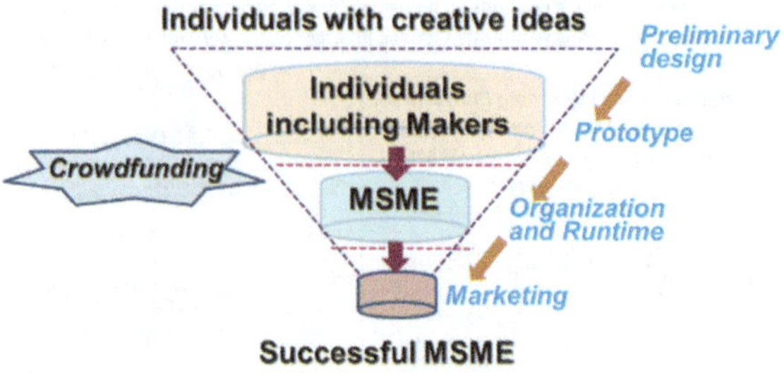

Fig. 12.12 Funnel model for individuals to reach crowdfunding-driven social manufacturing implementation

Here, either outsourcing or crowdsourcing mechanisms are used for determining the suitable producers based on both competition and collaboration so as to take a balance through sharing benefits related to production tasks. As soon as accomplishing the production tasks, most of these 3D printers are firstly delivered to the consumers who participated in the crowdfunding project. The rest of the 3D printers can also be sold to other consumers at a higher price. It should be mentioned that the correspondent PMC which uses *Taobao* e-commerce platform as its front-end and LMCs which utilize *WeChat* and limited interaction tools attached to the platform as their front-ends are concerned with to four steps of the manufacturing procedure shown in Fig. 12.11, which respectively are "*Detailed Design*", "*Team Expansion*", "*Batch Production*" and "*Marketing*".

In summary, this is a typical case that individual designers take advantage of SMRs to turn their ideas into available products under the context of social manufacturing. However, not all these practices can hold to the end due to many reasons, such as the ideas are not good enough, there are no enough team members, design team fails to acquire enough financial support, etc. So this kind of distributive implementation is to follow a "*Funnel model*" in which only a few individuals can eventually make their ideas into products among a huge number of individuals who have creative ideas too, Fig. 12.12 shows this situation.

12.4 Industrial Case: W-Company's Distributive Implementation with Completely-Centralized-Control Mechanism for X-Part Production in the Context of Elementary Social Manufacturing

Production outsourcing makes it possible for any core-enterprise to be able to transfer either its machining tasks of partial parts or its assembling tasks of its some assemblies and even products to partner co-enterprises in terms of using "*acquaintance model*" under the consideration of some factors like its own production capability, manufacturing costs, etc. Here, the core-enterprise and its partner co-enterprises consist of an enterprise alliance. Actually, the enterprise alliance is driven with acquaintance-model-based outsourcing mechanism and its value creations are mainly accomplished

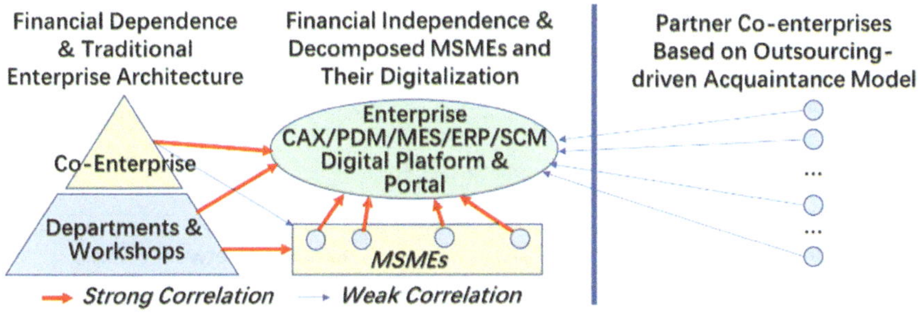

Fig. 12.13 Enterprise alliance and its changes in a digital and networking way

by the core enterprise. This kind of enterprise model has worked for decades even without the help of Internet. Currently, the core-enterprise has been changing its organizational structure and business flows in a digital and networking way, and has covered following key points (see Fig. 12.13):

- building a digital platform powered with CAX/PDM/MES/ERP/SCM systems and the correspondent networking portal for outside accesses, and
- decomposing workshops and some departments into financially-independent MSMEs.

The partner co-enterprises selected with the above acquaintance-model-based outsourcing mechanism can work together with the co-enterprise through a networking portal of the above digital platform.

This kind of new enterprise alliance at least is not completely self-organizational and virus-like propagated except including some online social connecting and communicating capabilities in business. We cannot call it as a kind of ideal social manufacturing paradigm. Fortunately, it is possible to let this enterprise alliance have the above two characteristics. This means that we can transfer traditional enterprise alliance to a kind of elementary social manufacturing paradigm driven with completely-centralized-control mechanism. In the following sub-sections, an industrial case which deals with X-part production will be discussed in detail too from the angle of how to use SMRs, how to shape an organizational structure and how to realize a runtime logic.

12.4.1 Limited Socialization of Manufacturing Resources

X-parts are a kind of key part used in many important products such as pumps, turbines, aircraft engines, windmill generators, etc. W-company takes the bigger market share for X-parts production in China, and also delivers its outsourced X-parts to overseas enterprises, and it has the capability of designing various X-parts.

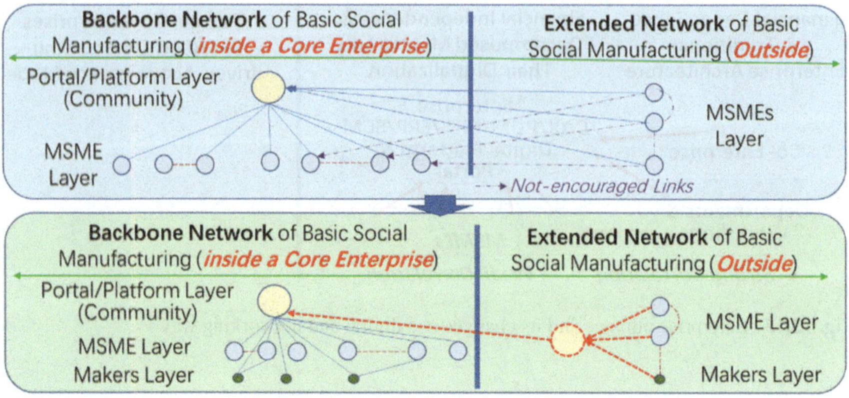

Fig. 12.14 Organizational structure changes from enterprise alliance to elementary social manufacturing paradigm

Because of a large number of outsourcing orders, W-company has developed a lot of MSMEs around it as its supplement of production capabilities in the form of partner co-enterprises. Actually, such MSMEs are also SMRs. In this way, a typical enterprise alliance based on "*acquaintance model*" can be constructed. At the same time, workshops and some departments inside the W-company are reconstructed as financially-independent MSMEs. A digital CAX/PDM/MES/ERP/SCM integrated platform and its networking portal are configured so as to support business interactions among this enterprise alliance.

In other words, it happens to use the limited socialization of manufacturing resources because the W-company:

- decomposes its workshops and some departments into financially-independent MSMEs, and
- integrates SMRs from the outside of the company as its supplement.

12.4.2 Organizational Architecture

In order to transfer this enterprise alliance to an elementary social manufacturing paradigm, referring to Fig. 12.14, the current organizational structure must be changed into a new one which presents partially self-organizational and virus-like propagated characteristics, extends its social connecting and communicating capabilities in business and works with both outsourcing and crowdsourcing mechanisms. It also means that not only "*acquaintance model*" bur also "*stranger model*" will play an important role in finishing production orders to produce a variety of X-parts. Such changes at least will deal with:

- enhancing the backbone network of elementary social manufacturing inside the W-company through increasing a maker layer in which at least designers inside the W-company may work as makers,
- strengthening the organizational robustness, pricing and quality controlling capability related to the extended network of elementary social manufacturing through setting up specific manufacturing communities which can integrate different types of MSMEs and makers from the outside and communicate with the W-company in a group not in an individual,
- using new elementary social manufacturing paradigm driven with completely-centralized-control mechanism to compete X-parts production orders, take the much more marketing share, and
- involving the design activities of various X-parts in the form of either outsourcing mechanism or crowdsourcing mechanism.

It can also be seen from Fig. 12.14 that the changes of organizational structure from an enterprise alliance to an elementary social manufacturing paradigm depends on using a combination of both backbone and extended networks for finishing production orders. Different from *Haier*'s paradigm, W-company only uses a digital platform instead of *Haier*'s a group of platforms as the portal of its community. It is because the W-company only focuses on a very specific market and what the company wants to do is to extend its business to various X-parts.

12.4.3 Runtime Logic

The runtime for the new organizational structure of elementary social manufacturing paradigm will be quite different from the original enterprise alliance. It's a little bit same as something shown in Fig. 12.6 in which there concurrently exist two situations dealing with either competing or assigning X-part production orders. Of course, the runtime depends respectively on outsourcing and crowdsourcing mechanisms, and has multiple paths to go. We can also learn such a fact from Fig. 12.15 that both "*acquaintance model*" and "*stranger model*" can be implemented through the manufacturing community which gathers a variety of SMRs from the outside of W-company. In fact, this manufacturing community is self-organized and virus-like propagated only focused on the X-part production field. Furthermore, W-company may use makers from both inside and outside of the company to participate a design competition especially in the form of crowdsourcing mechanism. It also presents characteristics of self-organization and virus-like propagation.

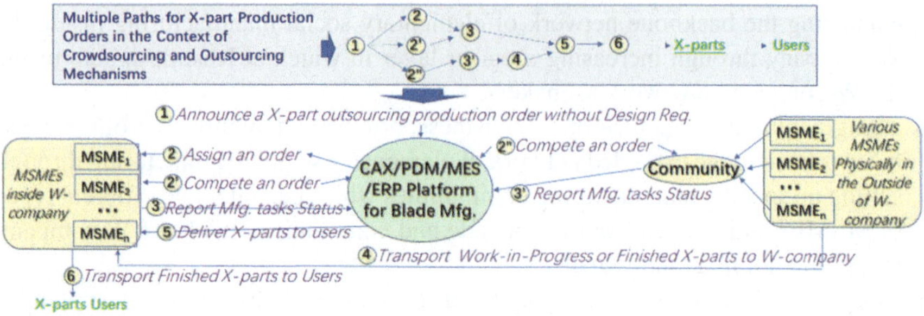

Fig. 12.15 Runtime related to elementary social manufacturing paradigm for handling X-part production orders without considering design requirements

12.4.4 Analysis of X-Parts Production

Different from *Haier*'s mode, W-company as an X-part supplier does its business mainly in a very specific part-level machining field. The business is extendable to designing and maintaining X-parts W-company produces. The case of W-company demonstrates such a fact that it is feasible to transfer an enterprise alliance to an elementary social manufacturing paradigm driven with completely-centralized-control mechanism through enabling the characteristics of self-organization and virus-like propagation on the basis of both outsourcing and crowdsourcing mechanisms.

From the angle of socializing manufacturing resources inside a core enterprise, on the one hand, W-company still must decompose its inner manufacturing resources into MSMEs and makers. Here, the role of makers is to increase two aspects of creative capabilities, that is, competitive design capability and maintaining capability. For example, maintaining capability may be concerned with either designing a scheme to enable the surface repairing of X-parts or designing and running a product-service-system-based workflow, etc. It is very important to let a mature maker develop into an MSME. This means that W-company should construct a makers' "*working space*" inside the company for researching, developing and testing X-parts. In addition, it is a basis to integrate different manufacturing resources by means of reconstructing these standalone CAX, PDM, MES and ERP systems into an integrated CAX/PDM/MES/ERP digital platform, and configuring the correspondent front-end networking portal. On the other hand, SMRs from the outside of W-company are geographically around the W-company and are presented as MSMEs and makers. They are basis of constructing an extension of the above paradigm. In a word, at least the following enabled technologies are needed for much better socialization of manufacturing resources:

- classifying and decomposing MSMEs and makers,
- designing an excitation mechanism to realize the transformation from a mature maker to an initial MSME, and

- implementing an integration among different standalone CAD, PDM, MES and ERP systems to generate a digital platform and developing a networking portal as the front-end of this digital platform.

From the angle of organizing these SMRs, on the one hand, W-company can organize their inner MSMEs and makers into a backbone production network and absolutely control it through the CAX/PDM/MES/ERP digital platform. On the other hand, MSMEs and makers from the outside of W-company are firstly organized into manufacturing communities for better pricing and quality controlling among W-company and them. And then, such manufacturing communities use the W-company's digital platform as their front-end although this extended production network is self-organized and virus-like propagated. However, it must be pointed out that the manufacturing communities only accept X-part production tasks from W-company and are under the control of it by using the digital platform mentioned above. This is also the reason we call this industrial case as a distributive implementation of social manufacturing paradigm driven with completely-centralized-control mechanism. In a word, at least the following enabled technologies are needed for a much better organization of manufacturing resources:

- constructing a social manufacturing network in terms of combining both backbone network and extended network, analyzing its static and dynamic performances, self-organized and virus-like propagated characteristics with the help of complex network theory and social computing,
- labeling datasets collected during social interactions inside or among manufacturing communities, and
- mining and identifying social business relationships by means of using deep learning algorithms, statistics, etc.

From the angle of runtime logic, a networking portal as the front-end of CAX/PDM/MES/ERP digital platform works for the connection between W-company and manufacturing communities where MSMEs and makers are gathered. Essentially, finishing X-part production orders is a basic task and taking a participation of competitively designing and maintaining various X-parts is an extended task. It is clear that crowdsourcing mechanism makes it possible for many MSMEs and makers to be able to help W-company to create the value. In a word, at least the following enabled technologies are needed for a much better runtime of this elementary social manufacturing paradigm:

- designing an order competitive mechanism of X-part productions referring to both outsourcing and crowdsourcing mechanisms,
- implementing production planning and scheduling, and
- tracking and tracing the machining quality during the whole procedure of X-part productions.

Compared with *Haier*'s paradigm, W-company faces to a very narrow marketing domain. Its socialized level is far lower than *Haier*'s one. But W-company is still in constructing a small ecological enterprise circle with the help of social manufacturing paradigm. This industrial case also provides a short-cut for a traditional core-enterprise to change its organizational structure from an enterprise alliance to an elementary social manufacturing paradigm.

12.5 Concluding Remarks

With the help of three industrial cases respectively from *Haier*, *RepRap* open-source 3D printers manufacturing, X-part productions, three types of distributive implementations for social manufacturing paradigm are described in detail from the angle of how to use SMRs, how to shape an organizational structure and how to realize a runtime logic. They are driven respectively with:

- partially-centralized-control mechanism,
- centerless-self-control mechanism, and
- completely-centralized-control mechanism.

Furthermore, their advantages and disadvantages are analyzed. At the same time, the correspondent key enabled technologies are also listed as references.

What we want to mention once again is that the fundamental starting point to create a social manufacturing application is to use Internet-based connecting and communicating behaviors in business well.

It is obvious that how to construct a sustainably ecological enterprise circle becomes one of the most important issues we discuss. It is also the amazing point social manufacturing paradigm possesses.

It is clear that the different decentralized implementations for social manufacturing demonstrate the future of manufacturing industry.

It can be said that social manufacturing paradigm will play a key role especially in enabling mass individualization in the near future [13]. Also along with the popularity of 3D printers, anyone can take a big opportunity to become a potential manufacturer for completely personal manufacturing which is actually an extreme shape of social manufacturing paradigm.

It is also urgently necessary to develop a new industrial software model so as to replace the current various "*platforms*" and public social media, and make the business procedures of social manufacturing work much better and easier.

We hope the tomorrow of social manufacturing paradigm would become better and better.

References

1. NN (2018) Introduction. Haier's Home Page, available at: http://www.haier.com
2. Jiang PY (2017) 2017 Annual report for project 'Industrial-cluster-driven innovation method applications in micro-and-small-scale manufacturing enterprises'. Project Documents Submitted to Ministry of Science & Technologies (MOST), China (in Chinese)
3. Zhang FQ, Jiang PY et al (2012) Modeling and analyzing of an enterprise collaboration network supported by service-oriented manufacturing. Proc IMechE, Part B: J Eng Manuf 226(9):1579–1593
4. Liu JJ, Jiang PY, Leng JW (2017) A framework of credit assurance mechanism for manufacturing services under social manufacturing context. In: The 13th IEEE International Conference on Automation Science and Engineering (IEEE CASE 2017), Xi'an, China, 20–23 August 2017
5. Ding K, Jiang PY et al (2016) Modeling and analyzing of an enterprise relationship network in the context of social manufacturing. Proc IMechE, Part B: J Eng Manuf 230(4):752–769
6. Zhang CY, Jiang PY et al (2016) Configuration design of the add-on cyber-physical system with CNC machine tools and its application perspectives. Procedia CIRP 56(2016):360–365
7. NN (2018) RepRap forums. Web Pages, available at: http://forums.reprap.org/
8. NN (2018) JingDong's crowdfunding platform. Web Pages, available at: https://z.jd.com/sceneIndex.html (in Chinese)
9. NN (2018) Kickstarter: a crowdfunding platform. Web Pages, available at: https://www.kickstarter.com
10. NN (2018) Alibaba's TaoBao network. Web Pages, available at: https://www.taobao.com (in Chinese)
11. NN (2018) Kickstarter. Web Pages, available at: https://www.amazon.com/
12. NN (2018) Athorbot project. Web Page, available at: https://z.jd.com/project/details/90916.html?from=jr_search&type=0, 2017 (in Chinese)
13. Jiang PY, Leng JW, Ding K, Gu PH, Koren Y (2016) Social manufacturing as a sustainable paradigm for mass individualization. Proc IMechE, Part B: J Eng Manuf 230(10):1961–1968

The manufacturer's authorised representative in the EU is Springer Nature Customer Service Centre GmbH, Europaplatz 3, 69115 Heidelberg, Germany. If you have any concerns regarding our products, please contact ProductSafety@springernature.com

Printed and bound by CPI Group (UK) Ltd, Croydon, CR0 4YY
30/09/2026
02225662-0008